FOODBORNE PARASITES

FOOD MICROBIOLOGY AND FOOD SAFETY SERIES

Food Microbiology and Food Safety publishes valuable, practical, and timely resources for professionals and researchers working on microbiological topics associated with foods, as well as food safety issues and problems.

Series Editor

Michael P. Doyle, *Regents Professor and Director of the Center for Food Safety, University of Georgia, Griffin, GA, USA*

Editorial Board

Francis F. Busta, *Director – National Center for Food Protection and Defense, University of Minnesota, Minneapolis, MN, USA*
Bruce R. Cords, *Vice President, Environment, Food Safety & Public Health, Ecolab Inc., St. Paul, MN, USA*
Catherine W. Donnelly, *Professor of Nutrition and Food Science, University of Vermont, Burlington, VT, USA*
Paul A. Hall, *Senior Director Microbiology & Food Safety, Kraft Foods North America, Glenview, IL, USA*
Ailsa D. Hocking, *Chief Research Scientist, CSIRO—Food Science Australia, North Ryde, Australia*
Thomas J. Montville, *Professor of Food Microbiology, Rutgers University, New Brunswick, NJ, USA*
R. Bruce Tompkin, *Formerly Vice President-Product Safety, ConAgra Refrigerated Prepared Foods, Downers Grove, IL, USA*

Titles

PCR Methods in Foods, John Maurer (Ed.) (2006)
Viruses in Foods, Sagar M. Goyal (Ed.) (2006)
Foodborne Parasites, Ynes R. Ortega (Ed.) (2006)

FOODBORNE PARASITES

Edited by

Ynes R. Ortega
University of Georgia, Griffin, GA, USA

Ynes R. Ortega
Center for Food Safety
Department of Food Science and Technology
University of Georgia
1109 Experiment St.
Griffin, GA 30223
Ortega@uga.edu

Library of Congress Control Number: 2006921159

ISBN-10: 0-387-30068-6   e-ISBN 0-387-31197-1   Printed on acid-free paper.
ISBN-13: 978-0387-30068-9

© 2006 Springer Science+Business Media, LLC
All rights reserved. This work may not be translated or copied in whole or in part without the written permission of the publisher (Springer Science+Business Media, LLC, 233 Spring Street, New York, NY 10013, USA), except for brief excerpts in connection with reviews or scholarly analysis. Use in connection with any form of information storage and retrieval, electronic adaptation, computer software, or by similar or dissimilar methodology now known or hereafter developed is forbidden.
The use in this publication of trade names, trademarks, service marks, and similar terms, even if they are not identified as such, is not to be taken as an expression of opinion as to whether or not they are subject to proprietary rights.

Printed in the United States of America.   (TB/EB)

9 8 7 6 5 4 3 2 1

springer.com

Preface

Microbiologists are being challenged as foodborne outbreaks are increasingly being observed worldwide. Most of these outbreaks are associated with viral and bacterial pathogens such as *Campylobacter*, *Salmonella*, and lately *Escherichia coli* O157:H7, which emerged in the 1990s. The role of food in transmission of parasites was not studied until later.

Although parasites have been evolving with man since antiquity, the control and eradication of these diseases is still far from being achieved and they are more frequently being reported in the literature as causative agents of food and waterborne illnesses.

Parasites have been consistently reported in developing and endemic countries. However, the presence of these parasites in the developed world has increased in the past two decades. For example, in the United States alone, foodborne outbreaks have multiplied by more than a factor of eight during the past 15 years.

Many factors can be attributed to the rise in parasitic foodborne outbreaks. The increase of international travel and population migration encourages rapid disease dissemination. Globalization of the food supply has introduced new challenges in the U.S. food safety arena. The food habits of Americans are changing, resulting in fresh foods and vegetables being consumed. Consequently, importation of foods is now necessary in order to satisfy the consumer demands for a year-round supply of certain commodities, such as exotic fruits and vegetables. Importations have also increased as the costs of particular products have been reduced. Unfortunately, transportation conditions such as controlled refrigeration favor survival of parasites on fruits and vegetables. In addition, restaurants presenting ethnic foods not traditionally consumed are revealing public health issues that have not been considered. For example, *Sarcocystis*, a nematode parasite, has been identified in meats served at Arab restaurants in Brazil where meats are consumed raw.

Moreover, populations are at increased risk for acquiring infection and developing gastrointestinal illness due to the prevalence of asymptomatic carriers and the zoonotic potential of some parasites that contribute to the spread of the pathogen. For example, inadequate composting and lack of agricultural practices has contributed to the exposure of food products to animal waste and hence, parasites.

Medical advances have contributed to changes in the US demographics, yet, the elderly, immunocompromised individuals and children are susceptible populations to parasitic infections.

Misdiagnosis of parasitic infections is not uncommon. Diagnostic laboratories and medical personnel may not be familiar with parasite identification and the risk factors for acquiring these infections. Excretion of parasites is intermittent, requiring that more than one stool sample be examined. In some instances, special staining procedures not routinely requested by medical personnel are required to identify parasites properly.

Outbreak investigations may take many days, by which time samples of the implicated product may not be available for investigation. Therefore, food scientists and the medical community need to be aware of the role of parasites as significant agents of foodborne outbreaks. Routine examination of clinical specimens of individuals with gastrointestinal illness does not search for parasites. In fact, testing for *Cyclospora* and *Cryptosporidium* must be specifically requested. Improvements in detection systems not only in the clinical settings, but also in the food microbiology laboratories have allowed for identification of infections that previously went undetected.

Molecular assays have been very helpful in foodborne outbreak investigations but are limited by isolation procedures for the parasites. Recovery is particularly important since parasites are generally inert in the environment, and enrichment procedures applied to bacterial contaminants are not an option.

In the 1980s, *Cryptosporidium* emerged as a significant pathogen responsible for high mortality in immunocompromised patients; particularly in those with AIDS. Although to date there is no effective therapy for cryptosporidiosis, a great deal of information was learned about this parasite, most notably that the cellular immune response plays a significant role in eradication. *Cyclospora*, another emerging parasite, has also proven to be a challenge to work with and has demonstrated to be highly resistant to environmental conditions. *Cyclospora* is a parasite that is associated with foodborne outbreaks in the United States and is caused by contaminated produce imported for human consumption. *Trichinella*, a nematode, is another emerging pathogen that is frequently reported in the United States and is commonly associated with the ingestion of game meats and sausages that are not properly cooked.

This book will review the two major parasite groups that are transmitted via water or foods: the protozoa, which are single celled organisms and the helminths. The helminths are classified into three sub groups, the cestodes (tapeworms), nematodes (round worms), and trematodes (flukes). To better understand their significance, each chapter will be covering the biology, mechanisms of pathogenesis, epidemiology, treatment, and inactivation of these parasites. A better understanding of the biology and control of parasitic infections is necessary to reduce and eliminate outbreaks in the United States and elsewhere.

PROTOZOA

Protozoa are single-celled organisms. They are eukaryotes which have compartmentalized organelles and the infectious stages are environmentally resistant. Protozoa can be blood borne or transmitted via a vector, such as a mosquito. The most well-known disease in this group is malaria. Other protozoa, acquired via water, are environmentally resistant, and can be responsible for worldwide human illness. A classic example of this group is *Cryptosporidium*. As we learn more of the molecular composition of these parasites, and the development of molecular tools for genotyping, new species are being described. Along with these, a reevaluation of these species and their significance in public health is changing. This is particularly true in the case of *Cryptosporidium*. The species *C. parvum* was considered to be unique, although there was significant evidence that the clinical presentation was variable as well as host susceptible. Now more genotypes are being identified in human and animal species. Although most of the outbreaks associated with *Cryptosporidium* have been waterborne, a few reports have been foodborne. The contamination has been associated with food preparation practices, hygienic conditions of the food handlers, or by fresh produce that may have been contaminated on the farms.

Most parasites are obligate intracellular organisms. In contrast with bacteria, parasites are inert and do not multiply in the environment. Therefore, studies of food matrices and determination of the safety of some products is challenging. Isolation and detection procedures are crucial because an enrichment process for parasites is not available. Although molecular assays overcome these difficulties, specific limitations of each organism will be discussed in the pertinent chapters.

Other protozoa can be acquired by more than one route. *Toxoplasma* is a good example of this group. Present in almost every region of the world, *Toxoplasma* can be acquired by most warm-blooded animals, by ingestion of oocyst (the environmentally resistant form) in water or fresh produce, or by ingestion of raw or inadequately cooked meats which contain tissue cysts of the parasite.

Most of the foodborne protozoa cause diarrheal illness in humans; however, *Toxoplasma* can have detrimental consequences for the fetus of pregnant women and in immunocompromised individuals. In immunocompetent individuals toxoplasmosis may be asymptomatic, but if the immune competency diminishes either by chemotherapy or by acquired deficiency, the individual can present symptomatic toxoplasmosis that is most commonly manifested as encephalitis.

HELMINTHS

This group is commonly identified as worms and is rapidly emerging in the United States and worldwide. Ingestion of raw or improperly cooked meats may be the source of infection with most of these parasites.

Helminths have very complex life cycles and in most instances humans are accidental hosts. However, in other cases the adult worms can find a home in the human body, start reproducing and shed eggs in the environment. These eggs could infect other susceptible individuals.

Adult nematodes, present as male and female worms, can be found in the gut of the infected individual. In contrast, cestodes and trematodes can be hermaphrodites and self fertilize their eggs; therefore, most of them do not require a male and female worm to complete their life cycle.

Some infections with helminths have been associated with immigrant populations that initiate an outbreak or start developing clinical signs infrequently observed by medical personnel. An example of this is cysticercosis. Outbreaks in the United States were described in an ethnic group that did not consume pork, and further investigation showed that infection was initiated by a foreign housekeeper who had teniasis.

Trematoda can also be acquired by ingestion of raw or undercooked fish. Although freezing can inactivate the cysts, the larval stages can remain viable at refrigeration temperatures. Many consumers prefer fresh fish due to the detrimental textural changes in the fish after freezing.

Ynes R. Ortega

Contributors

Ann M. Adams, US Food and Drug Administration, Kansas City District, Lenexa, KS 66214

Natalie Bowman, Columbia College of Physicians and Surgeons, Columbia University, New York, NY 10032

Vitaliano Cama, Department of International Health, Bloomberg School of Public Health, Johns Hopkins University, Baltimore, MD 21205

George D. Di Giovanni, Agricultural Research and Extension Center, Texas A&M University, El Paso, TX 79927

Joseph Donroe, Tufts University School of Medicine, Tufts University School of Public Health, Boston, MA 02111

Robert Gilman, Department of International Health, Bloomberg School of Public Health, Johns Hopkins University, Baltimore, MD 21205

Kristina D. Mena, School of Public Health, University of Texas Health Science Center, El Paso, Texas 79902

Ynes R. Ortega, Center for Food Safety, Department of Food Science and Technology, University of Georgia, Griffin, GA 30223

Charles R. Sterling, Department of Veterinary Science and Microbiology, University of Arizona, Tucson, AZ 85721

Gregory D. Sturbaum, CH Diagnostic and Consulting Service, Inc. Loveland, CO 80537

Irshad M. Sulaiman, Biotechnology Core Facility Branch, Scientific Resources Program, National Center for Infectious Diseases, Centers for Disease Control and Prevention, Atlanta, GA 30333

Lihua Xiao, Division of Parasitic Diseases, Centers for Disease Control and Prevention, Atlanta, GA 30341

Contents

Chapter 1. **Amoeba and Ciliates**
Ynes R. Ortega 1

1.1 Preface 1
1.2 Amoeba 1
    1.2.1 *Entamoeba histolytica* 2
1.3 *Dientamoeba fragilis* 6
    1.3.1 Morphology and Transmission 6
    1.3.2 Therapy 7
1.4 Nonpathogenic Amoeba 7
    1.4.1 *Entamoeba hartmanni* 7
    1.4.2 *Entamoeba coli* 7
    1.4.3 *Endolimax nana* 7
    1.4.4 *Iodamoeba butschlii* 7
1.5 Free-Living Amoebae 8
1.6 Ciliates 8
    1.6.1 Life Cycle and Morphology 9
    1.6.2 Clinical Significance 9
    1.6.3 Diagnosis and Treatment 11
    1.6.4 Epidemiology and Prevention 11
    References 11

Chapter 2. **The Biology of *Giardia* Parasites**
Irshad M. Sulaiman and Vitaliano Cama 15

2.1 Preface 15
2.2 Biology 16
2.3 Detection and Classification of *Giardia* 17
    2.3.1 Detection Methods 17
    2.3.2 Classification of *G. intestinalis* 19
    2.3.3 Genotyping of *G. intestinalis* 20
2.4 Transmission and Epidemiology 23
    2.4.1 Human 23
    2.4.2 Environmental 24
2.5 Control and Treatment 25
    References 28

Chapter 3. **Coccidian Parasites**
Vitaliano Cama 33

3.1 Preface 33
3.2 Background/History 33

3.3	Biology	36
3.4	Clinical Significance	38
3.5	Transmission and Epidemiology	39
	3.5.1 *Cyclospora*	39
	3.5.2 *Isospora*	42
	3.5.3 *Sarcocystis*	42
3.6	Diagnosis	45
3.7	Treatment and Control	46
	References	47

Chapter 4. *Cryptosporidium* and Cryptosporidiosis
Lihua Xiao and Vitaliano Cama 57

4.1	Preface	57
4.2	Taxonomy	57
4.3	Life Cycle and Developmental Biology	60
4.4	Epidemiology and Transmission	61
	4.4.1 Cryptosporidiosis in Immunocompetent Persons	61
	4.4.2 Cryptosporidiosis in Immunocompromised Persons	62
	4.4.3 Transmission Routes and Infection Sources: Anthroponotic Versus Zoonotic Transmission	63
	4.4.4 Waterborne Transmission	64
	4.4.5 Foodborne Transmission	66
4.5	Detection and Diagnosis	69
	4.5.1 Serologic Methods	69
	4.5.2 Methods for Detection of *Cryptosporidium* in Stool Specimens	69
	4.5.3 Methods for Detection of *Cryptosporidium* Oocysts in Environmental Samples	77
4.6	Treatment	81
4.7	Control of *Cryptosporidium* Contamination in Water and Food	82
	References	86

Chapter 5. Toxoplasmosis
Ynes R. Ortega 109

5.1	Preface	109
5.2	Parasite Description	109
5.3	Life Cycle	110
5.4	Transmission	112
5.5	Identification	112
	5.5.1 Molecular Assays	113
	5.5.2 Riboprinting	114
5.6	Pathogenicity	115

	5.7	Epidemiology 115
		5.7.1 Humans 115
		5.7.2 Swine 118
		5.7.3 Poultry 120
		5.7.4 Sheep and Goats 121
		5.7.5 Other Animal Species 122
	5.8	Treatment 123
	5.9	Inactivation 124
		References 125

Chapter 6. Food-Borne Nematode Infections
Charles R. Sterling 135

- 6.1 Preface 135
- 6.2 *Trichinella* spp. 135
  - 6.2.1 Background 135
  - 6.2.2 Speciation 136
  - 6.2.3 Life Cycle 136
  - 6.2.4 Epidemiology 137
  - 6.2.5 Human Trichinellosis–Epidemiology 140
  - 6.2.6 Clinical Manifestations 142
  - 6.2.7 Diagnosis and Treatment 143
  - 6.2.8 Prevention and Control 143
- 6.3 *Anisakis simplex* and
  Related Species 145
  - 6.3.1 Background 145
  - 6.3.2 Life Cycle 146
  - 6.3.3 Epidemiology 146
  - 6.3.4 Clinical Manifestations 147
  - 6.3.5 Diagnosis and Treatment 147
  - 6.3.6 Prevention and Control 148
- 6.4 *Angiostrongylus cantonensis* and
  Angiostrongylus costaricensis 148
- 6.5 *Gnathostoma* spp. 150
- 6.6 *Gongylonema* spp. 151
- 6.7 Other Nematode Infections with Food-Borne
  Associations 152
  References 153

Chapter 7. Foodborne Trematodes
Ann M. Adams 161

- 7.1 Preface 161
- 7.2 *Paragonimus* spp. 161
  - 7.2.1 Introduction 161
  - 7.2.2 Life Cycle 163

		7.2.3 Epidemiology	165
		7.2.4 Clinical Signs, Diagnosis, and Treatment	166
	7.3	*Clonorchis sinensis*	168
		7.3.1 Introduction	168
		7.3.2 Life Cycle	169
		7.3.3 Epidemiology	170
		7.3.4 Clinical Signs, Diagnosis, and Treatment	171
	7.4	*Opisthorchis viverrini* and *Opisthorchis felineus*	174
		7.4.1 Introduction	174
		7.4.2 Life Cycle	174
		7.4.3 Epidemiology	175
		7.4.4 Clinical Signs, Diagnosis, and Treatment	176
	7.5	*Nanophyetus salmincola*	177
		7.5.1 Introduction	177
		7.5.2 Life Cycle	177
		7.5.3 Salmon Poisoning Disease	179
		7.5.4 Clinical Signs, Diagnosis, and Treatment	180
	7.6	*Fasciola* spp.	182
		7.6.1 Introduction	182
		7.6.2 Life Cycle	182
		7.6.3 Epidemiology	184
		7.6.4 Clinical Signs, Diagnosis, and Treatment	184
	7.7	*Fasciolopsis buski*	186
		7.7.1 Introduction	186
		7.7.2 Life Cycle	187
		7.7.3 Epidemiology	188
		7.7.4 Clinical Signs, Diagnosis, and Treatment	188
	7.8	Prevention and Control	189
		References	191

Chapter 8. **Cestodes**
Natalie Bowman, Joseph Donroe, and Robert Gilman 197

	8.1	Preface	197
	8.2	Taenia	197
		8.2.1 *Taenia solium*	198
		8.2.2 *Taenia saginata*	207
		8.2.3 *Taenia asiatica*	209
	8.3	Diphyllobothrium	211
		8.3.1 *Diphyllobothrium latum*	211
		8.3.2 *Diphyllobothrium pacificum*	214
	8.4	Spirometra	215
		8.4.1 *Spirometra mansoides*	215
	8.5	Echinococcus	216
		8.5.1 *Echinococcus granulosus*	216
		8.5.2 *Echinococcus multilocularis*	221

8.6 Hymenolepis 223
    8.6.1 *Hymenolepis nana.* 223
    References 225

Chapter 9. Waterborne Parasites and Diagnostic Tools
Gregory D. Sturbaum and George D. Di Giovanni 231

9.1 Preface 231
9.2 Parasites 232
9.3 Concentration and Isolation Techniques 235
9.4 Detection Methodologies 237
    9.4.1 Microscopic Techniques 237
    9.4.2 Nucleic Acid Techniques 239
    9.4.3 Immunological-based Techniques 248
    9.4.4 Viability Techniques 250
9.5 Proper Evaluation (QA/QC) 253
    References 255

Chapter 10. Risk Assessment of Parasites in Food
Kristina D. Mena 275

10.1 Preface 275
10.2 The Risk Assessment Framework 276
    10.2.1 Defining the Hazard 276
    10.2.2 Exposure Assessment 277
    10.2.3 Hazard Characterization (Dose-response Assessment) 278
    10.2.4 Risk Characterization 280
    10.2.5 Assumptions, Assumptions, Assumptions 281
    10.2.6 Emerging Applications of Microbial Risk Assessment 282
    References 282

Index 285

CHAPTER 1

Amoeba and Ciliates

Ynes R. Ortega

1.1 PREFACE

Amoeba and ciliates are two groups of protozoan parasites that have long been known to infect humans. The amoeba are unicellular organisms which are characterized by the pseudopodia, which are cytoplasmic protrusions that provide motility to the organism. Amoeba are commonly found in the environment and a few are pathogenic to mammals.

The ciliates are protozoa, unicellular organisms that use the cilia on their surface for high motility. Ciliates are commonly found in environmental waters. The only species pathogenic to humans is *Balantidium coli*, which is also found to infect pigs and nonhuman primates.

Amoeba and ciliates can be acquired either by ingestion of contaminated water or food or by contamination of products or surfaces by food handlers.

1.2 AMOEBA

This group of parasites belong to the phylum Sarcomastigophora, subphylum Sarcodina (Bruckner, 1992). The cyst and trophozoite are the two morphological stages of the amoeba. Some amoeba (commensal) can infect humans, but do not cause illness. The free-living amoeba are frequently found in the environment, particularly water sources, but under certain circumstances they can infect humans. Three of public health relevance are *Acantamoeba*, *Naegleria*, and *Balamuthia*. Other amoeba can be identified as infecting humans, but not necessarily causing illness. Some of these commensal amoeba include the genera *Entamoeba*, *Endolimax*, and *Iodamoeba*. *Blastocystis* has been traditionally considered an amoeba, but has now been reclassified as fungi. The amoeba in humans is most often found to infect the buccal cavity or the gastrointestinal tract. *Entamoeba gingivalis* is commonly associated with gingivitis and is localized in the soft tartar between the teeth and the oral mucosa. It does not have a cyst stage and transmission is considered to be person to person or by contact with buccal secretions.

Nonpathogenic amoeba (commensal) that colonize the intestinal tract include *Entamoeba dispar*, *Entamoeba hartmanni*, *Entamoeba moshkovskii*, *Entamoeba polecki*, *Endolimax nana*, *Entamoeba chattoni*, *Entamoeba invadens*, *Iodamoeba butschlii*, and *Entamoeba coli*.

Amoeba can be identified by observing the morphology of trophozoites or cysts. Trophozoites can be observed only in fresh specimens from an infected individual. The arrangement, size, and pattern of the nuclear chromatin aid in the identification of the various species. The size and position of the karyosome also aid in the speciation of amoeba. The cytoplasm of the trophozoites may contain red blood cells, bacteria, yeasts, and molds. The number and size of the nuclei in the cyst are taken into consideration when identifying the genera and species of amoeba. Chromatoidal bodies and vacuoles present in the cytoplasm also aid in the identification of the species. All of these characteristics are not easily noted in fresh preparations, requiring permanent stains of fecal smears to be prepared and examined at 1000× magnification. Mixed infections are very common; therefore, observation of several parasitic structures is necessary for a conclusive diagnosis (Leber and Novak, 2005).

The pathogenic amoeba for humans is *Entamoeba histolytica*. It was described by Fedor Losch in 1875 from a Russian patient with dysenteric stools (Lösch, 1875). *E. histolytica* has been recovered worldwide and is more prevalent in the tropics and subtropics than in cooler climates. In areas of temperate and colder climates, it can be found in unsanitary conditions.

The pathogenicity of *Entamoeba* has been controversial. In some instances, *E. histolytica* may cause invasive disease and extraintestinal amebiasis, and in other instances, it may cause mild or asymptomatic infections. The host immune status, strain variability, environmental conditions, and the intestinal flora composition are factors that may influence the clinical presentation of the disease. Axenic cultivation of the amoeba has facilitated the study of isoenzyme profiles including glucophosphate isomerase, phosphoglucomutase, malate dehydrogenase, and hexokinase in various isolates. Sargeaunt concluded that *Entamoeba* could be characterized based on their isoenzyme analysis and characterized in various zymodemes. Of the amoeba that infects humans, *E. dispar* (nonpathogenic) and *E. histolytica* (pathogenic) are not only genotypically different, but phenotypically distinct, although they are morphologically similar.

The life cycle of amoeba starts when the cyst, which is the infectious form, is acquired by ingestion of contaminated materials, such as food and water, or by direct fecal-oral transmission. Once in the intestinal tract, excystation occurs, trophozoites are released and propagate via asexual multiplication. Cyst formation occurs in the colon where conditions are unfavorable for the trophozoite. Cysts are excreted in the feces and can remain viable in the environment for up to several weeks if protected from environmental conditions (Garcia, 1999).

1.2.1 *Entamoeba histolytica*

Entamoeba histolytica has been described worldwide. In areas of endemnicity up to 50% of the population may be infected. It ranks second in worldwide causes of morbidity by parasitic infections (Laughlin and Temesvari, 2005). Humans are the primary reservoirs of this parasite; however, it has been described as infecting non-human primates. This transmission can occur via person to person or by ingestion of cysts present in contaminated food or water. The cysts excyst in the intestine and trophozoites are released and start dividing. Some will encyst and be excreted with the feces. In invasive amebiasis, trophozoites may penetrate the bowel and disseminate to the liver, lungs, brain, pericardium, and other tissues. Invasive amebiasis tends to affect men predominantly, but asymptomatic infection is equally distributed among both genders (cuna-Soto *et al.*, 2000). Immigrants from South and Central America and Southeast Asia are two groups with a high incidence of amebiasis.

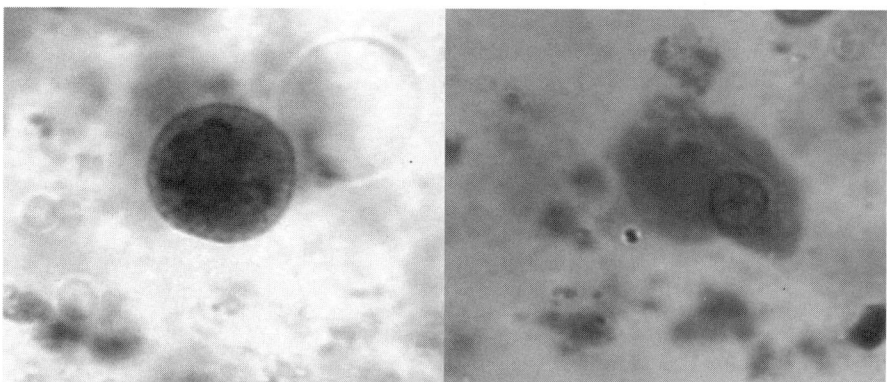

Figure 1.1. *Entamoeba histolytica* (left) cyst and (right) trophozoite.

Travelers are at high risk for acquiring the infection. In areas where *E. histolytica* and *E. dispar* are endemic, *E. histolytica* are more predominant in travelers and *E. dispar* are more predominant in residents. Amebiasis in homosexual males is frequently transmitted by sexual behavior. Asymptomatic presentation is up to 30% (Walderich *et al.*, 1997).

1.2.1.1 Morphology

Trophozoites are between 12 and 60 µm in diameter. The nucleus is characterized by evenly arranged chromatin on the nuclear membrane and a karyosome that is small, compact, and centrally located (Fig. 1.1). The cytoplasm is granular and has vacuoles containing bacteria or debris. In cases of dysentery, red blood cells may be present in the cytoplasm. Immature cysts are characterized by one to two nuclei, a glycogen mass, and chromatoidal bars with smooth round edges. Mature cysts have four nuclei, and the glycogen mass and the chromatoidal bars may disappear as the cyst matures. This process occurs while oocysts migrate in the intestine. The cyst measures 10–20 µm. Once the cyst is ingested by the new host, the gastric enzymes and neutral or alkaline pH in the intestine induce the trophozoites to become active, at which point they are liberated (Fig. 1.2).

1.2.1.2 Clinical Significance

The World Health Organization estimates 50 million infections and 100,000 deaths per year (Anonymous, 1997). The clinical presentation of *E. histolytica* can be asymptomatic, symptomatic without tissue invasion, and symptomatic with tissue invasion. Asymptomatic infection may be related to two genetically distinct invasive and noninvasive strains of *E. histolytica* (Zaki and Clark, 2001). Approximately 10% of infected individuals will have clinical symptoms such as dysentery, colitis, or in few instances, amebomas. Amebomas are localized granulomatous tissues with tumor-like lesions resulting from chronic ulceration. They may be mistaken for malignancy. Amoebic dysentery is characterized by diarrhea with cramping, lower abdominal pain, low fever, and the presence of blood and mucus in feces. Ulcers start at the surface of the epithelium that deepens into a classic flask-shaped ulcer.

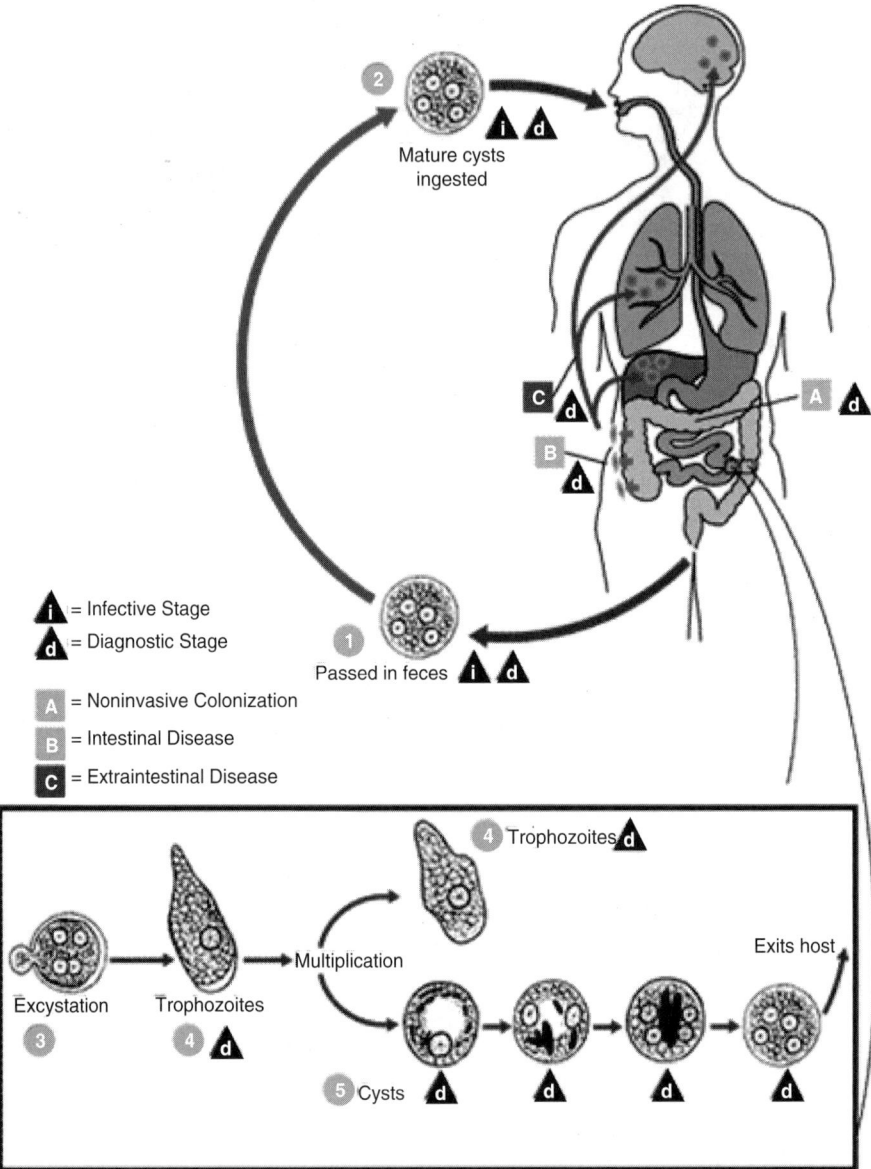

Figure 1.2. *Entamoeba histolytica* life cycle. Graph obtained from http://www.dpd.cdc.gov/dpdx/HTML/ImageLibrary

Abdominal perforation and peritonitis are rare, but can be serious complications. Amoebic colitis is characterized by intermittent diarrhea over a long period of time and can be misdiagnosed as ulcerative colitis or irritable bowel syndrome (Leber and Novak, 2005). Incubation period may vary from days to months.

If extraintestinal localization occurs, the liver is the most common site, but infection can also invade the lungs, pericardium, brain, etc. Symptoms may be acute or gradual and may include low-grade fever, pain in the right upper quadrant, and weight loss.

1.2.1.3 Pathogenesis and Immunity

Adhesins, amoebapores, and proteases have been associated with lysis of the colonic mucosa in intestinal amebiasis (Espinosa-Cantellano and Martinez-Palomo, 2000). *Entamoeba* has a cell surface protein that has a sensory activity and contributes to the surface adhesion of the trophozoite. The Gal/GalNAc lectin recognizes galactose and N-acetylgalactosamine found on the human colonic mucin glycoproteins. Interaction between this lectin and the host glycoproteins is required for adherence and contact-dependent cytolysis (Petri *et al.*, 1989). This lectin is unique in *E. histolytica* and has been used to develop the ELISA diagnostic assay produced by TechLab. The trophozoite moves forming the pseudopod in front, and the membrane moves to the uroid, which is a posterior foot. The amoeba collects surface antigens, including host antibodies, on the uroid. Membrane shedding is active at the uroid region, eliminating the accumulated ligands including antigens, Gal/GalNAc lectins, and the 96-kDa surface protein with the host antibodies. This process may contribute to the evasion of the host immune defenses. Amoeba with a defective cytoskeleton cannot form a cap or form uroids, and cannot cause cell cytolysis, suggesting that the cytoskeleton may play a role in contact-dependent cytolysis (Arhets *et al.*, 1998).

The cysteine proteinases are a major virulence factor. These proteinases can degrade elements of the extracellular matrix including fibronectin, laminin, and type I collagen (Keene *et al.*, 1986). These proteinases also interfere with the complement pathways and the humoral response of the human immune system. Gal/GalNAc lectin inhibits complement-mediated lysis because it mimics the CD59, a membrane inhibitor of C5b-9 in human blood cells. The proteinases can degrade and inactivate C3 and C5 to circumvent the host immune response, as well as degrade secretory IgG and IgA, limiting the host humoral immune response (Kelsall and Ravdin, 1993). The presence of IgA antilectin provides a marker of acquired immunity.

E. histolytica also secretes a pore-forming protein, the amoebapore containing three isoforms: A, B, and C. It works by inserting ion channels into artificial membranes and may be cytolytic to eukaryotic cells (Leippe *et al.*, 1994; Rosenberg *et al.*, 1989).

The mechanisms of host defense include production of mucin. The Gal/GalNac lectin binds to it. Whether it serves as defense or as an inducer for colonization needs to be determined (Petri *et al.*, 1989).

The inflammatory response provides another mechanism of defense. *In vitro* and *in vivo* studies demonstrated that the presence of trophozoites causes the expression of a variety of cytokines, including IL-1b and IL-8. This production occurred in regions other than those in direct contact with the parasite (Zhang *et al.*, 2000).

A subunit of the Gal/GalNAc lectin of 170 kDa induces production of IL-12 in human macrophages. The IL-12 promotes Th1 cytokine differentiation and, in turn, macrophage protection (Campbell *et al.*, 2000).

Diagnosis can be made by examination of fecal samples, material collected using a sigmoidoscope, tissue biopsy, and abscess aspirates. Serological testing can be used. Serum antibodies have been identified in 85% of patients with proven amebiasis (by histology) and in 99% of patients with extraintestinal amebiasis. Persons with *E. dispar* do not develop detectable levels of antibodies (Leber and Novak, 2005). Diagnosis is facilitated by the examination of permanently stained slides. Diagnostic assays specific for *E. histolytica* in clinical specimens are available on the market (TechLab, Blacksburg, VA) (Garcia *et al.*, 2000; Ong *et al.*, 1996; Pillai *et al.*, 1999). Zymodeme analysis has been used to differentiate between *E. histolytica* and *E. dispar*; which, although specific, is also expensive and time-consuming.

Molecular assays such as polymerase chain reaction (PCR) have been developed, but do not seem to be very sensitive when compared to conventional assays (Evangelopoulos *et al.*, 2000; Rivera *et al.*, 1996; Sanuki *et al.*, 1997; Zindrou *et al.*, 2001). Roy and collaborators compared a real-time PCR against the antigen detection tests and SS- rRNA and traditional PCR (72% sensitive and 99% specific). The real-time PCR was more sensitive (79% sensitive and 96% specific) than all the other assays and the specificity was higher by PCR. Using the TechLab antigen, detection kit detected only 49% of positive specimens (Roy *et al.*, 2005).

1.2.1.4 Therapy

If treating asymptomatic infection with cyst excretion, a luminal amoebicide such as iodoquinol or diloxanide furoate is recommended. If tissue invasion has occurred, tissue amoebicides such as metronidazole, chloroquine, or dehydroemetine are recommended. Follow-up stool examination is important, since these treatments may lead to drug resistance. The multidrug resistance gene *EhPgp1* is constitutively expressed in drug-resistant trophozoites (Ramirez *et al.*, 2005).

1.3 *DIENTAMOEBA FRAGILIS*

Dientamoeba, originally considered an amoeba, is now considered an amoeba-flagellate and is closely related to *Histomonas* and *Trichomonas* spp.

1.3.1 Morphology and Transmission

The trophozoite measures 5–15 μm and pseudopodia are angular. No flagella is present. The cytoplasm is highly granular and it is characterized as having one to two nuclei without peripheral chromatin and karyosome clusters of four to eight granules. Cysts have not been identified in *Dientamoeba fragilis*. This amoeba-flagellate does not have a cyst form and its transmission is less understood. However, transmission is suspected to be associated with helminth eggs such as *Acaris* and *Enterobius*. Higher incidences have been reported in mental institutions, missionaries, and Indians in Arizona. It has been reported in pediatric populations (Anonymous, 1993). Symptoms include fatigue, intermittent diarrhea, abdominal pain, anorexia, and nausea. It has been reported to cause noninvasive diarrheal illness. *Dientamoeba* colonizes the cecum and the proximal part of the colon. Reports of *Dientamoeba* are limited and this may be related to the difficulty in identifying the organisms. Asymptomatic cases of *D. fragilis* have been reported. This may be related to the description of two genetic variants using PCR-RFLP of the ribosomal genes (Johnson and Clark, 2000).

1.3.2 Therapy
Tetracycline or iodoquinol are recommended as the drug of choice for individuals with symptomatic infection. If co-infections include helminths such as *Enterobius*, mebendazole is usually included in the treatment (Butler, 1996).

1.4 NONPATHOGENIC AMOEBA

1.4.1 *Entamoeba hartmanni*
E. hartmanni is morphologically similar to *E. histolytica/E. dispar*. The trophozoite measures 5–12 μm and has one nucleus with a peripheral chromatin. The karyosome is small, compact, and centrally located. The cyst measures 5–10 μm. The mature cyst contains four nuclei. Chromatoidal bodies are like those of *E. histolytica*.

1.4.2 *Entamoeba coli*
It is commonly found in individuals in developing countries. It is characterized by having a cyst of 10–35 μm that may contain up to eight nuclei. Chromatoidal bars are splinter shaped and have rough pointed ends. The nuclei have distinctive characteristics, including the coarsely granular peripheral chromatin. The large karyosome is usually eccentric. The trophozoite measures between 15 and 50 μm and bacteria are usually present in the cytoplasm.

1.4.3 *Endolimax nana*
The trophozoite measures between 6 and 12 μm and has a granulated and vacuolar cytoplasm. The cyst measures between 5 and 10 μm. It is usually oval and when mature may have up to four nuclei. The nuclei have nonvisible peripheral chromatin and the karyosome is larger than the *Entamoeba*. Morphologically, it is very different from the *Entamoeba* species.

1.4.4 *Iodamoeba butschlii*
The trophozoite measures between 8 and 20 μm. The cytoplasm is granular and vacuolated. The cyst may be oval or round and measures between 5 and 20 μm. The mature cyst, contrary to the other amoeba, contains only one nucleus characterized by the absence of peripheral chromatin and a larger karyosome. It usually contains a large glycogen vacuole that stains brown when the sample is prepared using iodine. *Iodamoeba* can be easily differentiated from the other amoeba.

E. coli, E. nana, and *I. butschlii* can be easily differentiated from *E. histolytica* primarily by their size, followed by the nuclei characteristics and cytoplasmic inclusions.

1.5 FREE-LIVING AMOEBAE

Naegleria, *Acanthamoeba*, and *Balamuthia* have been identified in the central nervous system of humans and other animals. *Acanthamoeba* can also cause keratitis, and both *Acanthamoeba* and *Balamuthia madrillaris* may cause cutaneous infection in humans. *Naegleria fowleri* and *Acanthamoeba* spp. are commonly found in soil, water, sewage, and sludge. These amoebae feed on bacteria and multiply in the environment. They may harbor pathogenic bacteria to humans such as *Legionella*, *Mycobacterium avium*, *Listeria*, etc. Whether *Acanthamoeba* serves as a reservoir for human pathogens is unknown. Meningoencephalitis caused by *Naegleria* has been coined primary amebic meningoencephalitis. It is an acute and fulminant disease that can occur in previously healthy children and young adults who have been in contact with freshwater about 7–10 days prior to development of clinical signs. It is characterized by severe headache, spiking fever, stiff neck, photophobia, and coma, leading to death within 3–10 days after onset of symptoms. The amoeba find their way through the nostrils, to the olfactory lobes and cerebral cortex.

Acanthamoeba and *Balamuthia* encephalitis are found primarily in immunosuppressed individuals who have had exposure to recreational freshwater. Chronic granulomatous amoebic encephalitis (GAE) has an insidious onset and is usually chronic. The invasion and penetration to the central nervous system may be through the respiratory tract or the skin. These amoebae have been predominantly associated with waterborne transmission in recreational waters. Whether these amoebae are associated with foodborne transmission has not been determined.

N. fowleri is susceptible to amphotericin B alone or in combination with miconazole. Few patients infected with *Acanthamoeba* have survived when treated, but in most instances, patients with encephalitis have died. A successful recovery of a patient with GAE included surgery and treatment with sulfadiazine and fluconazole (Seijo *et al.*, 2000). Skin infections have a good prognosis and usually require topical treatment with 2% ketoconazole cream.

1.6 CILIATES

Ciliates are highly motile protozoa. They are characterized by cilia present on the surface. Free-living ciliates can be found in environmental waters. The only species pathogenic to humans is *B. coli*. It was initially identified in dysenteric stools of two patients and was later described by Leukhart in 1861 and Stein in 1862 (Diana, 2003). *Balantidium* can exist in reservoirs such as pigs and nonhuman primates. *B. coli* can be found in as many as 45% of pigs from intensive farming to 25% in wild boars (Solaymani-Mohammadi *et al.*, 2004; Weng *et al.*, 2005). In Denmark, 57% of suckling pigs had *Balantidium* (Hindsbo *et al.*, 2000); however, balantidiasis has not been reported in humans. In some regions of Venezuela, balantidiasis was observed to be 12% in humans and 33.3% in pigs. Nonhuman primates have also been reported carrying the infection in the tropics. Monkeys, chimpanzees, gibbons, macaques, and gorillas can harbor *Balantidium* (Nakauchi, 1999).

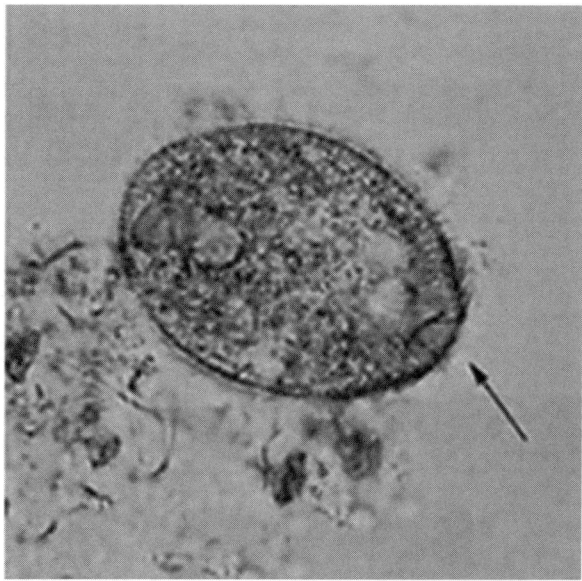

Figure 1.3. *Balantidium coli* trophozoite. Arrow points at the trophozoite cytostome. Picture obtained from http://www.dpd.cdc.gov/dpdx/HTML/ImageLibrary

Human infections can occur in warmer climates. Sporadic cases have been reported in cooler areas and in institutionalized groups with poor hygienic conditions. In the United States it is rarely found in clinical specimens. Deficient environmental sanitation favor dissemination of the infection (Devera *et al.*, 1999).

1.6.1 Life Cycle and Morphology
The trophozoite and the cyst are the only two stages of *B. coli*. The trophozoite is large and oval. It measures 50–100 μm long to 40–70 μm wide. The cyst measures 50–70 μm. The movement is rotary. The body is covered with longitudinal rows of cilia and they are longer near the cytostome. The trophozoite is pear shaped with an anterior end pointed and the posterior end broadly rounded. The cytoplasm contains vacuoles with ingested bacteria and cell debri. The trophozoite and cyst contain two nuclei: one large bean-shaped nucleus and a round micronucleus (Fig. 1.3). The cyst form is the infective stage. It has a thick cyst wall. Trophozoites secrete hyaluronidase, which aids in the invasion of the tissues. Cysts are formed as the trophozoite moves down the large intestine (Fig. 1.4).

1.6.2 Clinical Significance
Frequently, *Balantidium* infections can be asymptomatic; however, severe dysentery similar to those with amoebiasis may be present. Symptoms include diarrhea or dysentery, tenesmus, nausea, vomiting, anorexia, and headache. Insomnia, muscular weakness, and weight loss have also been reported. Diarrhea may persist for weeks or months prior to development of dysentery. Fluid loss is similar to that

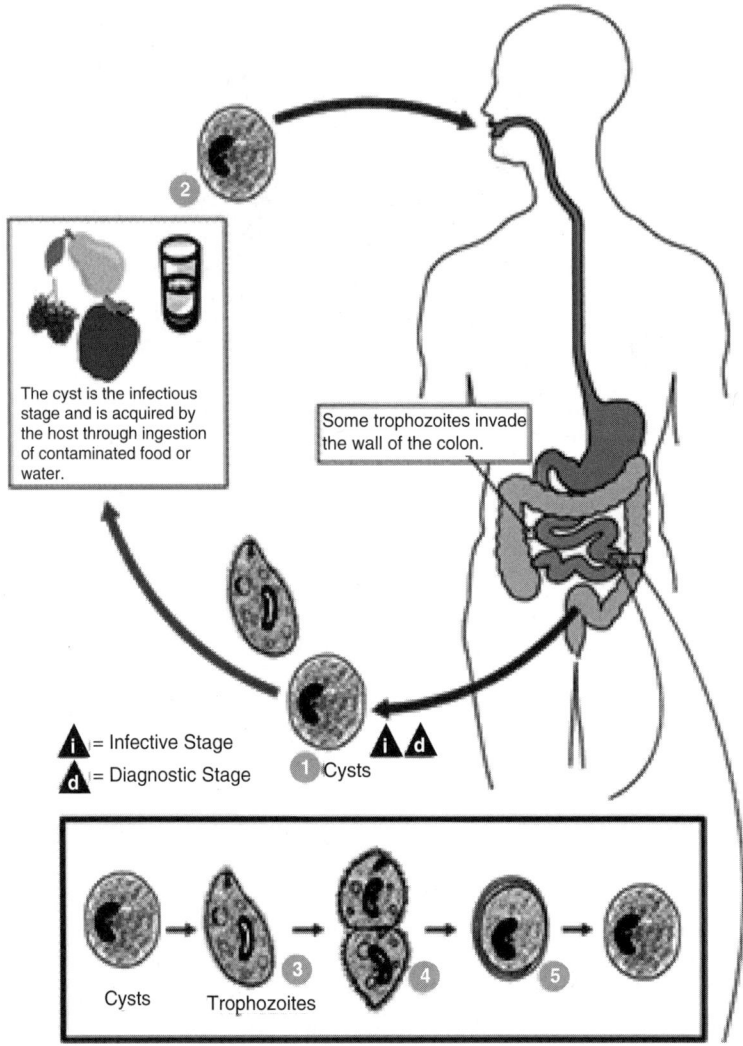

Figure 1.4. *Balantidium coli* life cycle diagram. Picture obtained from http://www.dpd.cdc.gov/dpdx/HTML/ImageLibrary

observed in cholera or cryptosporidiosis. Symptomatic infections can occur, resulting in bouts of dysentery similar to amebiasis. Colitis caused by *Balantidium* is often indistinguishable from *E. histolytica* (Castro *et al.*, 1983). Diarrhea, nausea, vomiting, headache, and anorexia are characteristic of balantidiasis.

The organism can invade the submucosa of the large bowel, causing ulcerative abscess and hemorrhagic lesions to occur. The shallow ulcers are prone to secondary infections by bacteria and can be problematic for the patient (Knight, 1978b). In few cases, extraintestinal disease such as peritonitis, urinary tract infection, and inflammatory vaginitis has been reported. *Balantidium* has been described in the urinary bladder of an infected individual (Knight, 1978a; Ladas *et al.*, 1989; Maleky, 1998). Pulmonary lesions can occur in immunocompromised patients without obvious contact with pigs, nor history of diarrhea prior to pulmonary infection (Anargyrou *et al.*, 2003). *Balantidium* pneumonia has been described in a 71-year-old woman suffering from anal cancer (Vasilakopoulou *et al.*, 2003). Chronic colitis and inflammatory polyposis of the rectum and sigmoid colon and an intrapulmonary mass have been described in a case with balantidiasis (Ladas *et al.*, 1989).

After ingestion, the trophozoite excretes hyaluronidase, which aids in the invasion of the tissue. On contact with mucosa, mucosal invasion is accomplished by cellular infiltration in the area of the developing ulcer. The organism can invade the submucosa of the large bowel. Ulcerative abscesses and hemorrhagic lesions can occur. Some of the abscess formations may extend to the muscular layer. The shallow ulcers and submucosal lesions are prone to secondary bacterial infection. Ulcers may vary in shape and the ulcer bed may be full of mucus and necrotic debris.

1.6.3 Diagnosis and Treatment

Wet preparation examinations of fresh and concentrated fecal material can determine the organism, as the shape and motility are characteristic of this ciliate. Tetracycline is the drug of choice, although it is considered an investigational drug for this infection. Iodoquinol or metronidazole may be used as alternative drugs.

1.6.4 Epidemiology and Prevention

Several studies have demonstrated the presence of *B. coli* in developing countries. Balantidiasis has been reported in 8% of children of the Bolivian Altiplano (Basset *et al.*, 1986).

Domestic hogs probably serve as the most important reservoir host for balantidiasis. In areas where pigs are the main domestic animal, the incidence of infection is high. Risk factors to acquire this infection include working at pig farms or slaughterhouses. Infection can turn into an epidemic if conditions favor propagation in the community. This has been observed in mental hospitals in the United States, where poor sanitary conditions are common. Preventive measures include increased attention to personal hygiene and sanitation measures, since the mechanisms of transmission are via contaminated water or foods with *Balantidium* cysts.

REFERENCES

Anargyrou, K., Petrikkos, G. L., Suller, M. T., Skiada, A., Siakantaris, M. P., Osuntoyinbo, R. T., Pangalis, G., and Vaiopoulos, G., 2003, Pulmonary *Balantidium coli* infection in a leukemic patient, *Am. J. Hematol.* **73**:180–183.

Anonymous, 1993, Intestinal protozoa: Flagellates and ciliates. In Garcia, L. S. and Bruckner, D. A. (eds), *Diagnositc Medical Parasitology*, Vol. 3, American Society for Microbiology, Washington, DC, pp. 31–44.

Anonymous, 1997, WHO/PAHO/UNESCO report. A consultation with experts on amoebiasis. Mexico City, Mexico 28–29 January, 1997, *Epidemiol. Bull.* **18**: 13–14.

Arhets, P., Olivo, J. C., Gounon, P., Sansonetti, P., and Guillen, N., 1998, Virulence and functions of myosin II are inhibited by overexpression of light meromyosin in *Entamoeba histolytica*, *Mol. Biol. Cell* **9**:1537–1547.

Basset, D., Gaumerais, H., and Basset-Pougnet, A., 1986, Intestinal parasitoses in children of an Indian community of Bolivian altiplano, *Bull. Soc. Pathol. Exot. Filiales.* **79**:237–246.

Bruckner, D. A., 1992, Amebiasis, *Clin. Microbiol. Rev.* **5**:356–369.

Butler, W. P., 1996, *Dientamoeba fragilis*. An unusual intestinal pathogen, *Dig. Dis. Sci.* **41**:1811–1813.

Campbell, D., Mann, B. J., and Chadee, K., 2000, A subunit vaccine candidate region of the *Entamoeba histolytica* galactose-adherence lectin promotes interleukin-12 gene transcription and protein production in human macrophages, *Eur. J. Immunol.* **30**:423–430.

Castro, J., Vazquez-Iglesias, J. L., and rnal-Monreal, F., 1983, Dysentery caused by *Balantidium coli*—report of two cases, *Endoscopy* **15**:272–274.

cuna-Soto, R., Maguire, J. H., and Wirth, D. F., 2000, Gender distribution in asymptomatic and invasive amebiasis, *Am. J. Gastroenterol.* **95**:1277–1283.

Devera, R., Requena, I., Velasquez, V., Castillo, H., Guevara, R., De, S. M., Marin, C., and Silva, M., 1999, Balantidiasis in a rural community from Bolivar State, Venezuela, *Bol. Chil. Parasitol.* **54**:7–12.

Diana, E., 2003, Intestinal parasitoses: A history of scientific progress and endemic disease in society. In Dionisio, D. (ed.), *Atlas of Intestinal Infections in AIDS*, Vol. 1, Springer-Verlag Italia, Milan, pp. 7–34.

Espinosa-Cantellano, M., and Martinez-Palomo, A., 2000, Pathogenesis of intestinal amebiasis: From molecules to disease, *Clin. Microbiol. Rev.* **13**:318–331.

Evangelopoulos, A., Spanakos, G., Patsoula, E., Vakalis, N., and Legakis, N., 2000, A nested, multiplex, PCR assay for the simultaneous detection and differentiation of *Entamoeba histolytica* and *Entamoeba dispar* in faeces, *Ann. Trop. Med. Parasitol.* **94**:233–240.

Garcia, L. S., 1999, Flagellates and ciliates, *Clin. Lab Med.* **19**:621–638, vii.

Garcia, L. S., Shimizu, R. Y., and Bernard, C. N., 2000, Detection of *Giardia lamblia*, *Entamoeba histolytica/Entamoeba dispar*, and *Cryptosporidium parvum* antigens in human fecal specimens using the triage parasite panel enzyme immunoassay, *J. Clin. Microbiol.* **38**:3337–3340.

Hindsbo, O., Nielsen, C. V., Andreassen, J., Willingham, A. L., Bendixen, M., Nielsen, M. A., and Nielsen, N. O., 2000, Age-dependent occurrence of the intestinal ciliate *Balantidium coli* in pigs at a Danish research farm, *Acta Vet. Scand.* **41**:79–83.

Johnson, J. A., and Clark, C. G., 2000, Cryptic genetic diversity in *Dientamoeba fragilis*, *J. Clin. Microbiol.* **38**:4653–4654.

Keene, W. E., Petitt, M. G., Allen, S., and McKerrow, J. H., 1986, The major neutral proteinase of *Entamoeba histolytica*, *J. Exp. Med.* **163**:536–549.

Kelsall, B. L., and Ravdin, J. I., 1993, Degradation of human IgA by *Entamoeba histolytica*, *J. Infect. Dis.* **168**:1319–1322.

Knight, R., 1978a, Giardiasis, isosporiasis, and balantidiasis, *Clin. Gastroenterol.* **7**:31–47.

Knight, R., 1978b, Giardiasis, isosporiasis, and balantidiasis, *Clin. Gastroenterol.* **7**:31–47.

Ladas, S. D., Savva, S., Frydas, A., Kaloviduris, A., Hatzioannou, J., and Raptis, S., 1989, Invasive balantidiasis presented as chronic colitis and lung involvement, *Dig. Dis. Sci.* **34**:1621–1623.

Laughlin, R. C., and Temesvari, L. A., 2005, Cellular and molecular mechanisms that underlie *Entamoeba histolytica* pathogenesis: Prospects for intervention, *Expert. Rev. Mol. Med.* **7**:1–19.

Leber, A. L., and Novak, S. M., 2005, Intestinal and urogenital parasites. In Murray, P. R., Baron, E. J., Jorgensen, J. H., Pfaller, M. A., and Yolken, R. H. (eds), *Manual of Clinical Microbiology*, Vol. 133, ASM Press, Washington, pp. 1990–2007.

Leippe, M., Andra, J., Nickel, R., Tannich, E., and Muller-Eberhard, H. J., 1994, Amoebapores, a family of membranolytic peptides from cytoplasmic granules of *Entamoeba histolytica*: Isolation, primary structure, and pore formation in bacterial cytoplasmic membranes, *Mol. Microbiol.* **14**:895–904.

Lösch, F., 1875, Massenhafte entwicklung von amobën im dickdarm. *Archiv für pathologische anatomie und physiologie und für klinsche medicin, von Rudolf Virchow* **65**:196–211.

Maleky, F., 1998, Case report of *Balantidium coli* in human from south of Tehran, Iran, *Indian J. Med. Sci.* **52**:201–202.

Nakauchi, K., 1999, The prevalence of *Balantidium coli* infection in fifty-six mammalian species, *J. Vet. Med. Sci.* **61**:63–65.

Ong, S. J., Cheng, M. Y., Liu, K. H., and Horng, C. B., 1996, Use of the ProSpecT microplate enzyme immunoassay for the detection of pathogenic and non-pathogenic *Entamoeba histolytica* in faecal specimens, *Trans. R. Soc. Trop. Med. Hyg.* **90**:248–249.

Petri, W. A., Jr., Chapman, M. D., Snodgrass, T., Mann, B. J., Broman, J., and Ravdin, J. I., 1989b, Subunit structure of the galactose and N-acetyl-D-galactosamine-inhibitable adherence lectin of *Entamoeba histolytica*, *J. Biol. Chem.* **264**:3007–3012.

Pillai, D. R., Keystone, J. S., Sheppard, D. C., MacLean, J. D., MacPherson, D. W., and Kain, K. C., 1999, *Entamoeba histolytica* and *Entamoeba dispar*: Epidemiology and comparison of diagnostic methods in a setting of nonendemicity, *Clin. Infect. Dis.* **29**:1315–1318.

Ramirez, M. E., Perez, D. G., Nader, E., and Gomez, C., 2005, *Entamoeba histolytica*: Functional characterization of the -234 to -196 bp promoter region of the multidrug resistance *EhPgp1* gene, *Exp. Parasitol.* **110**:238–243.

Rivera, W. L., Tachibana, H., Silva-Tahat, M. R., Uemura, H., and Kanbara, H., 1996, Differentiation of *Entamoeba histolytica* and *E. dispar* DNA from cysts present in stool specimens by polymerase chain reaction: Its field application in the Philippines, *Parasitol. Res.* **82**:585–589.

Rosenberg, I., Bach, D., Loew, L. M., and Gitler, C., 1989, Isolation, characterization and partial purification of a transferable membrane channel (amoebapore) produced by *Entamoeba histolytica*, *Mol. Biochem. Parasitol.* **33**:237–247.

Roy, S., Kabir, M., Mondal, D., Ali, I. K., Petri, W. A., Jr., and Haque, R., 2005, Real-time-PCR assay for diagnosis of *Entamoeba histolytica* infection, *J. Clin. Microbiol.* **43**:2168–2172.

Sanuki, J., Asai, T., Okuzawa, E., Kobayashi, S., and Takeuchi, T., 1997, Identification of *Entamoeba histolytica* and *E. dispar* cysts in stool by polymerase chain reaction, *Parasitol. Res.* **83**:96–98.

Seijo, M. M., Gonzalez-Mediero, G., Santiago, P., Rodriguez De, L. A., Diz, J., Conde, C., and Visvesvara, G. S., 2000, Granulomatous amebic encephalitis in a patient with AIDS: Isolation of *Acanthamoeba* sp. Group II from brain tissue and successful treatment with sulfadiazine and fluconazole, *J. Clin. Microbiol.* **38**:3892–3895.

Solaymani-Mohammadi, S., Rezaian, M., Hooshyar, H., Mowlavi, G. R., Babaei, Z., and Anwar, M. A., 2004, Intestinal protozoa in wild boars (Sus scrofa) in western Iran, *J. Wildl. Dis.* **40**:801–803.

Vasilakopoulou, A., Dimarongona, K., Samakovli, A., Papadimitris, K., and Avlami, A., 2003, *Balantidium coli* pneumonia in an immunocompromised patient, *Scand. J. Infect. Dis.* **35**:144–146.

Walderich, B., Weber, A., and Knobloch, J., 1997, Differentiation of *Entamoeba histolytica* and *Entamoeba dispar* from German travelers and residents of endemic areas, *Am. J. Trop. Med. Hyg.* **57**:70–74.

Weng, Y. B., Hu, Y. J., Li, Y., Li, B. S., Lin, R. Q., Xie, D. H., Gasser, R. B., and Zhu, X. Q., 2005, Survey of intestinal parasites in pigs from intensive farms in Guangdong Province, People's Republic of China, *Vet. Parasitol.* **127**:333–336.

Zaki, M., and Clark, C. G., 2001, Isolation and characterization of polymorphic DNA from *Entamoeba histolytica*, *J. Clin. Microbiol.* **39**:897–905.

Zhang, Z., Wang, L., Seydel, K. B., Li, E., Ankri, S., Mirelman, D., and Stanley, S. L., Jr., 2000, *Entamoeba histolytica* cysteine proteinases with interleukin-1 beta converting enzyme (ICE) activity cause intestinal inflammation and tissue damage in amoëbiasis, *Mol. Microbiol.* **37**:542–548.

Zindrou, S., Orozco, E., Linder, E., Tellez, A., and Bjorkman, A., 2001, Specific detection of *Entamoeba histolytica* DNA by hemolysin gene targeted PCR, *Acta Trop.* **78**:117–125.

CHAPTER 2

The Biology of *Giardia* Parasites

Irshad M. Sulaiman and Vitaliano Cama

2.1 PREFACE

Giardia intestinalis (syn. *G. lamblia* or *G. duodenalis*) is one of the ten major enteric parasites affecting humans worldwide. It is also considered the most common intestinal pathogenic protozoa of humans and is recognized as a recurrent parasite of other nonhuman species, including cattle, beavers, and domestic dogs. Even though *Giardia* was first observed by Van Leeuwenhoek in 1681, in the past it was debated whether *Giardia* was a pathogen. However, now it is accepted that *Giardia* can cause intestinal disease in humans and in a wide range of domestic and wild mammals. The clinical presentations of giardiasis range from an asymptomatic cyst excreting state to diarrhea, which can be acute, chronic, or intermittent (Karanis *et al.*, 1996; Kirkpatrick and Benson, 1987; Monzingo and Hibler, 1987; Nizeyi *et al.*, 1999; Olson *et al.*, 1995; Pacha *et al.*, 1985; Patton and Rabinowitz, 1994; Rickard *et al.*, 1999; Sulaiman *et al.*, 2003; Thompson *et al.*, 2000; Wallis *et al.*, 1984; Xiao, 1994). Recently, *Giardia* infections have been associated with growth faltering due to nutrient malabsorption (Berkman *et al.*, 2002).

Giardia intestinalis is a parasite of public health importance as it can be transmitted through several routes, including water (drinking as well as recreational) and fresh food products (Nichols, 2000). Since foodborne outbreaks occur more frequently on a smaller scale than waterborne outbreaks, they are identified less frequently. Nonetheless, there are well-documented outbreaks implicating *G. intestinalis* as the causative agent (Anonymous, 1989; Mintz *et al.*, 1993).

In developed countries, *Giardia* is currently referred as a reemerging infectious agent because of its increasing role in outbreaks of diarrhea in day-care centers, and water and foodborne outbreaks affecting the general population. However, in developing countries located in Asia, Africa, and Latin America, approximately 200 million people per year experience symptomatic giardiasis (Thompson *et al.*, 2000).

There are multiple recognized species of the genus *Giardia*, although only *G. intestinalis* is found to be pathogenic to humans. Recent genetic studies have identified distinct groups within this species (Andrews *et al.*, 1989; Meloni *et al.*, 1995), and several researchers now regard *G. intestinalis* as a species-complex (Andrews *et al.*, 1989; Ey *et al.*, 1997; Monis *et al.*, 1998) (Table 2.1). Furthermore, the zoonotic potential of some animal isolates of *G. intestinalis* has recently been suggested (Sulaiman *et al.*, 2003). These findings plus their impact on the public health emphasize the importance to understand the biology, epidemiology, transmission, control, and treatment of *G. intestinalis* parasites.

Table 2.1. *Giardia intestinalis* assemblages and host range of isolates.

Assemblage (Genotype)	Host range	Reference
A	Human, cat, dog, calf, horse, pig, deer, lemur, beaver, slow loris, guinea pig	Homan *et al.*, 1992; Maryhofer *et al.*, 1995; Meloni *et al.*, 1995; Monis *et al.*, 1996; Ey *et al.*, 1997; Karanis and Ey 1998; Monis *et al.*, 1999; Trout *et al.*, 2003; Sulaiman *et al.*, 2003
B	Human, dog, monkey, beaver, muskrat, chinchilla, guinea pig, rabbit	Homan *et al.*, 1992; Maryhofer *et al.*, 1995; Meloni *et al.*, 1995; Ey *et al.*, 1997; Monis *et al.*, 1999; Sulaiman *et al.*, 2003
C, D	Dog	Maryhofer *et al.*, 1995; Hopkins *et al.*,1997; Monis *et al.*, 1998; Monis *et al.*, 1999; Sulaiman *et al.*, 2003
E	Cow, goat, sheep, pig	Ey *et al.*, 1997; Sulaiman *et al.*, 2003
F	Cat	Maryhofer *et al.*, 1995; Meloni *et al.*, 1995; Hopkins *et al.*, 1999; Monis *et al.*, 1999
G	Rat	Monis *et al.*, 1999; Sulaiman *et al.*, 2003

2.2 BIOLOGY

Giardia species are flagellated unicellular enteric protozoan parasites, inhabiting the intestinal tracts of almost every group of vertebrates, causing giardiasis. Some of the infections can be asymptomatic, although diarrhea and other discomforts are commonly observed in a large number of infected animals, children, and adults. The incubation or pre-patent period is longer than bacterial or viral infections, usually lasting from 12 to 19 days (Jokipii *et al.*, 1985).

The vegetative trophozoites and the environmentally resistant cysts are the two major stages in the life cycle of *Giardia*. Infection occurs when viable cysts are ingested by a susceptible host either through contaminated water, food, or contact with contaminated materials. Within a few hours, excystation or cyst opening occurs in the proximal part of the intestine where two newly formed trophozoites are released and proceed to infect intestinal cells. The parasite adheres and multiplies on the luminal lining of the small intestine, leading to diarrhea which may interfere with nutrient absorption. After multiplication, the trophozoites pass to the terminal region of the intestine and form new cysts that are excreted in stools. The cysts are released in variable quantities, and can survive well in the environment, allowing the parasite to reach other susceptible hosts and cause new infections.

Giardia is regarded as one of the most primitive eukaryotes in existence, and is considered to be the missing evolutionary link between eukaryotes and prokaryotes by a number of evolutionary biologists (Nasmuth, 1966; Sogin, 1991). As a typical eukaryotic organism, *Giardia* parasites contain distinct nuclei with a nuclear membrane and cytoskeleton. However, some other organelles commonly present in eukaryotes, such as the nucleoli, peroxisomes, and mitochondria are absent (Adam, 2001). Additionally, *Giardia* does not reproduce sexually. These characteristics, plus the lack of fossil records and similarities in microscopic characteristics (including morphology and ultrastructural features), make it difficult to define a species structure of *Giardia*, which can truly reflect its biologic characteristics and evolutionary relationships (Barta, 1997; Monis, 1999).

The taxonomy of *Giardia* at the species level was confused and complicated through the first half of the twentieth century. Historically, 41 species of *Giardia* with uncertain validity can be found in the literature. These reports istakenly named *Giardia* species based on the host of origin (Campbell *et al.*, 1990; van Keulen *et al.*, 1993).

At present, only six species of the genus *Giardia* are considered valid. Filice (1952) classified *Giardia* using morphological characteristics and recognized three distinct species: *G. duodenlalis* infecting a wide range of mammals, including humans, livestock, and companion animals; *G. agilis* in amphibians; and *G. muris* in rodents. In 1987, the species *G. ardeae* and *G. psittaci* were described in birds (Erlandsen and Bemrick, 1987; Erlandsen *et al.*, 1990), and shortly thereafter *G. microti* was described in muskrats and voles. This last species of *Giardia* was identified on the basis of cyst morphology and rRNA nucleotide sequence analysis (Feely, 1988; van Keulen *et al.*, 1998). Thus, *G. microti* was the first species of *Giardia* to be determined using both morphological and molecular features. This species was further validated by the characterization of nucleotide sequences of the triose phosphate isomerase (TPI) gene isolated from muskrats (Sulaiman *et al.*, 2003).

It has been recently postulated that the taxonomy of *Giardia* at the species level is still unresolved, as there are biological and molecular differences between isolates within a species; although their cysts or trophozoites morphologically identical. This is further supported by the observations of significant diversity among *G. intestinalis* isolates in host-infectivity assays (Visvesvara *et al.*, 1988), metabolism (Hall *et al.*, 1992), and *in vitro* and *in vivo* growth requirement (Andrews *et al.*, 1992; Binz *et al.*, 1992).

2.3 DETECTION AND CLASSIFICATION OF *GIARDIA*

2.3.1 Detection Methods

The ability to accurately identify the sources of foodborne or waterborne human infections requires efficient detection methodologies. In clinical laboratories, microscopy is usually the first choice for the diagnosis of parasites. *G. intestinalis* can be detected by trained microscopists while performing routine ova and parasite examination of human clinical specimens, or stool samples of domestic animals. The cysts of *G. intestinalis* are oval shaped, and measure approximately 8–12 μm long and 7–10 μm wide.

The trophozoite has unique morphological characteristics that are easily visualized. It is shaped like a split pear, rounded on the anterior end and pointed posteriorly, convex on its dorsal side and the ventral region resembles a concave disk. Two nuclei are symmetrically located on each side of the midline of the trophozoite, giving the appearance of a face. Four pairs of flagella originate from the midline between the nuclei and are directed backward. The trophozoites of *G. intestinalis* are very fragile

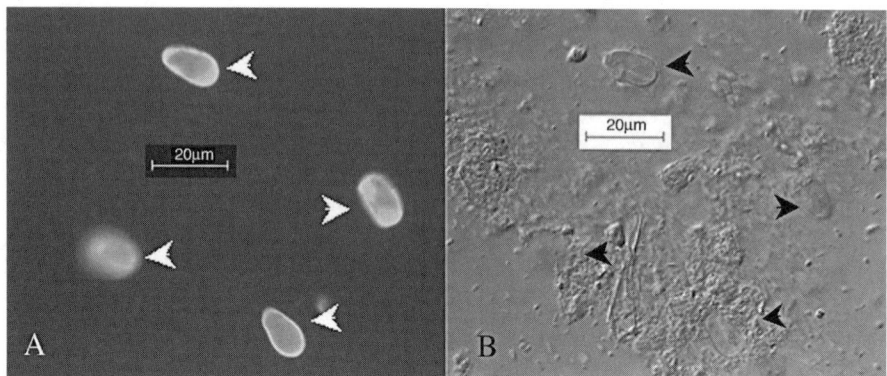

Figure 2.1. *Giardia intestinalis*, 400×magnification. A. Immunofluorescent microscopy. B. Nomarski microscopy.

and highly susceptible to changes in osmotic pressures, thus can be easily destroyed during sample processing. Whenever this stage is sought, microscopic examination of direct wet mount slides is the method of choice.

For the clinica diagnosis of *Giardia*, it is recommended to examine the stool specimens obtained from three consecutive days. Otherwise, diagnostic results cannot be assured. Despite these guidelines, it is estimated that about 30% of infections still remain undetected (Wolfe, 1990). An advantage of direct microscopy is its simplicity, low cost, and high specificity at the genus level. Since a positive microscopy result is usually correct, negative results, however, may require additional testing, especially if the gastrointestinal symptoms persist and no other etiologies are identified. The detection of *G. intestinalis* has been significantly improved with the development of immunofluorescent antibody microscopy (IFA) and enzyme immunoassays (EIA). The IFA method uses fluorescent-labeled antibodies that react with *Giardia* cysts, enhancing and simplifying the visualization of the parasite (Fig. 2.1). This method is of significant help for the untrained clinical microscopist, although its contribution has been more significant in the detection of parasites in environmental and food samples.

Clinical laboratories use either microscopy or EIA methods for detecting *G. intestinalis* in human specimens. Regardless of the diagnostic approach, it is imperative to test at least three stool samples per patient. When comparing these methods, neither is substantially better than the other. Light microscopy does not require any special reagents, is simple, and has a very high specificity. Meanwhile, EIA methods are better suited for processing large batches of samples, some of which can be quantitative and can be automated. In these settings, EIAs are more widely used, especially under the format of enzyme-linked immunosorbent assays (ELISA) that can detect soluble *Giardia* antigens in stools (Ungar *et al.*, 1984).

In contrast to the clinical specimens, the microscopic detection of *Giardia* cysts in source or finished water and in fresh produce is still a significant challenge. Currently, testing for the occurrence of *Giardia* in surface water to be used by drinking water treatment plants is mandatory under the 1996 amendment of the *Safe*

Drinking Water Act. To meet this mandate, in 1999 the US Environmental Protection Agency (EPA) implemented a validated procedure that involved immunomagnetic recovery of *Giardia* cysts followed by IFA microscopy. The test is known as US EPA Method 1623 for the detection of *Giardia* and *Cryptosporidium*. As a consequence, there has been an increase in the availability of validated tests for the detection of *Giardia* cysts and *Cryptosporidium* oocysts. While these tests were developed for water testing, several of these methods have been adapted for detecting both parasites in food matrices.

In summary, microscopy and ELISA/EIA are the methods most commonly used in clinical laboratories. Meanwhile, immunoassays and IFA are most frequently used for detecting *Giardia* in environmental and food samples.

2.3.2 Classification of *G. intestinalis*

Serology or typing based on immunological reactivity has been commonly used for typing the bacteria. This method, however, has demonstrated poor diagnostic differentiation for most parasites, including the *G. intestinalis*.

Zymodeme analysis, more accurately described as multilocus enzyme electrophoresis, identifies differences in enzymes that are the result of amino acid substitutions. Panels of 15–20 different enzymes are tested to identify differences in the charges of the native enzymes and can usually yield at least one with a difference. Nonetheless, this method is limited to selected research laboratories since it requires a large number of organisms.

In recent years, differences in nucleotide sequences have been more widely used to understand the diversity of an organism. One alternative is by identification of specific gene sequences that can be visualized in the form of restriction fragment length polymorphisms (RFLPs). This approach has been used for classification of bacteria and has proven very useful with some protozoa. One of the methods for detecting RFLPs is the digestion of DNA by an infrequently digesting restriction enzyme, followed by electrophoretic separation of the digested fragments in agarose gels, resulting in patterns that can be used for genotyping.

Pulsed field gel electrophoresis (PFGE) has been used for epidemiologic investigation of outbreaks of bacterial infections (Lipuma, 1998; Maslow *et al.*, 1993). However, PFGE requires relatively large amounts of DNA, and thus it has limited application in the typing of noncultured protozoa. Nonetheless, PFGE without enzyme digestion has been successfully used with *G. intestinalis* (Adam *et al.*, 1988) and *Cryptosporidium parvum* (Blunt *et al.*, 1997; Caccio *et al.*, 1998; Hays *et al.*, 1995; Mead *et al.*, 1988). Despite the vast amounts of DNA required, its results do not always correlate well with genotypes.

Recently, a polymerase chain reaction (PCR) followed by RFLP was used for the detection and subsequent genotype determination of *Giardia*. A significant advantage is that small amounts of *Giardia* DNA are required, while the PCR-amplified products can be analyzed using RFLP (Caccio *et al.*, 2002). Another DNA amplification method, such as random amplified polymorphic DNA, also requires little DNA and does not require prior knowledge of DNA sequence.

Most molecular methods based on PCR require a very little amount of DNA and can be quantitative in their comparisons. During the past few years, the cost of these methods, including sequencing, have drastically decreased while the quality has improved, making this a highly attractive method of classification and even detection. As with all DNA-based methods, a potential limitation is finding adequate gene targets that are easily amplifiable by PCR. Additionally, these PCR products should have sequence polymorphisms that can be utilized to differentiate closely related organisms, but most importantly, those with human or animal infectious potential.

To date, the isolates of *G. intestinalis* have been classified into two major genotypes or assemblages, A and B (Adam, 2000). Assemblage A comprises groups 1 (genotype A-1) and 2 (genotype A-2) previously described by Nash. These two groups can be differentiated readily by molecular techniques (Baruch *et al.*, 1996; Lu *et al.*, 1998). Genotype B (Nash group 3) is very different from genotype A, having 19% nucleotide divergence for the TPI gene (Lu *et al.*, 1998) and 13% difference for the ADP-ribosylating factor gene (Murtagh *et al.*, 1992).

2.3.3 Genotyping of *G. intestinalis*

The use of molecular tools not only confirms the validity of a species, but can identify specific subpopulations, also described as genotypes or assemblages. Recent molecular studies have confirmed distinct genotypes within *G. intestinalis* (Homan *et al.*, 1992; Maryhofer *et al.*, 1995; Monis *et al.*, 1999; Nash and Mowatt, 1992; Sulaiman *et al.*, 2003; Thompson *et al.*, 2000) (Table 2.1), and two major groups of human-pathogenic *G. intestinalis* have been recognized worldwide (Homan *et al.*, 1992; Maryhofer *et al.*, 1995; Nash *et al.*, 1985; Nash and Mowatt, 1992). Furthermore, various phylogenetic studies based on the characterization of nucleotide sequences of glutamate dehydrogenase (GDH), elongation factor 1α (EF1α), SSU rRNA, and TPI genes have demonstrated the existence of five to seven defined lineages of *G. intestinalis* (van Keulen *et al.*, 1991; Mowatt *et al.*, 1994; Monis *et al.*, 1996, Monis *et al.*, 1999; Sulaiman *et al.*, 2003).

Among the genes analyzed thus far for *Giardia* isolates from various hosts, the highest degree of polymorphism was observed at the TPI locus in a study conducted on 4 human isolates, 2 mice isolates, and 1 isolate each from cats, dogs, pigs, rats, and blue herons (Monis *et al.*, 1999). Later, a new TPI-based nested-PCR protocol was developed to amplify the TPI fragment from various *Giardia* isolates using primers complementary to the conserved published TPI nucleotide sequences of various *Giardia* parasites downloaded from the GenBank (Sulaiman *et al.*, 2003). Using this new nested protocol, a larger sample including 37 human isolates, 15 dog isolates, 8 muskrat isolates, 7 isolates each from cattle and beavers, and 1 isolate each from a rat and a rabbit (Sulaiman *et al.*, 2003) revealed that phylogenetic differences at the TPI gene was largely in agreement with all the previous results based on other genes (Monis *et al.*, 1996, 1999; Mowatt *et al.*, 1994; van Keulen *et al.*, 1991). This TPI-based phylogenetic study revealed the following groupings of *G. intestinalis* parasites: (i) a group containing only human isolates (assemblage A); (ii) a major group containing human muskrat, beavers, and rabbit isolates (assemblage B); (iii) a group containing isolates from cattle and pigs (assemblage E or the Hoofed livestock genotype); (iv) a group containing isolates from dogs (assemblage C); (v) a cat genotype; and (vi) a genotype from rats (Monis *et al.*, 1999; Sulaiman *et al.*, 2003).

Table 2.2. Distribution of assemblage A and B in humans reported in some previous studies.

Location	Number of positive samples examined	Assemblage of samples (%)			Reference
		A*	B**	Mixed	
Australia	13	100	0	0	Hopkins et al., 1997
Australia	11	36	64	0	Andrews et al., 1998
Netherlands	24	50	50	0	Homan et al., 1998
Germany	12	92	8	0	Karanis and Ey, 1998
China	3	0	67	33	Lu et al., 1998
Australia	4	50	50	0	Monis et al., 1999
China	8	50	50	0	Yong et al., 2000
Korea	7	100	0	0	Yong et al., 2000
The Netherlands	18	50	50	0	Homan and Mank, 2001
China, Cambodia, Australia, USA	9	56	44	0	Lu et al., 2002
United Kingdom	33	27	64	9	Amar et al., 2002
Italy	30	80	20	0	Caccio et al., 2002
Mexico	22	100	0	0	Ponce-Macotela et al., 2002
Mexico	26	100	0	0	Cedillo-Rivera et al., 2003
New Zealand	5	100	0	0	Learmonth et al., 2003
India	10	0	100	0	Sulaiman et al., 2003
Peru	25	24	76	0	Sulaiman et al., 2003
United States	2	0	100	0	Sulaiman et al., 2003

*(group I, II or Polish),
**(group III, IV or Belgian)

Remarkably, these findings have demonstrated that assemblage B contains host-adapted *Giardia* genotypes that infect humans and other mammalian species, thus suggesting the zoonotic potential of this genotype/assemblage. There is still some controversy regarding the transmission potential of *Giardia* from animals to humans (Thompson, 2000; Thompson *et al.*, 2000), since identical genotypes have been reported in humans and animals, supporting the zoonotic potential of this parasite.

Even though two major molecular groups of *G. intestinalis* have been recognized infecting humans worldwide (Table 2.2), there is no agreement in naming these genotypes, and the nomenclature varies by the location of the scientist involved. The following genotypes are reported in the literature: (i) Polish and Belgian genotypes in Europe (Homan *et al.*, 1992), (ii) Groups 1, 2, and 3 in North America (Nash *et al.*, 1985; Nash and Mowatt, 1992), and (iii) assemblages A and B in Australia (Maryhofer *et al.*, 1995). It may be several years before a consensus may be reached

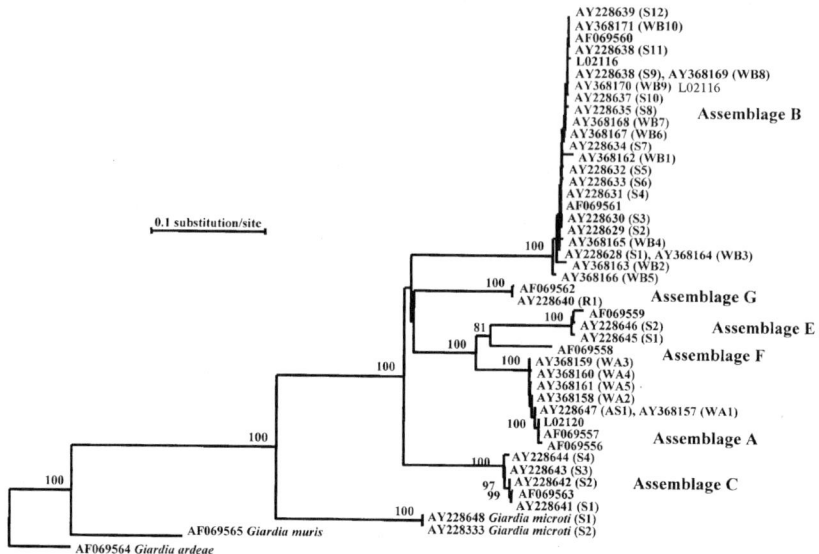

Figure 2.2. Phylogenetic relationships of *Giardia* parasites inferred by the neighbor-joining analysis of the TPI nucleotide sequences.

in naming *Giardia* genotypes; however, the term "assemblage" has been increasingly used by scientists worldwide, and will be used to describe the importance of genotyping.

Major advances in understanding the genetic diversity within *G. intestinalis* have been accomplished by the development of various PCR-based methods and protocols. Thus far, the TPI gene has been found to exhibit the highest degree of polymorphism in *G. intestinalis* at both inter- as well as intra-genotype levels, and the TPI-based genotyping has proven to be very useful in epidemiological investigations of human giardiasis (Monis *et al.*, 1999; Sulaiman *et al.*, 2003, 2004).

An interesting fact of *G. intestinalis* genotyping is that the phylogenetic distance separating assemblages A and B is greater than typically used to differentiate the two protozoan species (Fig. 2.2) (Monis *et al.*, 1996, 1998; Sulaiman *et al.*, 2003; Van Keulen *et al.*, 1998), reinforcing the notion that *G. intestinalis* may be a species complex. Several studies compared isolates of *Giardia* from different hosts as well as geographic regions and identified various genetically identical groups and subgroups within the *G. intestinalis* assemblages around the world (Monis *et al.*, 1999; Sulaiman *et al.*, 2003).

Both assemblages A and B of *G. intestinalis* have been reported worldwide, and these parasites have been found to be human-pathogenic in most continents (Table 2.2). The reported distribution of human-infectious genotypes, however, differed by study and geographical location. For example, a study in Mexico reported that all 22 isolates and derived clones from human specimens belonged to assemblage A subtype II, with a complete lack of assemblage B. The predominance of this subtype of assemblage A was attributed to biologic or geographic factors related to the study area (Ponce-Macoteala *et al.*, 2002). Another study conducted in Mexico showed 26 human isolates belonging to assemblage A (Cedillo-Rivera *et al.*, 2003). Similarly, all seven human isolates in Korea characterized at 16S rDNA locus belonged to assemblage A of *G. intestinalis* (Yong *et al.*, 2000). However, in Peru and United Kingdom, assemblages A and B are found in human stools, although assemblage A was responsible for more human infections than assemblage B (Amar *et al.*, 2002; Sulaiman *et al.*, 2003).

A study in the United Kingdom of sporadic cases of humans giardiasis successfully used a TPI-based PCR-RFLP genotyping tool. Of the 33 TPI-PCR positive infected patients, 21 (64%) contained assemblage B, 9 (27%) had assemblage A, and 3 (9%) samples showed mixed infection of assemblages A and B (Amar *et al.*, 2002). Similar results were obtained with samples from a nursery outbreak, in which 88% (21 of 24) samples were shown to be *G. duodenlais* assemblage B parasites and rest of them to be assemblage A parasites (Amar *et al.*, 2002). Recently, the intra-genotypic variations of TPI in assemblage B were also considered as useful markers in subgenotyping outbreak isolates in a study conducted in the United States (Sulaiman *et al.*, 2003).

2.4 TRANSMISSION AND EPIDEMIOLOGY

Giardia species colonize the intestine of almost every group of vertebrates. *G. intestinalis* is the species that infects humans, domestic pets, farm animals, and wild mammals (Xiao *et al.*, 1994; Olson *et al.*, 1995; Thompson *et al.*, 2000; Sulaiman *et al.*, 2003).

2.4.1 Human

In the United States, giardiasis is the most frequently detected parasitic disease. Results from a surveillance study from January 1992 through December 1997 revealed the occurrence of *Giardia* infections in all the major geographic areas of the United States, with an estimated incidence of 2.5 million cases per year (Furness *et al.*, 2000).

Recent molecular characterization studies indicate that there are many host-adapted *G. intestinalis* genotypes, and with the use of these molecular tools, it is now possible to assess the human infective potentials of the *Giardia* cysts found in the water system. Additionally, these methods would be beneficial for tracking sources of contaminations (Sulaiman *et al.*, 2003).

Although most human infections are thought to occur due to anthroponotic activities, drinking and recreational water, fresh foods, and certain mammals may also play significant roles in the transmission of *G. intestinalis* (Mead *et al.*, 1999; Nichols, 2000). Additionally, misting during commercialization or industrial washing is another way in which fresh produce can be contaminated with *G. intestinalis*, and other parasites (Amahmid *et al.*, 1999).

The identification of *Giardia* as the causative agent in a foodborne outbreak is a daunting task. The long incubation time, usually between 2 and 3 weeks, the variable percentage of symptomatic infections, and the use of detection methods that do not allow trace-back investigations, recall bias from the affected people and the lack of samples suspect food products at the time infections are almost invariably observed in a suspected foodborne outbreak of giardiasis. Nonetheless, there are outbreaks where fresh produce or uncooked foods were confirmed as the vehicles of infection (Schantz, 1991).

Giardia intestinalis has a ubiquitous presence in the environment and is detected worldwide. A review by CDC researchers (Mead *et al.*, 1999) estimated that every year 225,000 cases of foodborne giardiasis occur in the United States. Because giardiasis is frequently considered a water- or foodborne illness, the identification of *Giardia* at the genus and species levels is insufficient to establish its role in outbreaks.

The accurate identification and molecular characterization of foodborne parasites is critical for successful epidemiological and trace-back studies. Findings from these investigations can lead to the identification of sources or points of contamination, the magnitude of parasite dissemination, and to the elaboration of appropriate interventions to prevent future outbreaks. Additionally, it can also be used to evaluate the efficacy of current or new preventative measures to prevent future transmission.

2.4.2 Environmental

Waterborne outbreaks of giardiasis are a major public health problem in many industrialized nations, including the United Kingdom, Sweden, Canada, and the United States (Ljungstrom and Castor, 1992; Moore *et al.*, 1993). Human sewage has been considered a source of *Giardia* cysts contamination in water. In Canada and Italy, a high prevalence (73–100%) of *Giardia* cysts was reported in raw sewage samples (Caccio *et al.*, 2003; Heitman *et al.*, 2002; Wallis *et al.*, 1996). The public health importance and contamination sources of *Giardia* cysts found in water, however, are largely unclear, because very few studies have been carried out to genetically characterize the *Giardia* cysts in water. Nevertheless, *G. intestinalis* cysts of assemblage A have been identified in a few clams collected from the Rhode River, a Chesapeake Bay subestuary in Maryland (Graczyk *et al.*, 1999).

Molecular characterization of *Giardia* species in wastewater provides a valuable tool for community-wide surveillance of human giardiasis. Several attempts have been made to detect and differentiate *Giardia* species in environmental samples using PCR techniques for the detection and differentiation of *Giardia* parasites (Caccio *et al.*, 2003; van Keulen *et al.*, 2002; Wallis *et al.*, 1996). The distribution of *Giardia* species in environmental samples is likely dependent on human, agricultural, and wildlife activities. Two human-pathogenic genotypes of *G. intestinalis* (assemblage A in 10 and assemblage B in 4) were identified in 14 environmental samples (Wallis *et al.*, 1996) from Canada using a SSU rRNA-based PCR-RFLP protocol (van Keulen *et al.*, 2002). In Italy, 16 samples from four wastewater plants were analyzed by beta-giardin-based PCR-RFLP method. Assemblage A was found in eight of the samples, whereas both assemblages A and B were found in the remaining eight samples (Caccio *et al.*, 2003).

In a recent *Giardia* surveillance study, conducted on a much larger sample size of wastewater collected over a three-year period from an urban area of Milwaukee,

WI, both genotypes of human-pathogenic *G. intestinalis* (assemblages A and B) were found in the 131 samples (Sulaiman *et al.*, 2004). The majority (84.7%) of the wastewater samples (111 samples) belonged to *G. intestinalis* assemblage A, which had five distinct subtypes (WA1–WA5). However, one subtype (WA1) accounted for most of the assemblage A isolates (107 of 111), indicating humans in Milwaukee were infected with subtype WA1. This subtype was identical to a sequence previously reported in an axenically cultured strain from humans (AF069557, assemblage A, group II) in Australia (Monis *et al.*, 1999), and in six fecal samples from humans (AY228647) in Peru (Sulaiman *et al.*, 2003). The significance of a predominant subtype in Milwaukee is not clear. It is tempting to conclude that a common source of human infection was responsible for the wide occurrence of subtype WA1. However, the 20 wastewater samples that formed assemblage B had a high genetic diversity (with 10 distinct subtypes), indicating that it is unlikely the transmission of *Giardia* infection in Milwaukee is restricted to one source. It is possible that subtype WA1 of assemblage A is more infectious than other *Giardia* parasites (Sulaiman *et al.*, 2004). Recently, phenotypic differences between assemblages A and B have been observed; assemblage B was seen in patients with persistent diarrhea, whereas assemblage A was seen mostly in patients with intermittent diarrhea (Homan and Mank, 2001).

Phylogenetic analysis based on nucleotide sequencing has been successful not only in understanding the species structure of various microorganisms of public health importance, but also in identifying their transmission routes. Phylogenetic analysis based on all published *Giardia* TPI sequences from various previous studies, including *Giardia* isolates from humans, animals, and environmental samples, clearly demonstrated only assemblages A and B to be human-pathogenic parasites (Monis *et al.*, 1999; Sulaiman *et al.*, 2003, 2004) (Fig. 2.2). Thus, results of the phylogenetic analysis could be useful in understanding the public health importance of some *G. intestinalis* parasites. It is important to mention that the human *G. intestinalis* parasites belonged to only two distinct lineages (assemblages A and B), whereas four lineages contained the *G. intestinalis* from animals (assemblages C and E, and cat and rat genotypes). Assemblage B, however, also contains various animal isolates, such as all the beaver isolates and some isolates from muskrats, rabbit, and mice, strongly indicating that these animal *G. intestinalis* isolates have the potential to infect humans. Previously, it was suggested that *Giardia* parasites from beavers could be a source of infection among hikers and some waterborne outbreaks of giardiasis (Monzingo and Hibler, 1987;Thompson, 2000; Thompson *et al.*, 2000; Wallis *et al.*, 1984). Results of these recent studies have provided genetic evidence to substantiate these claims.

2.5 CONTROL AND TREATMENT

Since *Giardia* is primarily transmitted through the fecal-oral route, one of the major vehicles for transmission is contaminated drinking water. The water supply systems may become contaminated by the introduction of sewage or animal activity in their watershed. *Giardia* cysts can be effectively removed and inactivated in water supplies by a combination of filtration and disinfection.

The filtration of drinking water supplies is accomplished by removing particulate matter from water by passage through porous media. A large number of filtration technologies are utilized for this purpose, including diatomaceous earth filtration, slow sand filtration, and coagulation filtration. There are several types of coagulation-filtration practices that include conventional filtration, direct filtration, and in-line filtration. It has been estimated that all of the above filtration methods can remove $\geq 99\%$ of the *Giardia* cysts from raw water, provided they are operated and maintained properly (Logsdon, 1988). Since filtration can reduce, but not necessarily eliminate the levels of contaminants in water, disinfection is an additional process needed to ensure microbiologically safe water. This is particularly true when the source of drinking water is surface water.

It is indeed important to mention that approximately 14–15 million households in the United States depend on a personal household well for drinking water each year, and more than 90,000 new wells are drilled all over the United States each year (www.cdc.gov/ncidod/dpd/healthywater/privatewell.htm). During 1999–2000, contaminated private well water caused 26% of the drinking water-borne outbreaks, making people ill (www.cdc.gov/ncidod/dpd/healthywater/factsheets/wellwater.htm). Therefore, the contamination of a private well is not only the concern of the household served by the well, but also nearby households using other water supplies and the aquifer that the water is drawn from. Rules from the US EPA that protect public drinking water systems do not apply to privately owned wells. Most states of the United States have rules for private wells, but these rules may not completely protect the private wells. A guideline (as fact sheets) has also been posted in the above Web site of Division of Parasitic Disease, Centers for Disease Control and Prevention, Atlanta, providing information on contaminants that can be found in well water, and information on making well water safe to drink.

Giardia cysts from human or animal sources that contaminate surface water can survive well at temperatures below 20°C, and still can be infective to humans. Previously described filtration protocols can remove 99% or more of *Giardia* cysts, and various chemical disinfectants (namely chlorine, chlorine dioxide, chloramines, and ozone) can further reduce the cyst burden in drinking water. However, the success of these methods are dependent on several factors such as pH, concentration, and contact time.

In a recent review, 21 waterborne outbreaks on cruise ships were examined (Rooney *et al.*, 2004). During the period of January 1970 to June 2003, more than 6400 people were found affected. Importantly, many outbreaks were neither reported nor published. The study also revealed that the above outbreaks could have been prevented if water had been uplifted from reliable sources, and extra treatment (such as filtration and disinfection) as well as routine monitoring of residual disinfectants in distribution systems was accomplished on a regular basis (Rooney *et al.*, 2004).

Giardia continues to be an emerging pathogen in various epidemiologic settings. The PCR-based molecular tools are valuable in identifying routes and dynamics of transmission, and species structure. Data from molecular epidemiologic studies using modern diagnostic tools will help prevent future outbreaks.

Control measures to prevent human giardiasis should be integrated with programs to prevent waterborne and foodborne pathogens. Personal control measures include paying special attention to personal hygiene, in order to prevent exposure to infectious feces. In the United States and other developed countries this is fairly simple to accomplish. Nonetheless, special attention must be given to day-care centers. Control and prevention of giardiasis in these settings require proper training and education of care providers as well as careful supervision of the disposal of fecal contamination. In places where an optimal sanitary infrastructure is not widely available, including areas in developing countries, environmental contamination with fecal material poses a significant challenge. In these circumstances, the best alternative is to educate people in personal hygiene habits and ways to protect their food and water used for their consumption. People facing those conditions, as well as campers and hikers, should boil, treat, or filter their drinking water prior to consumption.

Giardia cysts are very resistant to conventional water treatment, such as chlorination and ultraviolet irradiation. For large water systems, sand filtration or a similar method for physical removal, in addition to an effective disinfection treatment, can be a successful water treatment option.

For individual water supplies it is advisable to have the water tested prior to selecting a treatment option. Frequently used are combinations of filtration disinfection, and occasionally, reverse osmosis. To verify the efficacy of water treatment systems or point of use devices, visit the Web site of the National Sanitation Foundation (www.nsf.org).

In the event of infections, several prescription drugs are available to treat giardiasis. The cure rates of antigiardia compounds vary by the study, and range from 80 to 100%. Metronidazole (commercial name Flagyl) is recommended by the World Health Organization for chemotherapy. It is very effective and is the most commonly used drug in adult patients in the United States; however it has an unpleasant flavor and children do not tolerate it very well. As tolerance may be a problem in pregnant mothers and young children, furazolidone is recommended in expectant women (DuPont, 1989), although nausea or vomiting may still occur (Altamirano and Bondani, 1989). Furazolidone or nitoxozanide may be used for treating pediatric giardiasis. Additional drugs include tinidazole, a compound chemically related to metronidazole, and quinacrine. These two products are not available in the United States, although tinidazole is frequently used in other countries. Quinacrine is useful in the management of difficult cases and can be obtained through Panorama Pharmacy, Panorama City, CA (Nash, 2001).

Other products used to treat giardiasis include paramomycin and albendazole and metronidazole-related compounds such as ornidazole and secnidazole (Gardner and Hill, 2001). A more complete and up to date information regarding drugs against *Giardia* can always be found in the electronic versions of The Medical Letter (http://www.medletter.com) or the website of the Division of Parasitic Diseases, Centers for Disease Control and Prevention (www.cdc.gov/ncidod/dpd/parasites/default.htm).

REFERENCES

Adam, R. A., 2001, Biology of *Giardia lamblia*, *Clin. Microbiol. Rev.* **14**:447–475.
Adam, R. D., Nash, T. E., and Wellems, T. E., 1988, The *Giardia lamblia* trophozoite contains sets of closely related chromosomes, *Nucl. Acids Res.* **16**:4555–4567.
Adam, R. D., 2000. The *Giardia lamblia* genome. *Int. J. Parasitol.* **30**:475–484.
Altamirano, A., and Bondani, A., 1989, Adverse reactions to furazolidone and other drugs: A comparative review, *Scand. J. Gastroenterol. Suppl.* **169**:70–80.
Amahmid, O., Asmama, S., and Bouhoum, K., 1999, The effect of waste water reuse in irrigation on the contamination level of food crops by *Giardia* cysts and *Ascaris* eggs, *Int. J. Food. Microbiol.* **49**:19–26.
Amar, C. F., Dear, P. H., Pedraza-Diaz, S., Looker, N., Linnane, E., and McLauchlin, J., 2002, Sensitive PCR-restriction fragment length polymorphism assay for detection and genotyping of *Giardia intestinalis* in human feces, *J. Clin. Microbiol.* **40**:446–452.
Andrews, R. H., Adams, M., Boreham, P. F., Maryhofer, G., and Meloni, B. P., 1989, *Giardia intestinalis*: Electrophoretic evidence for a species-complex, *Int. J. Parasitol.* **19**:183–190.
Andrews, R. H., Chilton, N. B., and Maryhofer, G., 1992, Selection of specific genotypes of *Giardia intestinalis* by growth *in vitro* and *in vivo*, *Parasitology* **105**:375–386.
Andrews, R. H., Monis, P. T., Ey, P. L., and Maryhofer, G., 1998, Comparison of the levels of intra-specific genetic variation within *Giardia muris* and *Giardia intestinalis*, *Int. J. Parasitol.* **8**:1179–1185.
Anonymous, 1989, Common-source outbreak of giardiasis—New Mexico. *MMWR* **16**:405–407.
Barta, J. R., 1997, Investigating phylogenetic relationships within the Apicomplexa using sequence data: The search of homology, *Methods* **13**:81–88.
Baruch, A. C., Isaac-Renton, J., and Adam, R. D., 1996, The molecular epidemiology of *Giardia lamblia*: A sequence-based approach, *J. Infect. Dis.* **174**:233–236.
Berkman, D. S., Lescano, A. G., Gilman, R. H., Lopez, S. L., and Black, M. M., 2002, Effects of stunting, diarrhoeal disease, and parasitic infection during infancy on cognition in late childhood: A follow-up study, *Lancet* **16**:564–571.
Binz, N., Thompson, R. C. A., Lymbery, A. J., and Hobbs, R. P., 1992, Comparative studies on the growth dynamics of two genetically distinct isolates of *Giardia intestinalis in vitro*, *Int. J. Parasitol.* **22**:195–202.
Blunt, D. S., Khramtsov, N. V., Upton, S. J., and Montelone, B. A., 1997, Molecular karyotype analysis of *Cryptosporidium parvum*: Evidence for eight chromosomes and a low-molecular-size molecule, *Clin. Diagn. Lab. Immunol.* **4**:11–13.
Caccio, S., Camilli, R., La Rosa, G., and Pozio, E., 1998, Establishing the *Cryptosporidium parvum* karyotype by NotI and SfiI restriction analysis and Southern hybridization, *Gene* **219**:73–79.
Caccio, S. M., De Giacomo, M., and Pozio, E., 2002, Sequence analysis of the beta-giardin gene and development of a polymerase chain reaction-restriction fragment length polymorphism assay to genotype *Giardia intestinalis* cysts from human fecal samples. *Int. J. Parasitol.* **32**:1023–1030.
Caccio, S. M., De Giacomo, M., Aulicino, F. A., and Pozio, E., 2003, *Giardia* cysts in wastewater treatment plants in Italy, *Appl. Environ. Microbiol.* **69**:3393–3398.
Campbell, S. R., van Keulen, H., Erlandsen, S. L., Senturia, J. B., and Jarrol, J. L., 1990, *Giardia* sp: Comparison of electrophoretic karyotypes, *Exp. Parasitol.* **71**:470–482.
Cedillo-Rivera, R., Darby, J. M., Enciso-Moreno, J. A., Ortega-Pierres, G., and Ey, P. L., 2003, Genetic homogeneity of axenic isolates of *Giardia intestinalis* derived from acute and chronically infected individuals in Mexico, *Parasitol. Res.* **90**:119–123.

DuPont, H. L., 1989, Progress in therapy for infectious diarrhea, *Scand. J. Gastroenterol. Suppl.* **169**:1–3.
Erlandsen, S. L., and Bemrick, W. L., 1987, SEM evidences for a new species *Giardia psittaci*, *J. Parasitol.* **73**:623–629.
Erlandsen, S. L., Bemrick, W. J., Wellis, C. L., Feely, D. E., Kundson, L., Cambell, S. R., van Keulen, H., and Jarrol, E. L., 1990, Axenic culture and characterization of *Giardia ardea* from the great blue heron (*Ardea herodias*), *J. Parasitol.* **76**:717–724.
Ey, P. L., Andrews, R. H., and Maryhofer, G., 1993, Differentiation of major genotypes of *Giardia intestinalis* by polymerase chain reaction analysis of a gene encoding a tropozoite surface antigen, *Parasitol.* **106**:347–356.
Feely, D. E., 1988, Morphology of the cyst of *Giardia microti* by light and electron microscopy, *J. Protozool.* **35**:52–54.
Filice, F. P., 1952, Studies on the cytology and life history of a *Giardia* from the laboratory rat, *Univ. Calif. Publ. Zool.* **57**:53–1146.
Furness, B. W., Beach, M. J., and Roberts, J. M., 2000, Giardiasis surveillance-Unites States, 1992–1997, *MMWR* **29**:1–13.
Gardner, T. B., and Hill, D. R., 2001, Treatment of giardiasis, *Clin. Microbiol. Rev.* **14**:114–128.
Graczyk, T. K., Thompson, R. C., Fayer, R., Adams, P., Morgan, U. M., and Lewis, E. J., 1999, *Giardia intestinalis* cysts of genotype A recovered from clams in the Chesapeake Bay subestuary, Rhode River, *Am. J. Trop. Med. Hyg.* **61**:526–529.
Hall, M. L., Costa, N. D., Thompson, R. C. A., Lymbery, A. J., Meloni, B. P., and Wales, R. G., 1992, Genetic variants of *Giardia intestinalis* differ in their metabolism, *Parasitol. Res.* **78**:712–714.
Hays, M. P., Mosier, D. A, and Oberst, R. D., 1995, Enhanced karyotype resolution of *Cryptosporidium parvum* by contour-clamped homogeneous electric fields, *Vet. Parasitol.* **58**:273–280.
Heitman, T. L., Frederick, L. M., Viste, J. R., Guselle, N. J., Morgan, U. M., Thompson, R. C., and Olson, M. E., 2002, Prevalence of *Giardia* and *Cryptosporidium* and characterization of *Cryptosporidium* spp. isolated from wildlife, human, and agricultural sources in the North Saskatchewan River Basin in Alberta, Canada, *Can. J. Microbiol.* **48**:530–541.
Homan, W. L., and Mank, T. G., 2001, Human giardiasis: Genotype linked differences in clinical symptomatology, *Int. J. Parasitol.* **31**:822–826.
Homan, W. L., van Enckevort, F. H. J., Limper, L., van Eys, G. J. J. M., Schoone, G. J., Kasprzak, W., Majewska, A. C., and van Knapen, F., 1992, Comparison of *Giardia* isolates from different laboratories by isoenzyme analysis and recombinant DNA probes. *Parasitol. Res.* **78**:316–323.
Homan, W. L., Gilsing, M., Bentala, H., Limper, L., and van Knapen, F., 1998, Characterization of *Giardia intestinalis* by polymerase-chain-reaction fingerprinting, *Parasitol. Res.* **84**:707–714.
Hopkins, R. M., Meloni, B. P., Groth, D. M., Wetherall, J. D., Reynoldson, J. A., and Thompson, R. C., 1997, Ribosomal RNA sequencing reveals differences between the genotypes of *Giardia* isolates recovered from humans and dogs living in the same locality, *J. Parasitol.* **83**:44–51.
Jokipii, A. M., Hemila, M., and Jokipii, L., 1985, Prospective study of acquisition of *Cryptosporidium*, *Giardia lamblia*, and gastrointestinal illness, *Lancet* **31**:487–489.
Karanis P., and Ey, P. L., 1998, Characterization of axenic isolates of *Giardia intestinalis* established from humans and animals in Germany, *Parasitol. Res.* **84**:442–449.
Karanis, P., Opiela, K., Renoth, S., and Seith, H. M., 1996, Possible contamination of surface waters with *Giardia* spp. through muskrats, *Zent. Bakteriol.* **284**:302–306.

Kirkpatrick, C. E., and Benson, C. E., 1987, Presence of *Giardia* spp. and absence of *Salmonella* spp. in New Jersey muskrats (*Ondarta zibethicus*), *Appl. Environ. Microbiol.* **53**:1790–1792.

Learmonth, J. J., Ionas, G., Pita, A. B., and Cowie, R. S., 2003, Identification and genetic characterisation of *Giardia* and *Cryptosporidium* strains in humans and dairy cattle in the Waikato Region of New Zealand, *Water. Sci. Technol.* **47**:21–26.

Lipuma, J. J., 1998, Burkholderia cepacia epidemiology and pathogenesis: Implications for infection control. *Curr. Opin. Pulm. Med.* **4**:337–341.

Ljungstrom, I., and Castor, B., 1992, Immune response to *Giardia lamblia* in a water-borne outbreak of giardiasis in Sweden, *J. Med. Microbiol.* **36**:347–352.

Logsdon, G. S., 1988, Comparison of some filtration processes appropriate for *Giardia* cyst removal. In: Wallis, P.M. and Hammond, B.R. (eds), Advances in *Giardia* Research University of Calgary Press, Calgary, Alberta, Canada, pp. 95–102.

Lu, S. Q., Baruch, A. C., and Adam, R. D., 1998, Molecular comparison of *Giardia lamblia* isolates, *Int. J. Parasitol.* **28**:1341–1345.

Lu, S., Wen, J., Li, J., and Wang, F., 2002, DNA sequence analysis of the triose phosphate isomerase gene from isolates of *Giardia lamblia*, *Chin. Med. J.* **115**:99–102.

Maryhofer, G., Andrews, R. H., Ey, P. L., and Chilton, N. B., 1995, Division of *Giardia* isolates from humans into two genetically distinct assemblages by electrophoretic analysis of enzymes coded at 27 loci and comparison with *Giardia muris*, *Parasitology* **111**:11–17.

Maslow, J. N., Mulligan, M. E., and Arbeit, R. D., 1993, Molecular epidemiology: Application of contemporary techniques to the typing of microorganisms, *Clin. Infect. Dis.* **17**:153–162.

Mead, J. R., Arrowood, M. J., Current, W. L., and Sterling, C. R., 1988, Field inversion gel electrophoretic separation of *Cryptosporidium* spp. chromosome-sized DNA, *J. Parasitol.* **74**:366–369.

Mead, P. S., Slutsker, L., Dietz, V., McCaig, L. F., Bresee, J. S., Shapiro, C., Griffin, P. M., and Tauxe, R. V., 1999, Food-related illness and death in the United States, *Emerg. Infect. Dis.* **5**:607–625.

Meloni, B. P., Lymbery, A. J., and Thompson, R. C. A., 1995, Genetic characterization of isolates of *Giardia intestinalis* by enzyme electrophoresis: Implication of reproductive biology population structure taxonomy and epidemiology, *J. Parasitol.* **81**:368–383.

Mintz, E. D., Hudson-Wragg, M., Mshar, P., Cartter, M. L., and Hadler, J. L., 1993, Foodborne giardiasis in a corporate office setting. *J. Infect. Dis.* **167**:250–253.

Monis, P. T., Maryhofer, G., Andrews, R. H., Homan, W. L., Limper, L., and Ey, P. L., 1996, Molecular genetic analysis of *Giardia intestinalis* isolates at the glutamate dehydrogenase locus, *Parasitology* **112**:1–12.

Monis, P. T., Andrews, R. H., Maryhofer, G., Kulda, J., Isaac-renton, J. L., and Ey, P. L., 1998. Novel lineages of *Giardia intestinalis* identified by genetic analysis of organisms isolated from dogs in Australia, *Parasitology* **116**:7–19.

Monis, P. T., Andrews, R. H., Maryhofer, G., and Ey, P. L., 1999, Molecular systematics of the parasitic protozoan *Giardia intestinalis*, *Mol. Biol. Evol.* **16**:1135–1144.

Monzingo, D. L., and Hibler, C. P., 1987, Prevalence of *Giardia* sp. in a beaver colony and the resulting environmental contamination, *J. Wildl. Dis.* **23**:576–585.

Moore, A. C., Herwaldt, B. L., Craun, G. F., Calderon, R. L., Highsmith, A. K., and Juranek, D. D., 1993, Surveillance for waterborne disease outbreaks—United States, 1991–1992, *MMWR* **42**:1–22.

Mowatt, M. R., Weinbach, E. C., Howard, T. C., and Nash, T. T., 1994, Complementation of *Escherichia coli* glycolysis mutant by *Giardia lamblia* triosephosphate isomerase, *Exp. Parasitol.* **78**:85–92.

Murtagh, J. J. Jr., Mowatt, M. R., Lee, C. M., Lee, F. J., Mishima, K., Nash, T. E., Moss, J., and Vaughan, M., 1992, Guanine nucleotide-binding proteins in the intestinal parasite *Giardia lamblia*. Isolation of a gene encoding an approximately 20-kDa ADP-ribosylation factor, *J. Biol. Chem.* **267**:9654–9662.

Nash, T. E., 2001, Treatment of *Giardia lamblia* infections, *Pediatr. Infect. Dis. J.* **20**:193–195.

Nash, T. E., and Mowatt, M. R., 1992, Identification and characterization of a *Giardia lamblia* group-specific gene, *Exp. Parasitol.* **75**:369–378.

Nash, T. E., McCutchan, T., Keister, D., Dame, J. B., Conard, J. D., and Gillin, F. D., 1985, Restriction endonuclease analysis of DNA from 15 *Giardia* isolates obtained from humans and animals, *J. Infect. Dis.* **152**:64–73.

Nasmuth, K., 1996, A homage to *Giardia*, *Curr. Biol.* **6**:1042.

Nichols, G. L., 2000, Food-borne protozoa, *Br. Med. Bull.* **56**:209–235.

Nizeyi, J. B., Mwebe, R., Nanteza, A., Cranfield, M. R., Kalema, G. R., and Graczyk, T. K., 1999, *Cryptosporidium* sp. and *Giardia* sp. infectious in mountain gorillas (*Gorilla gorilla beringi*) of the Bwindi Impenetrable National Park, Uganda, *J. Parasitol.* **85**:1084–1088.

Olson, M. E, McAllister, T. A., Deselliers, L., Morck, D. W., Cheng, K. J., Buret, A. G., and Ceri, H., 1995, Effects of giardiasis on production in a domestic ruminant (lamb) model, *A. J. Vet. Res.* **56**:1470–1474.

Pacha, R. E., Clark, G. W., and Williams, E. A., 1985, Occurence of *Campylobacter jejuni* and *Giardia* species in muskrats (*Ondarta zibethica*), *Appl. Environ. Microbiol.* **50**:177–178.

Patton, S., and Rabinowitz, A. R., 1994, Parasites of wild felidae in Thailand: A coprological survey, *J. Wildl. Dis.* **30**:472–475.

Ponce-Macotela, M., Martinez-Gordillo, M. N., Bermudez-Cruz, R. M., Salazar-Schettino, P. M., Ortega-Pierres, G., and Ey, P. L., 2002, Unusual prevalence of the *Giardia intestinalis* A-II subtype amongst isolates from humans and domestic animals in Mexico, *Int. J. Parasitol.* **32**:1201–1202.

Rickard, L. G., Siefker, C., Boyle, C. R., and Gentz, E. J. 1999, The prevalence of *Cryptosporidium* and *Giardia* spp. in fecal samples from free-ranging white tailed deer (*Odocoileus virginianus*) in the southeastern United States. *J. Vet. Diagn Invest.* **11**:65–72.

Rooney, R. M., Bartram, J. K., Cramer, E. H., Mantha, S., Nichols, G., Suraj, R., and Todd, E. C., 2004, A review of outbreaks of waterborne disease associated with ships: Evidence for risk management, *Pub. Heal. Rep.* **119**:435–442.

Schantz, P. M., 1991, Parasitic Zoonoses in Perspective, *Int. J. Parasitol.* **21**:161–170.

Sogin, M. L., 1991, Early evolution and the origin of eukaryotes. Current Opinion in *Genet. Develop.* **1**:457–463.

Sulaiman, I. M., Fayer, R., Bern, C., Gilman, R. H., Trout, J. M., Schantz, P. M., Das, P., Lal, A. A., and Xiao, L., 2003, Triosephosphate isomerase gene characterization and potential zoonotic transmission of multispecies *Giardia intestinalis*, *Emerg. Infect. Dis.* **9**:1444–1452.

Sulaiman, I. M., Jiang, J., Singh, A., and Xiao, L., 2004, Distribution of *Giardia intestinalis* genotypes and subgenotypes in raw urban wastewater in Milwaukee, Wisconsin, *Appl. Environ. Microbiol.* **9**:3776–3780.

Thompson, R. C. A., 2000, Giardiasis as a re-emerging infectious disease and its zoonotic potential, *Int. J. Parasitol.* **30**:1259–1267.

Thompson, R. C. A., Hopkins, R. A., and Homan, W. L., 2000, Nomenclature and genetic groupings of *Giardia* infecting mammals, *Parasitology* **16**:210–218.

Trout, J. M., Santin, M., and Fayer, R., 2003, Identification of assemblage A *Giardia* in white-tailed deer, *J. Parasitol.* **89**:1254–1255.

Ungar, B. L. P., Yolken, R. H., Nash, T., and Quinn, T. C., 1984, Enzyme linked immunosorbent assay (ELISA) for the detection of *Giardia lamblia* in fecal specimens, *J. Infect. Dis.* **149**:90–97.

van Keulen, H., Gutell, R., Gates, M., Campbell, S., Erlandsen, S. L., Jarrol, E. L., Kulda, J., and Meyer, E. A., 1993, Unique phylogenetic position of Diplomonadida based on the complete small subunit ribosomal RNA sequence of *Giardia ardeae, Giardia muris, Giardia intestinalis*, and *Hexamita* sp. *FASEB J.* **7**:223–231.

van Keulen, H., Feely, D. E., Macechko, T., Jarrol, E. L., and Erlandsen, S. L., 1998, The sequence of *Giardia* small subunit rRNA shows that voles and muskrats are parasitized by a unique species *Giardia microti, J. Parasitol.* **84**:294–300.

van Keulen, H., Macechko, P. T., Wade, S., Schaaf, S., Wallis, P. M., and Erlandsen, S. L., 2002, Presence of human *Giardia* in domestic, farm and wild animals, and environmental samples suggests a zoonotic potential for giardiasis, *Vet. Parasitol.* **108**:97–107.

Visvesvara, G. S., Dickerson, J. W., and Healy, G. R., 1988, Variable infectivity of human-derived *Giardia lamblia* cysts for Mongolian gerbils (*Meriones unguiculatus*). *J. Clin. Microbiol.* **26**:837–841.

Wallis, P. M., Buchanan-Mappin, J. M., Faubert, G. M., Belosevic, M., 1984, Reservoirs of *Giardia* spp. in southwestern Alberta. *J. Wildl. Dis.* **20**:279–283.

Wallis, P. M., Erlandsen, S. L., Isaac-Renton, J. L., Olson, M. E., Robertson, W. J., and van Keulen, H., 1996, Prevalence of *Giardia* cysts and *Cryptosporidium* oocysts and characterization of *Giardia* spp. isolated from drinking water in Canada. *Appl. Environ. Microbiol.* **62**:2789–2797.

Wolfe, M. S., 1990, Clinical Symptoms and diagnosis by traditional methods. In Meyer, E. A. (ed), *Giardiasis*, Human Parasitic Diseases, Vol. 3, Elsevier, Amsterdam, New York, pp. 176–185.

Xiao, L., 1994, *Giardia* infection in farm animals. *Parasitol. Today* **10**:436–438.

Yong, T. S., Park, S. J., Hwang, U. W., Yang, H. W., Lee, K. W., Min, D.Y., Rim, H. J., Wang, Y., and Zheng, F., 2000, Genotyping of *Giardia lamblia* isolates from humans in China and Korea using ribosomal DNA Sequences. *J. Parasitol.* **86**:887–891.

CHAPTER 3

Coccidian Parasites
Cyclospora cayetanensis, Isospora belli, Sarcocystis hominis/suihominis

Vitaliano Cama

3.1 PREFACE

Cyclospora cayetanensis, *Isospora belli*, and the *Sarcocystis* spp. *Sarcocystis hominis* and *Sarcocystis suihominis* are parasites that infect the enteric tract of humans (Beck *et al.*, 1955; Frenkel *et al.*, 1979; Ortega *et al.*, 1993). These parasites cause disease when infectious oocysts are ingested by humans. The routes of transmission can be direct human to human contact or through contaminated food (Connor and Shlim, 1995; Fayer *et al.*, 1979) or water (Wright and Collins, 1997). Taxonomically, these parasites are very distinct: *Cyclospora* and *Isospora* belong to the Eimeriidaes, whereas *Sarcocystis* belongs to the Sarcocystidae. Nonetheless, some similarities are noteworthy: infectious stages of these parasites have morphological similarities, they have been reported to be food-borne, they cause infections of the intestinal tract of humans, and their clinical presentations have similarities (Mansfield and Gajadhar, 2004). Thus, relatedness and differences between several aspects of cyclosporiasis, isosporosis, and sarcocystosis will be covered in this chapter.

3.2 BACKGROUND/HISTORY

Cyclospora is probably the most important foodborne pathogen of the three parasites. It is endemic in several regions of the world, primarily in developing countries (Markus and Frean, 1993; Ortega *et al.*, 1993), whereas in the developed world it has been associated with important foodborne outbreaks (Charatan, 1996; Herwaldt, 2000; Herwaldt and Beach, 1999). It has also been reported in travelers returning from endemic areas (Gascon *et al.*, 1995; Soave *et al.*, 1998).

Isospora belli is an infrequent parasite of humans, with most cases reported from tropical areas. In the immunocompetent population, *Isospora* infections are usually asymptomatic (Teschareon *et al.*, 1983). Immunocompromised patients, however, can suffer *I. belli* infections associated with severe clinical disease (Soave, 1988).

Human sarcocystiosis is even more infrequent, and infectious parasites are incidentally identified in feces (Bunyaratvej *et al.*, 1982; Fayer, 2004). The clinical significance of the detection of oocysts or sporocysts in feces is unknown in most cases (Dubey, 1993).

Cyclospora cayetanensis was fully recognized as a parasite less than 15 years ago (Ortega *et al.*, 1993), and gained significant recognition because of its association with several food-borne outbreaks in the United States. The name and description of human infectious *Cyclospora* dates back to 1992, however, previous studies described organisms similar to *Cyclospora*. One of the oldest, if not the first documented report was published in 1979 (Ashford, 1979), describing a parasite with microscopic characteristics of immature (nonsporulated) *C. cayetanensis* oocysts. This report concluded that the described organism probably was a new species of *Isospora* infecting humans (Ashford, 1979).

In the following years, other investigators reported the organisms autofluoresced when observed under UV light (Long *et al.*, 1991). This finding led some investigators to believe it was a blue-green algae or a cyanobacterium. Thus, the term cyanobacterium-like body or CLB was coined and used for some years. Other terms found in the literature that also referred to *Cyclospora* were coccidian-like body (also CLBs) (Hoge *et al.*, 1993) or big *Cryptosporidium*, as reported by Naranjo *et al.* in the 1989 Annual Meeting of the American Society of Tropical Medicine and Hygiene. Findings in this report were on the basis of microscopy of acid-fast stained organisms and transmission electron micrographs that showed sections of organisms resembling encysted flagellates.

Shortly thereafter, studies led by Ortega demonstrated that CLBs were indeed a coccidian parasite, with each oocyst containing two sporocysts, and each of them containing two sporozoites. These findings were first presented at the Annual Meeting of the American Society of Tropical Medicine and Hygiene in November 1992. The morphological characteristics of the parasite allowed its classification as a protozoan parasite belonging to the genus *Cyclospora* (Ortega *et al.*, 1993). Shortly thereafter, the parasite was named *C. cayetanensis*, in honor of Universidad Peruana Cayetano Heredia, a Peruvian University where preliminary studies were conducted (Ortega *et al.*, 1994). Thus far, it is the most recently recognized coccidian parasite affecting humans.

In contrast to *Cyclospora*, *Sarcocystis*, and *Isospora* have been known to infect humans for several decades. *I. belli* is an enteric parasite exclusively from humans (Ferreira *et al.*, 1962; Panosian, 1988). Although it infects, replicates, and completes its life cycle in enteric epithelial cells, it does not cause any significant disease in the immune-competent population. People with impaired immunity, however, may develop extra-intestinal infections and isosporiasis may pose a risk to their lives (Arnaud-Battandier, 1985; Figueroa *et al.*, 1985; Furio and Wordell, 1985; Henry *et al.*, 1986; Jonas *et al.*, 1984; Kobayashi *et al.*, 1985; Ma *et al.*, 1983; Modigliani *et al.*, 1985; Ng *et al.*, 1984; Ros *et al.*, 1987; Shein and Gelb, 1984; Whiteside *et al.*, 1984) (Table 3.1).

Sarcocystis spp. were first reported by Miescher in 1843, who described white milky lesions in the muscles of house mice. Since the initial report, multiple authors described similar lesions and infections in the flesh of various animal species. The transmission cycle, however was not yet elucidated. In 1943, Scott demonstrated that there was no direct transmission from infected to healthy sheep. Despite the negative results, it gave an indication that there was no direct transmission between members of the same species. It took several decades until the life cycle of *S. hominis* and *S. suihominis* were finally described in the 1970s (Heydorn, 1977; Rommel and Heydorn, 1972), when it was demonstrated that both of these parasites required two

Table 3.1. Similarities and differences between *Cyclospora*, *Sarcocystis*, and *Isospora* species that infect humans.

	Cyclospora cayetanesis	*Sarcocystis hominis, S. suihominis*	*Isospora belli*
Life cycle			
Number of hosts	1 host	2 host	1 host
Type of cycle	Anthroponotic	Prey-predator	anthroponotic
Transmission	Fecal–oral	Ingestion of contaminated meat	Fecal oral
Epidemiology			
Developed countries	Epidemic (infrequent)	Very rare.	Opportunistic infection in immune impaired people
Developing countries	Endemic in certain areas	Endemic in areas where undercooked beef/pork is consumed	Opportunistic infection in immune impaired people. Infrequent in healthy people.
Symptoms			
Immune competent	Diarrhea	Mild diarrhea	Asymptomatic
Immune impaired	Diarrhea	Mild diarrhea	Chronic diarrhea, ocular infections
Oocysts			
Shape	Round	Ellipsoidal	Ellipsoidal
Size (μm)	8–10	13×10	30×12
Infectious stage	Sporulated oocysts	Sporocyst	Oocyst or sporocyst
No. of sporocysts	2	2	2
Shape sporocysts	Round 4	Ellipsoidal	ellipsoidal
Sporozoites per sporocyst		2	2
Schematic representation	(image)	(image)	(image)
Sporulation	Outside host	Intestine of host	In or outside host
Stage found in stools	Unsporulated oocysts	Sporulated sporocysts	Unsporulated oocysts

hosts: a definitive host; also described by others as the predator species, and an intermediate host or prey. Infections in the definitive hosts usually infect the enteric tract, where asexual and sexual stages develop. In contrast, in intermediate hosts the parasites develop sexual stages only, forming white milky threads (sarco = flesh and kystis = bladder) in their muscles. Shortly thereafter in 1975, it was demonstrated that cattle and sheep could be the intermediate hosts for more than two species of *Sarcocystis*, and that simultaneous infections may occur (Heydorn *et al.*, 1975).

3.3 BIOLOGY

The oocysts of *C. cayetanensis* are spherical, with 8–10 μm in diameter, and although its life cycle has not been fully demonstrated, evidence so far supports the notion that this is a parasite that only infects humans (Ashford, 1979; Ortega *et al.*, 1994; Ortega *et al.*, 1993). *Cyclospora* has been reported in feces of some animal species, including ducks (Zerpa *et al.*, 1995), chickens (Garcia-Lopez *et al.*, 1996), dogs (Carollo *et al.*, 2001; Yai *et al.*, 1997), and monkeys (Chu *et al.*, 2004); however, no evidence of tissue infections was described in these animals. *Cyclospora* spp. were observed in non-human primates and genetic analysis based on the small subunit RNA gene revealed significant host specificity. *Cyclospora papionis* was exclusively found in baboons, vervet monkeys were only infected with *Cyclospora cercopitheci*, and only colobus monkeys had infections with *Cyclospora collobi* (Eberhard *et al.*, 2001). Animal infectivity studies using nine strains of mice, including adult and neonatal immunocompetent and immune-deficient inbred and outbred strains, rats, sandrats, chickens, ducks, rabbits, jirds, hamsters, ferrets, pigs, dogs, owl monkeys, rhesus monkeys, and cynomolgus monkeys, were unsuccessful. These findings suggest that *C. cayetanensis* is a species specific parasite that only infects humans (Eberhard *et al.*, 2000).

There are two species of *Sarcocystis* that can infect humans, *S. hominis* and *S. suihominis*. As their names suggest, these are two-host parasites and man becomes infected by ingesting viable cysts in the meat of bovids or swine, respectively. Oocysts of these parasites are indistinguishable by microscopy; elliptical in shape, measuring 13 by 10 μm. Two sporocysts develop and in contrast to *Cyclospora*, the oocyst wall is fragile and usually ruptures, so that the sporocysts are more commonly found in clinical specimens and in the environment.

Isospora oocysts are elliptical, measuring 20–30 μm by 10–20 μm. These oocysts are shed unsporulated into the environment; thus, clinical specimens from patients will contain eplitical red structures when stained with acid fast. In environmental samples, however, the sporulated forms are likely to be found. Within the oocysts wall, two spherical structures can be observed. Initially they are called spheroblasts, and when mature, sporocysts. Each sporocyst will be spherical in shape, measuring 10–12 μm in diameter and will contain four sporozoites each.

Cyclospora and *Isospora* are believed to be monoxenus parasites, infecting only humans. Infectious stages of *Cyclospora* have been identified in the small intestine, although the complete life cycle has only been postulated.

Sarcocystis are commonly found in several species of farm animals. Of public health importance are *S. hominis* and *S. suihominis*. People acquire these parasites by ingesting raw or undercooked infected beef or pork.

The life cycle of these parasites is very different. While *Cyclospora* and *Isospora* are reported to be anthroponotic, meaning these parasites complete their life cycles exclusively in humans, *Sarcocystis* parasites have a more complex life cycle and require two different hosts: an intermediate bovid in the case of *S. hominis*, and pigs for *S. suihominis*. In both cases, humans are the definitive host.

Human infections of these parasites affect the digestive tract. Cyclosporiasis starts after a susceptible host ingests sporulated oocysts, the infectious stage. After ingestion, oocysts are ruptured in the upper gastrointestinal tract and the two sporocysts are liberated. Shortly thereafter, two sporozoites are released from each sporocyst and proceed to infect epithelial cells of the intestine. The extent of infection in the human gastrointestinal tract has not been clearly determined. Ortega and collaborators (Ortega et al., 1997a) have described intracellular stages from histological sections of upper gastrointestinal biopsies from 17 Peruvian patients. Meronts, macrogametocytes, and microgametocytes were identified in the cytoplasm of jejunal enterocytes, demonstrating that oocysts could be formed in this section of the enteric lumen. The extent of intestinal infection or preferential site of infection has yet to be reported. This apparent lack of information is in part due to the absence of an animal model to study human cyclosporiasis (Sadaka and Zoheir, 2001). Laboratory studies have attempted to infect immunocompetent and immunodeficient animals which were unsuccessful, and none of these species developed clinical symptoms or patent infections (Eberhard et al., 2000).

There are anecdotic reports of *Cyclospora* in ducks (Zerpa et al., 1995), dogs (Yai et al., 1997), chickens, and monkeys found naturally infected with *Cyclospora* (Chu et al., 2004). Observed using acid fast stain, coccidian-like bodies were described in the stools of a duck owned by a *Cyclospora*-infected patient, while fecal droppings from ducks of noninfected people were *Cyclospora* negative (Zerpa et al., 1995). Although two dogs in Sao Paulo, Brazil were reported positive for *Cyclospora*, a follow-up study of 140 dogs from the same city did not find any infected dogs (Carollo et al., 2001). Recently, dogs, chickens, and monkeys from Nepal were found microscopy positive for *Cyclospora* and these samples were further confirmed by polymerase chain reaction (PCR). The authors concluded that infections in these species needed further confirmation including histological demonstration of infectious stages (Chu et al., 2004).

Seventeen Peruvian patients positive for *Cyclospora* organisms were surveyed and underwent endoscopy, and their symptoms were recorded. Patients presented with gastrointestinal symptoms, including diarrhea, flatulence, weight loss, abdominal discomfort, and nausea. Jejunal biopsies showed an altered mucosal architecture with shortening and widening of the intestinal villi due to diffuse edema and infiltration by a mixed inflammatory cell infiltrate. There was reactive hyperemia with vascular dilatation and congestion of villous capillaries. Parasitophorous vacuoles contained sexual and asexual forms. Type I and II meronts, with 8–12 and 4 fully differentiated merozoites, respectively, were found at the luminal end of epithelial cells. These findings demonstrated the complete developmental cycle associated with host changes due to *C. cayetanensis* (Ortega et al., 1997a).

Microscopy studies of human tissues infected with sarcocysts revealed the complexity of this parasite. Lesions from 40 patients were studied by histopathology, revealing seven morphological types of the parasite. Each type represented one to several different species, all of which were zoonotic. Among four types of sarcocysts found in skeletal muscle, three closely resembled a corresponding species found commonly in monkeys: one from a man in Uganda, another from an Indian patient, and one from a patient in Southeast Asia. Among three types of sarcocysts found in the human heart, one resembled a species commonly seen in the heart of cattle. Of the 40 *Sarcocystis* infections in man, 13 probably were acquired in Southeast Asia, 8 in India, 5 in Central or South America, 4 each in Africa and Europe, 3 in US, 1 in China and 2 from unknown localities. Symptoms associated with these infections included muscle soreness or weakness, subcutaneous swellings, eosinophilia, and periarteritis or polyarteritis nodosa. Nonetheless, there is no conclusive evidence of pathogenicity of the mature sarcocyst (Beaver *et al.*, 1979).

Occasionally, local infections may present complications due to large localized inflammatory reactions. Eight out of 11 presented with infections of the tongue and nasopharynx (Pathmanathan and Kan, 1992). Potential association with vasculitis has also been proposed (McLeod *et al.*, 1980).

Infections with *I. belli* are found worldwide, but are more frequent in tropical and subtropical areas. Immunocompetent people usually have asymptomatic disease, while immunocompromised persons develop chronic diarrhea and extraintestinal infections. The life cycle of this organism stars with the excertion of unsporulated or partially sporulated occysts. Upon sporulation, which can occur as fast as 24 hours, the oocysts become infectious. The cycle continues when a person ingests infectious oocysts and the sporozoites are released in the small intestine. There, *Isospora* infects enterocytes, undergoes asexual and sexual reproduction. Leading to the production and shedding of *I. belli* unsporualted or partially sporulated oocysts, also known as sporoblasts (Lindsay *et al.*, 1997).

3.4 CLINICAL SIGNIFICANCE

Diarrhea, malaise, lack of energy, and appetite are symptoms associated with several gastrointestinal pathogens, and are present in patients with cyclosporiasis in developed countries. In endemic areas, clinical infections are usually detected in children 4–10 years of age. About 50% of these children present with diarrhea and other gastrointestinal discomforts, including malaise, bloating, and anorexia. Infections may resolve spontaneously, suggesting that immunity may play a role in clearance of infections.

In the case of human isosporiasis, severe symptoms have been reported in AIDS patients or people with other forms of immune-suppression. The most frequently clinical symptoms are fever, malaise, chronic and persistent diarrhea, steatorrhea, and loss of weight. The infection can even cause death.

Clinical disease associated with human sarcocystosis is usually limited (Dissanaike, 1994; Pozio, 1991), and most human infection studies demonstrated spontaneous resolution within a month postinfection (Chen *et al.*, 1999). Infection studies in humans (Lian *et al.*, 1990) reported clinical symptoms such as anemia, abdominal pain, diarrhea, fatigue, and dizziness on day 3 postinfection. Sporocysts and oocysts were found in the feces on day 8. The patent period of sporocyst excretion was more than 42 days. In addition to digestive malaise, another symptom associated with *Sarocysctis* infections may be muscle aches (Pamphlett and O'Donoghue, 1990).

Experimental infections using human volunteers have been reported in the literature. Eight medical students ate raw meat from a pig experimentally infected with *S*.

suihominis. Six to twenty-four hours after the meal all persons suffered from acute clinical symptoms, particularly diarrhea and vomiting, and coldness and sweating which decreased by the second day (Piekarski *et al.*, 1978). At the Institute of Medical Parasitology, University of Bonn, eleven medical students and six members of the institute participated in a meal with raw pork of an experimentally *S. suihominis* infected pig. Only individuals who ingested excessively high quantities of infected meat suffered severe symptoms (Kimmig *et al.*, 1979).

No serological associations were found between heart conditions and *Sarcocystis* infections in Egypt (Azab and el-Shennawy, 1992), but serological evidence associates this disease with chronic fatigue syndrome (Pamphlett and O'Donoghue, 1992).

Twenty-two intestinal specimens surgically resected due to segmental enterocolitis were classified into three groups: (1) acute inflammation with hemorrhage and necrosis; (2) constrictive lesion; and (3) false diverticulum with perforation. The predominant finding was unisegmental involvement, distributed in jejunum, ileum, and ileocolon. Microscopically, small parasitic structures, interpreted to be an unconventional excystation stage of *S. hominis*, were present on the luminal border and within the crypt-lining epithelial cells (Dubey, 1976). At the ulcerated area, tissue invasion by Gram-positive bacteria were consistently seen and considered as second pathogen (Bunyaratvej and Unpunyo, 1992).

3.5 TRANSMISSION AND EPIDEMIOLOGY

3.5.1 *Cyclospora*

Foodborne outbreaks in the United States have increased in absolute numbers over the past 30 years. Data analysis from the Foodborne Outbreak Surveillance System revealed 190 outbreaks between 1973 and 1997. During this period, *Cyclospora* and *E. coli* O157:H7 were newly recognized causes of foodborne illness (Sivapalasingam *et al.*, 2004).

Cyclospora cayetanensis infections, thus far, have only been confirmed in humans. Currently, no animal species is considered to be an intermediate or accidental host, or a reservoir of *C. cayetanensis*. Epidemiological studies have demonstrated transmission through the foodborne route. Surveys of fresh produce were conducted in an endemic area, detecting *Cyclospora* in 1.8% of the samples (Ortega *et al.*, 1997b).

Results from studies to determine waterborne transmission of *Cyclospora* are controversial. The first report came from Nepal, where contaminated chlorinated water was identified as the source of *Cyclospora* infections between expatriates in Nepal (Rabold *et al.*, 1994). A retrospective epidemiological investigation of an outbreak within a hospital in IL identified drinking water as the source of infection (Huang *et al.*, 1995). A prospective epidemiological study in Haiti (Lopez *et al.*, 2003) reported variations in the percentage of cases of *Cyclospora* infections, from 12 to 1.1 % between February and April 2001. One artesian well was positive prior to this study; however, none of the wells was positive thereafter. Therefore, no epidemiological associations were established between infections and water sources.

Molecular investigations have reported the detection of parasite DNA in surface water from California (Dowd et al., 2003; Shields and Olson, 2003). These findings indicate that *Cyclospora* may indeed be present in water and that water borne transmission needs to be further studied.

Currently, *Cyclospora* is primarily considered a food-borne parasite, although before 1996, cyclosporiasis was originally considered a diarrheal disease affecting returning travelers.

This concept changed in the spring of 1996, when a large food-borne outbreak of cyclosporiasis occurred in North America, affecting 1465 people from 20 states, the District of Columbia, and two provinces (Herwaldt and Ackers, 1997). Epidemiological investigations confirmed the vast majority of cases were associated with 55 events that served raspberries. This investigations later demonstrated a significant association between cyclosporiasis and consumption of raspberries imported from Guatemala (Herwaldt and Ackers, 1997). Detailed epidemiological descriptions of specific events included a wedding reception in Massachusetts (Fleming et al., 1998), and a luncheon in Charleston, S.C. where 38 of 64 attendees met the case definition of cyclosporiasis (Caceres et al., 1998). Cluster investigations in Florida showed that raspberries were the only food common to this outbreak and confirmed them to be imported from Guatemala as well (Katz et al., 1999).

In the spring of 1997, another large outbreak of cyclosporiasis affected the USA and Canada. Epidemiological and trace-back investigations identified 41 infection clusters that comprised 762 cases of acute cyclosporiasis and 250 sporadic cases of cyclosporiasis. Similar to the 1996 outbreak, there were significant associations between *Cyclospora* infections and consumption of Guatemalan raspberries. As a consequence of this second outbreak, exportation of Guatemalan raspberries was voluntarily suspended in May 1997 (Herwaldt and Beach, 1999).

Other fresh produce has been implicated in the transmission of *Cyclospora*. Basil was directly implicated in a 1999 outbreak in Missouri, where 62 cases were identified. All of these people had previously eaten either pasta chicken salad at one event or tomato basil salad at another event (Lopez et al., 2001). European countries also have documented cases of foodborne cyclosporiasis. Although a specific vegetable was not identified, the foods associated with disease risk was lettuce imported from southern Europe that was spiced with fresh green herbs (Doller et al., 2002).

Recently, snow peas have been implicated as sources of transmission of cyclosporiasis. A recent outbreak found that 50 potential cases of cyclosporiasis were linked to the consumption of snow peas imported from Guatemala (Anonymous, 2004).

Cyclospora infections had been previously reported in travelers returning from developing countries. Although infections have been reported from Southeast Asia, Papua New Guinea, Indonesia, India, Pakistan, Nepal, the Middle East, North Africa, the United Kingdom, the Caribbean, the United States, Central America, and South America, the true prevalence of this parasite in any population is unknown. Infections were reported in travelers returning from South America (Drenaggi et al., 1998). In 1997, *Cyclospora* was reported in 5 of 469 returning travelers with diarrhea (Jelinek et al., 1997). Among expatriots living in endemic areas, a study in Nepal reported a higher risk of diarrhea among foreigners during the first 2 years of residence (Shlim et al., 1999).

Endemic cyclosporiasis has been reported in several areas of the world, mainly in developing countries. Although there are socioeconomical and geographical similarities with other related pathogens including *Cryptosporidium, Giardia*, and bacterial and viral disease, there are also marked differences. In general, *Cyclospora* has been reported affecting children in areas where access to clean water or sanitation is marginal or suboptimal. A prospective study in Peru found that children between three and six years of age were more frequently affected (Bern et al., 2002), and infections were rare after 10 years of age. These findings suggest that humans living in endemic areas where exposure to the parasite is frequent can develop protective immunity.

These observations, however, are limited to specific areas where the parasite is endemic. Another study on upper-middle-class Peruvians, showed that this population presents sporadic disease which closely resembles cyclosporiasis in developed countries (Ortega et al., 1997a). Although the study sites were only a few miles apart, the sanitary infrastructure was markedly different, emphasizing the role of sanitation as an important factor for cyclosporiasis.

Endemic cyclosporiasis has been reported around the world. The first report suggested that *Cyclospora* infections were more commonly found in children, (Ortega et al., 1993) especially those under 5 years of age (Hoge et al., 1995). In Latin America, *Cyclospora* infections have been reported in children with diarrhea in Brazil (2/315) (Ribes et al., 2004) and 6.1% of people living in impoverished areas in Venezuela (Chacin-Bonilla et al., 2003). A study conducted with 36 case and 37 control Egyptian children reported *Cyclospora* infections in 5.6% of malnourished children compared to 2.8% of the controls (Rizk and Soliman, 2001). *Cyclospora* is not considered an HIV opportunistic agent, with similar incidence rates among immunocompromised and immunocompetent people. Nonetheless, a recent study in Egypt reported 6% *Cyclospora* among Hodgkins Lymphoma patients receiving chemotherapy (Rizk and Soliman, 2001). Additionally, *Cyclospora* infections were reported in 7 of 71 Venezuelan HIV infected patients and 7 of 132 otherwise normal children ages 1–12, while the highest frequency was observed in children 2–5 years of age (Chacin-Bonilla et al., 2001).

An epidemiologic study among Guatemalan children reported significant differences in the epidemiology of *Cyclospora* in Guatemala. *Cyclospora* was detected in 117 (2.1%) of 5520 specimens, mainly in children 5 years of age. *Cyclospora* infections were more strongly associated with diarrhea than *Cryptosporidium* infections (Bern et al., 2000). *Cyclospora* was also reported in 1.5% of Guatemalan people, however, none of the infected cases were raspberry farm workers (Pratdesaba et al., 2001).

Findings from a three-year longitudinal study (1995–1998) in children living in an impoverished area of Peru, showed an incidence rate of 0.20 cases per child-year, which was constant among children 1–9 years of age. Infections were more frequent during the warmer months, December to May, showing a seasonal pattern (Bern et al., 2002). *C. cayetanensis* oocysts were also detected in 1.1% of 5836 Peruvian children studied over 2 years (Madico et al., 1997). A longitudinal study in several areas of Nepal from April 1995 to November 2000 found marked seasonality, with highest infection rates occurring during the summer and rainy season of the year (Sherchand and Cross, 2001).

Additional studies have reported endemic cyclosporiasis on all continents, primarily affecting children. A study in Lagos, Nigeria, reported an overall prevalence of 0.9% (Alakpa et al., 2002, 2003). In Yunnan, China, 5.29% of pediatric diarrhea cases had *Cyclospora (Zhang et al., 2002)*. In 2002, Uga found *Cyclospora* in the Bekasi District in West Java, Indonesia (Uga et al., 2002). *Cyclospora* infections have been reported in pediatric and adult patients with diarrhea in Tanzania (Cegielski et al., 1999) and the sub-Saharan region (Markus and Frean, 1993).

3.5.2 *Isospora*

Isospora belli infections are infrequently reported in humans, however, patients with impaired immune systems may develop chronic life-threatening diarrhea. Within this population, there has been a sharp drop in the incidence of isosporiasis due to the prophylactic use of trimethropin sulfamethoxazole. The clinical, transmission, and epidemiological features of human isosporiasis are usually found in the form of case reports.

Human isosporiasis affecting immunocompetent people is very rare. In the United States, a review of human stools received in Kentucky clinics from March to September 2003 did not detect *Isospora* positive specimens (Ribes et al., 2004). Similar findings were reported from Japan, where a 7-year study of 4273 specimens collected from patients with infectious enteritis and admitted to hospitals found only three positive samples (Obana et al., 2002). In developing countries, however, *I. belli* has been reported in children with diarrhea (Tavarez et al., 1991).

In patients with AIDS-associated complications, *I. belli* infections accompanied by chronic diarrhea and wasting were reported in people not receiving or with subtherapeutic plasma levels of antiretroviral medications (Brantley et al., 2003; Maiga et al., 2002). In developing countries, the prevalence of *I. belli* infections in HIV-infected patients varies from 1.5 to 17% (Brandonisio et al., 1999; Cimerman et al., 1999; Cranendonk et al., 2003; Ferreira, 2000; Gassama et al., 2001; Joshi et al., 2002; Lainson and da Silva, 1999; Lebbad et al., 2001; Mohandas et al., 2002; Wiwanitkit, 2001).

Severe *Isospora* infections have also been reported in patients with neoplastic diseases (Makni et al., 2000; Resiere et al., 2003) and chronic renal failure (Ali et al., 2000).

3.5.3 *Sarcocystis*

Despite its biological similarities with *Toxoplasma* and *Isospora*, *Sarcocystis* is not considered an AIDS-opportunistic infection (Dionisio et al., 1992). Most reports of cases of *Sarcocystis* in animals and humans are from Asia.

3.5.3.1 Human Infections

Reports of human infections with *Sarcocystis* are limited. Most of them are from Asia. In Thailand, six patients aged 3 to 70 years presented with acute enteritis associated with *Sarcocystis* infection (Bunyaratvej et al., 1982).

In Tibet, fecal specimens of 926 persons from Duilongdeqing, Milin, and Linzhi counties were examined. The prevalences of *S. hominis* in the three counties were 20.5, 22.5, and 22.9% respectively, with an average of 21.8%. *S. suihominis* prevalence was 0, 0.6, and 7.0% respectively. No significant difference in infection rate was found between different age or sex groups. *Sarcocystis* was detected in 42.9% of beef specimens from markets. The infected cases were generally asymptomatic (Yu, 1991).

In Malaysia, *Sarcocystis* cysts have been reported from domestic and wild animals, including domestic and field rats, moonrats, bandicoots, slow loris, buffalo, and monkey. The overall seroprevalence in humans was 19.8% among the main racial groups in Malaysia (Kan and Pathmanathan, 1991).

Sarcocystis sp. was identified in 14 Vietnamese individuals from a total of 1228 examined (1.1%) who came to Central Slovakia in the course of 18 months in 1987–1989. The subjects were from the north eastern part of the country from Hanoi-Haiphong areas (Straka *et al.*, 1991).

Three cases of muscular sarcocystosis from West Malaysia were reported. Eight of 11 cases were associated with malignancies, especially of the tongue and nasopharynx (Pathmanathan and Kan, 1992).

The prevalence of human skeletal muscle sarcocystosis in Malaysia was determined by examination of tongue tissues from autopsies of subjects aged 12 years or more. Of 100 tongues examined, 21% were found to contain *Sarcocystis*. The number of cysts per case varied from 1 to 13. The age range of positive cases was from 16 to 57 years (mean 37.7 years). Prevalence did not differ with regard to race, sex, or occupation (Wong and Pathmanathan, 1992).

Sarcocystis sp. was identified in 23.2% of 362 Thai laborers who were going abroad for work. *Sarcocystis* were frequently found in male laborers (83.3%) ($p < 0.01$). The laborers from northeastern Thailand ($n = 278$) had a higher prevalence (26.6%) of *Sarcocystis* infection ($p < 0.01$) (Wilairatana *et al.*, 1996).

Fifty samples of raw kibbe from 25 Arabian restaurants in the city of Sao Paulo, Brazil, were examined for the presence of *Sarcocystis*. Sarcocysts was found in all 50 samples. Based on cyst wall structure, *S. hominis* (94%), *S. hirsuta* (70%), and *S. cruzi* (92%) were identified; mostly as mixed infections (Pena *et al.*, 2001).

A patient from West Malaysia presented with *Sarcocystis* in the larynx (Kutty and Dissanaike, 1975). *Sarcocystis* was identified in biopsy specimens from two adults in Singapore, one in Bombay, one in Uganda, and in the heart of a child in Costa Rica. Among the sarcocysts seen in the 40 cases (35 old and 5 new), seven morphological types were recognized, each representing one to several different species, all of which are zoonotic. Of the 40 *Sarcocystis* infections in man, 13 probably were acquired from Southeast Asia, 8 from India, 5 from Central or South America, 4 each from Africa and Europe, 3 from USA, 1 from China, and 2 from unknown localities (Beaver *et al.*, 1979; Gut, 1982).

3.5.3.2 *Sarcocystis* in Animals

In Czechoslovakia, *Sarcocystis cruzi*, *Sarcocystis hirsuta* and *S. hominis* were detected in 87% of 200 cattle and *Sarcocystis ovicanis* and *Sarcocystis tenella* in 92% of 100 sheep examined. All of 200 pigs examined were negative (Gut, 1982).

In New Zealand, muscle tissue from the oesophagus and diaphragm of 500 beef cattle were examined. All cattle were infected with *Sarcocystis*: 98% had *S. cruzi* and 79.8% had *S. hirsuta/S. hominis* (Bottner et al., 1987).

In Belgium, muscle tissue from the oesophagus, diaphragm, and heart of 100 cattle were examined for *Sarcocystis* infection. Of these, 97% were positive. Thick-walled cysts were recovered from 56% of animals, but these could not be identified as *S. hirsuta* and/or *S. hominis* on morphological grounds (Vercruysse et al., 1989).

In India, muscle samples from 890 slaughtered pigs were examined for the presence of sarcocysts. The prevalence rate was 67.98%, of which 43.14% was *Sarcocystis miescheriana* and 47.11% was *S. suihominis* (Saleque and Bhatia, 1991).

The prevalence of *Sarcocystis* in muscle of 36 caribou examined in Newfoundland, Canada, was 53%. Infected animals were more frequently found in the central part of the island (Khan and Fong, 1991).

In Ethiopia, *Sarcocystis* was identified in 93% of sheep, 82% of cattle, 81% of goats, 16.6% of donkeys, and 6.6% of chickens, from a total of 671 animals. None of the 40 heart muscles from bovine, ovine, caprine, and donkey fetuses examined harbored *Sarcocystis* (Woldemeskel and Gebreab, 1996).

In Fars Province of Iran, 786 (57.7%) of 1362 animals examined were positive for *Sarcocystis*. The prevalence was significantly higher ($p < 0.05$) in animals owned by nomadic Assyrians (67.95%) than in those owned by local people (41.86%). Animals older than 2 years of age (69.98%) were more infected compared to young ones (30.02%). *Sarcocystis gigantea* was predominantly identified in the oesophagus, *Sarcocystis medusiformis* mainly in the diaphragm, *S. tenella* in the oesophagus, diaphragm, tongue, and heart, and *Sarcocystis arieticanis* in the oesophagus, tongue, and occasionally in the diaphragm (Oryan et al., 1996).

In Japan, *S. suihominis* was detected for the first time in the heart and diaphragm of 5 out of 600 older culled breeding pigs slaughtered in the Saitama Prefecture (Saito et al., 1998).

A survey was carried out to investigate the occurrence of *Sarcocystis* infection in the loin of 482 Japanese and imported beef. The prevalence of *Sarcocystis* was lower in Japanese beef (total 6.31%: 0% in Holstein castrated, 12.96% in Holstein milk cow, 3.33% in Japanese shorthorn, and 11.58% in Japanese black cattle) than in beef imported from America (36.78%) or Australia (29.49%). All detected cysts except one were identified as *S. cruzi*. One thick walled cyst was found in Australian beef, but it could not be distinguished as either *S. hirsuta* or *S. hominis* (Ono and Ohsumi, 1999).

In the Upper East Region (UER) of Ghana a cross-sectional study was carried out to estimate the prevalence of parasitic infections in local cross-bred pigs. Ten out of 60 villages with a human population of 200–1000 inhabitants were randomly selected for the study. The number of pigs varied from 50 to 200 pigs per village. *Sarcocystis* sp. was observed in 28.3% animals (Permin et al., 1999).

A survey of *Sarcocystis* infection was conducted in Mongolia between June 1998 and July 1999. The prevalence of infection was cattle 90.0% (27/30), yak 93.3% (28/30), hainag 100% (30/30), sheep 96.9% (753/777), horses 75% (3/4), and camels 100% (5/5). Heart was most commonly infected in cattle (100%), yak

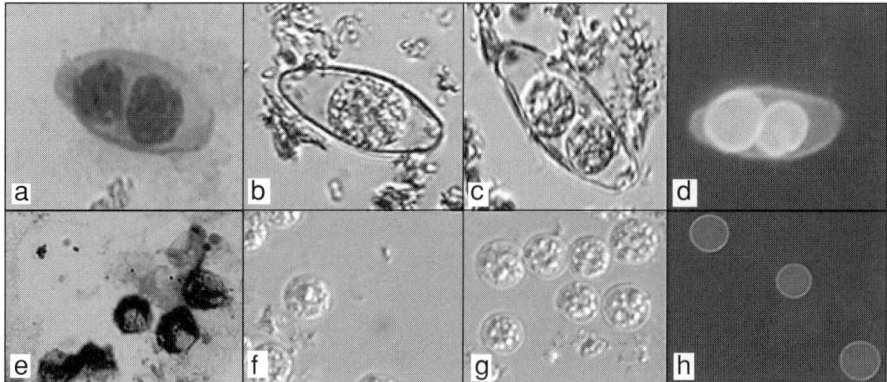

Figure 3.1. *Isospora* (a) modified acid-fast stain, (b, c) unsporulated and sporulated oocyst, (d) autofluorescence. *Cyclospora* oocysts (e) modified acid-fast stain, (f, g) unsporulated oocyst, and (g) autofluorescence.

(86.7%), and hainag (100%); tongue was most commonly infected in sheep (100%) and horses (100%) (Fukuyo *et al.*, 2002).

Domestic animals have also been reported harboring *Sarcocystis*. In Rio de Janeiro, Brazil, 0.8% of fecal samples had *Sarcocystis* sp. in a survey of 131 domesticated and stray cats (Serra *et al.*, 2003).

3.6 DIAGNOSIS

Diagnosis of these parasites is done by the identification of the oocysts or sporocysts in the fecal samples of the infected individuals. Oocysts can be observed using phase contrast or bright field or DIC microscopy (Fig. 3.1). *Cyclospora* and *Isospora* oocysts and *Sarcocystis* sporocysts autofluoresce; therefore, a fecal suspension is observed using a epifuorescence microscope with an excytation filter of 360/40 nm, a long pass dichroic mirror of 400 nm and a 420 nm emission filter (Fig. 3.1) (Lindquist *et al.*, 2003). A Kinyoun, Ziehl Nielsen, or carbolfucsin modified acid fast stain can be used to stain oocysts, however *Cyclospora* oocysts take the stain variably; therefore, if a sample contains a low number of oocysts they could be missed. A safranin stain has also been reported (Fig. 3.1). This stain works better than the previously described because oocysts stain bright pink and most oocysts take the stain (Visvesvara *et al.*, 1997).

These procedures are not very practical when examining environmental samples, because oocysts are usually in small numbers and many structures may look alike. It is then recommended to use molecular tools to identify these parasites.

In 1996, a PCR assay for *Cyclospora* was developed using clinical samples (Relman *et al.*, 1996). It was later determined that this PCR, although sensitive, was not specific for *Cyclospora*. When working with environmental samples, other coccidian parasites noninfectious to humans were also amplifying a product of the same size as *Cyclospora*. Jineman then developed a PCR-RFLP which could differentiate among these coccidia (Jinneman *et al.*, 1998). This was followed by new strategies to develop more sensitive and specific assays. These included the oligonucleotide-ligation assay (OLA) (Jinneman *et al.*, 1999) and real-time PCR (Varma *et al.*, 2003). The sensitivity of these assays could be improved by optimizing the DNA extraction procedures and eliminating PCR inhibitors. Some of these strategies included the use of FTA filters (Orlandi and Lampel, 2000) and inclusion of resin matrix treatment during oocyst disruption (Jinneman *et al.*, 1998). PCR amplification of the *Isospora* SSU-rRNA gene has also been described Li *et al.*, 2002; Muller *et al.*, 2000; Yang *et al.*, 2002).

Sarcocystis hirsuta isolates from cattle, *Sarcocystis hominis*-like isolates and a *Sarcocystis cruzi* isolate were examined by PCR amplification of the 18S rRNA DNA. These species could be fast and easily differentiated by sequencing of the amplified products and PCR-RFLP (Fischer and Odening, 1998; Yang *et al.*, 2002).

An IFA-test (Indirect Fluorescent Antibody Test) for the diagnosis of sarcocystosis in the intermediate host was identified in mice inoculated experimentally with *Sarcocystis dispersa*; however, cross reaction was observed with *Frenkelia* (Cerna and Kolarova, 1978). Habeeb studied the humoral response in patients with idiopathic cardiac diseases and rheumatic diseases. The enzyme-linked immunosorbent assay (ELISA) and indirect fluorescent antibody technique (IFAT) using *Sarcocystis fusiformis* antigen could specifically identify cases with *Sarcocystis* and idiopathic cardiac diseases (Cerna and Kolarova, 1978; Habeeb *et al.*, 1996).

Histopathological descriptions of these infections have been reported. Microscopic examination of the excised tissue of cases that presented with lumps, pain in the limbs, or a discharging sinus showed characteristic cysts of *Sarcocystis* (Mehrotra *et al.*, 1996). Parasitic vacuoles of patients with cyclosporiasis were observed more commonly in individuals that were immunocompromised. Immunocompetent individuals rarely presented parasitic vacuoles; however, the epithelial disarray and inflammation was consistently present (Ortega *et al.*, 1997a). *Isospora* intracellular stages have been observed in intestinal biopsies of infected individuals, particularly in AIDS patients.

3.7 TREATMENT AND CONTROL

Treatment of choice for cyclosporiasis is trimethoprim sulfamethoxazole (TMP-SMX) (Madico *et al.*, 1993; Pape *et al.*, 1994). In a randomized control trial, ciprofloxacin was suggested as an alternative treatment for patients with cyclosporiasis who are allergic to sulfa drugs (Verdier *et al.*, 2000).

Trimethoprim sulfamethoxazole (TMP/SMX) is the drug of choice for treatment of isosporiasis. However, patients allergic or intolerant to TMP/SMX may take Ciprofloxacine as an alternative treatment. (Verdier *et al.*, 2000). Patients with AIDS have a high rate of adverse reactions to this therapy. Two patients with AIDS and isosporiasis that were sulfonamide allergic were treated successfully with pyrimethamine alone, 75 mg/d, and recurrence was prevented with daily pyrimethamine therapy, 25 mg/d (Weiss *et al.*, 1988). Doxycycline and nifuroxazide has also been studied in AIDS patients. Relapsed *Isospora* infections were long-term treated with doxycycline (Meyohas *et al.*, 1990).

Sarcocystis infections in humans responded to a one month treatment with sulfadiazine or finidazole (Yu, 1991).

Because the mechanisms of transmission of *Cyclospora* and *Isospora* are associated with the consumption of contaminated water or foods, control of these infections can be achieved by avoiding fresh produce in areas of endemnicity and drinking purified water (by filtration or boiling). Since *Cyclospora* is highly resistant to chlorination, this procedure should not be considered as an alternative water treatment. *Sarcocystis* infections can be prevented by eating well-cooked meats, particularly game meats, and pork (Yu, 1991).

The methods currently used in the sanitation of fresh food products are ineffective against *Cyclospora, Isospora* and *Sarcocystis*. These parasites are particularly resistant to sanitizers and disinfectants. Therefore, prevention is the best strategy to reduce the risk of infections and can be accomplished by the implementation of good agricultural practices during all stages of production, harvesting and commercialization.

REFERENCES

Anonymous, 2004, Outbreak of cyclosporiasis associated with snow peas—Pennsylvania, 2004, *Morb. Mortal. Wkly. Rep.* **53**: 876–878.

Alakpa, G., Fagbenro-Beyioku, A. F., and Clarke, S. C., 2002, *Cyclospora cayetanensis* in stools submitted to hospitals in Lagos, Nigeria, *Int. J. Infect. Dis.* **6**:314–318.

Alakpa, G. E., Clarke, S. C., and Fagbenro-Beyioku, A. F., 2003, *Cyclospora cayetanensis* infection in Lagos, Nigeria, *Clin. Microbiol. Infect.* **9**:731–733.

Ali, M. S., Mahmoud, L. A., Abaza, B. E., and Ramadan, M. A., 2000, Intestinal spore-forming protozoa among patients suffering from chronic renal failure, *J. Egypt. Soc. Parasitol.* **30**:93–100.

Arnaud-Battandier, F., 1985, Cryptosporidiosis, isosporiasis, lambliasis in immunologic deficiencies, *Arch. Fr. Pediatr.* **42**(Suppl 2):959–963.

Ashford, R. W., 1979, Occurrence of an undescribed coccidian in man in Papua New Guinea, *Ann. Trop. Med. Parasitol.* **73**:497–500.

Azab, M. E., and el-Shennawy, S. F., 1992, Investigation of *Sarcocystis* as a causative agent in cardiac disease, *J. Egypt. Soc. Parasitol.* **22**:611–616.

Beaver, P. C., Gadgil, K., and Morera, P., 1979, *Sarcocystis* in man: A review and report of five cases, *Am. J. Trop. Med. Hyg.* **28**:819–844.

Beck, J. W., Stanton, R. L., and Langford, G. C., Jr., 1955, Human infection with *Isospora belli*; report of a case in Florida, *Am. J. Clin. Pathol.* **25**:648–651.

Bern, C., Hernandez, B., Lopez, M. B., Arrowood, M. J., de Merida, A. M., and Klein, R. E., 2000, The contrasting epidemiology of *Cyclospora* and *Cryptosporidium* among outpatients in Guatemala, *Am. J. Trop. Med. Hyg.* **63**:231–235.

Bern, C., Ortega, Y., Checkley, W., Roberts, J. M., Lescano, A. G., Cabrera, L., Verastegui, M., Black, R. E., Sterling, C., and Gilman, R. H., 2002, Epidemiologic differences between cyclosporiasis and cryptosporidiosis in Peruvian children, *Emerg. Infect. Dis.* **8**:581–585.

Bottner, A., Charleston, W. A., Pomroy, W. E., and Rommel, M., 1987, The prevalence and identity of *Sarcocystis* in beef cattle in New Zealand, *Vet. Parasitol.* **24**:157–168.

Brandonisio, O., Maggi, P., Panaro, M. A., Lisi, S., Andriola, A., Acquafredda, A., and Angarano, G., 1999, Intestinal protozoa in HIV-infected patients in Apulia, South Italy, *Epidemiol. Infect.* **123**:457–462.

Brantley, R. K., Williams, K. R., Silva, T. M., Sistrom, M., Thielman, N. M., Ward, H., Lima, A. A., and Guerrant, R. L., 2003, AIDS-associated diarrhea and wasting in Northeast Brazil is associated with subtherapeutic plasma levels of antiretroviral medications and with both bovine and human subtypes of *Cryptosporidium parvum*, *Braz. J. Infect. Dis.* **7**:16–22.

Bunyaratvej, S., and Unpunyo, P., 1992, Combined *Sarcocystis* and gram-positive bacterial infections. A possible cause of segmental enterocolitis in Thailand, *J. Med. Assoc. Thai.* **75**(Suppl 1):38–44.

Bunyaratvej, S., Bunyawongwiroj, P., and Nitiyanant, P., 1982, Human intestinal sarcosporidiosis: Report of six cases, *Am. J. Trop. Med. Hyg.* **31**:36–41.

Caceres, V. M., Ball, R. T., Somerfeldt, S. A., Mackey, R. L., Nichols, S. E., MacKenzie, W. R., and Herwaldt, B. L., 1998, A foodborne outbreak of cyclosporiasis caused by imported raspberries, *J. Fam. Pract.* **47**:231–234.

Carollo, M. C., Amato Neto, V., Braz, L. M., and Kim, D. W., 2001, Detection of *Cyclospora* sp. oocysts in the feces of stray dogs in Greater Sao Paulo (Sao Paulo State, Brazil), *Rev. Soc. Bras. Med. Trop.* **34**:597–598.

Cegielski, J. P., Ortega, Y. R., McKee, S., Madden, J. F., Gaido, L., Schwartz, D. A., Manji, K., Jorgensen, A. F., Miller, S. E., Pulipaka, U. P., Msengi, A. E., Mwakyusa, D. H., Sterling, C. R., and Reller, L. B., 1999, *Cryptosporidium*, enterocytozoon, and *cyclospora* infections in pediatric and adult patients with diarrhea in Tanzania, *Clin. Infect. Dis.* **28**:314–321.

Cerna, Z., and Kolarova, I., 1978, Contribution to the serological diagnosis of sarcocystosis, *Folia Parasitol. (Praha)* **25**:289–292.

Chacin-Bonilla, L., Estevez, J., Monsalve, F., and Quijada, L., 2001, *Cyclospora cayetanensis* infections among diarrheal patients from Venezuela, *Am. J. Trop. Med. Hyg.* **65**:351–354.

Chacin-Bonilla, L., Mejia de Young, M., and Estevez, J., 2003, Prevalence and pathogenic role of *Cyclospora cayetanensis* in a Venezuelan community, *Am. J. Trop. Med. Hyg.* **68**:304–306.

Charatan, F. B., 1996, *Cyclospora* outbreak in US, *BMJ* **313**:71.

Chen, X., Zuo, Y., and Zuo, W., 1999, Observation on the clinical symptoms and sporocyst excretion in human volunteers experimentally infected with *Sarcocystis hominis*, *Zhongguo Ji Sheng Chong Xue Yu Ji Sheng Chong Bing Za Zhi* **17**:25–27.

Chu, D. M., Sherchand, J. B., Cross, J. H., and Orlandi, P. A., 2004, Detection of *Cyclospora cayetanensis* in animal fecal isolates from Nepal using an FTA filter-base polymerase chain reaction method, *Am. J. Trop. Med. Hyg.* **71**:373–379.

Cimerman, S., Cimerman, B., and Lewi, D. S., 1999, Prevalence of intestinal parasitic infections in patients with acquired immunodeficiency syndrome in Brazil, *Int. J. Infect. Dis.* **3**:203–206.

Connor, B. A., and Shlim, D. R., 1995, Foodborne transmission of *Cyclospora*, *Lancet* **346**:1634.

Cranendonk, R. J., Kodde, C. J., Chipeta, D., Zijlstra, E. E., and Sluiters, J. F., 2003, *Cryptosporidium parvum* and *Isospora belli* infections among patients with and without diarrhoea, *East Afr. Med. J.* **80**:398–401.

Dionisio, D., Santucci, M., Comin, C. E., Di Lollo, S., Orsi, A., Gabbrielli, M., Milo, D., Rogasi, P. G., Meli, M., and Vigano, S., 1992, Isosporiasis and sarcocystosis: The current findings, *Recenti. Prog. Med.* **83**:719–725.

Dissanaike, A. S., 1994, Human *Sarcocystis* infection, *Trans. R. Soc. Trop. Med. Hyg.* **88**:364.

Doller, P. C., Dietrich, K., Filipp, N., Brockmann, S., Dreweck, C., Vonthein, R., Wagner-Wiening, C., and Wiedenmann, A., 2002, Cyclosporiasis outbreak in Germany associated with the consumption of salad, *Emerg. Infect. Dis.* **8**:992–994.

Dowd, S. E., John, D., Eliopolus, J., Gerba, C. P., Naranjo, J., Klein, R., Lopez, B., de Mejia, M., Mendoza, C. E., and Pepper, I. L., 2003, Confirmed detection of *Cyclospora cayetanesis*, Encephalitozoon intestinalis and *Cryptosporidium parvum* in water used for drinking, *J. Water Health* **1**:117–123.

Drenaggi, D., Cirioni, O., Giacometti, A., Fiorentini, A., and Scalise, G., 1998, Cyclosporiasis in a traveler returning from South America, *J. Travel Med.* **5**:153–155.

Dubey, J. P., 1976, A review of *Sarcocystis* of domestic animals and of other coccidia of cats and dogs, *J. Am. Vet. Med. Assoc.* **169**:1061–1078.

Dubey, J. P., 1993, *Toxoplasma, Neospora, Sarcocystis*, and other Tissue Cyst-Forming Coccidia of Humans and Animals. In Kreier, J. P. and Baker, J. R. (eds), *Pathogenic Protozoa*, Academic Press, Inc., San Diego, California, pp. 1–156.

Eberhard, M. L., Ortega, Y. R., Hanes, D. E., Nace, E. K., Do, R. Q., Robl, M. G., Won, K. Y., Gavidia, C., Sass, N. L., Mansfield, K., Gozalo, A., Griffiths, J., Gilman, R., Sterling, C. R., and Arrowood, M. J., 2000, Attempts to establish experimental *Cyclospora cayetanensis* infection in laboratory animals, *J. Parasitol.* **86**:577–582.

Eberhard, M. L., Njenga, M. N., DaSilva, A. J., Owino, D., Nace, E. K., Won, K. Y., and Mwenda, J. M., 2001, A survey for *Cyclospora* spp. in Kenyan primates, with some notes on its biology, *J. Parasitol.* **87**:1394–1397.

Fayer, R., 2004, *Sarcocystis* spp. in human infections, *Clin. Microbiol. Rev.* **17**:894–902.

Fayer, R., Heydorn, A. O., Johnson, A. J., and Leek, R. G., 1979, Transmission of *Sarcocystis suihominis* from humans to swine to nonhuman primates (Pan troglodytes, Macaca mulatta, Macaca irus), *Z. Parasitenkd.* **59**:15–20.

Ferreira, L. F., Coutinho, S. G., Argento, C. A., and da, S. J., 1962, Experimental human coccidial enteritis by *Isospora belli* Wenyon, 1923: A study based on the infection of 5 volunteers, *Hospital (Rio J)* **62**:795–804.

Ferreira, M. S., 2000, Infections by protozoa in immunocompromised hosts, *Mem. Inst. Oswaldo Cruz* **95**(Suppl 1):159–162.

Figueroa, F., Palacios, A., Rivero, S., Oddo, D., Roa, I., Honeyman, J., Gatica, M. A., and Acuna, G., 1985, Chronic diarrhea due to *Isospora belli* and Kaposi's sarcoma in a male homosexual. Report of the 1st case of acquired immunodeficiency syndrome in Chile, *Rev. Med. Chil.* **113**:772–779.

Fischer, S., and Odening, K., 1998, Characterization of bovine *Sarcocystis* species by analysis of their 18S ribosomal DNA sequences, *J. Parasitol.* **84**:50–54.

Fleming, C. A., Caron, D., Gunn, J. E., and Barry, M. A., 1998, A foodborne outbreak of *Cyclospora cayetanensis* at a wedding: Clinical features and risk factors for illness, *Arch. Intern. Med.* **158**:1121–1125.

Frenkel, J. K., Heydorn, A. O., Mehlhorn, H., and Rommel, M., 1979, Sarcocystinae: Nomina dubia and available names, *Z. Parasitenkd.* **58**:115–139.

Fukuyo, M., Battsetseg, G., and Byambaa, B., 2002, Prevalence of *Sarcocystis* infection in meat-producing animals in Mongolia, *Southeast Asian J. Trop. Med. Public Health* **33**:490–495.

Furio, M. M., and Wordell, C. J., 1985, Treatment of infectious complications of acquired immunodeficiency syndrome, *Clin. Pharm.* **4**:539–554.

Garcia-Lopez, H. L., Rodriguez-Tovar, L. E., and Medina-De la Garza, C. E., 1996, Identification of *Cyclospora* in poultry, *Emerg. Infect. Dis.* **2**:356–357.

Gascon, J., Corachan, M., Bombi, J. A., Valls, M. E., and Bordes, J. M., 1995, *Cyclospora* in patients with traveller's diarrhea, *Scand. J. Infect. Dis.* **27**:511–514.

Gassama, A., Sow, P. S., Fall, F., Camara, P., Gueye-N'diaye, A., Seng, R., Samb, B., M'Boup, S., and Aidara-Kane, A., 2001, Ordinary and opportunistic enteropathogens associated with diarrhea in Senegalese adults in relation to human immunodeficiency virus serostatus, *Int. J. Infect. Dis.* **5**:192–198.

Gut, J., 1982, Effectiveness of methods used for the detection of sarcosporidiosis in farm animals, *Folia Parasitol. (Praha)* **29**:289–295.

Habeeb, Y. S., Selim, M. A., Ali, M. S., Mahmoud, L. A., Abdel Hadi, A. M., and Shafei, A., 1996, Serological diagnosis of extraintestinal Sarcocystosis, *J. Egypt. Soc. Parasitol.* **26**:393–400.

Henry, M. C., De Clercq, D., Lokombe, B., Kayembe, K., Kapita, B., Mamba, K., Mbendi, N., and Mazebo, P., 1986, Parasitological observations of chronic diarrhoea in suspected AIDS adult patients in Kinshasa (Zaire), *Trans. R. Soc. Trop. Med. Hyg.* **80**:309–310.

Herwaldt, B. L., 2000, *Cyclospora cayetanensis*: A review, focusing on the outbreaks of cyclosporiasis in the 1990s, *Clin. Infect. Dis.* **31**:1040–1057.

Herwaldt, B. L., and Ackers, M. L., 1997, An outbreak in 1996 of cyclosporiasis associated with imported raspberries. The *Cyclospora* Working Group, *N. Engl. J. Med.* **336**:1548–1556.

Herwaldt, B. L., and Beach, M. J., 1999, The return of *Cyclospora* in 1997: Another outbreak of cyclosporiasis in North America associated with imported raspberries. *Cyclospora* Working Group, *Ann. Intern. Med.* **130**:210–220.

Heydorn, A. O., 1977, Life-cycle of Sarcosporidia. IX. Developmental cyclus of *Sarcocystis suihominis* n. spec., *Berl. Munch. Tierarztl. Wochenschr.* **90**:218–224.

Heydorn, A. O., Mehlhorn, H., and Gestrich, R., 1975, Light and electron microscopic studies on cysts of *Sarcocystis fusiformis* in the muscles of calves infected experimentally with oocytes and sporocysts of the large form of *Isospora* bigemina from dogs. 2. The fine structure of cyst stages., *Zentralbl. Bakteriol. [Orig A]* **233**:123–137.

Hoge, C. W., Shlim, D. R., Rajah, R., Triplett, J., Shear, M., Rabold, J. G., and Echeverria, P., 1993, Epidemiology of diarrhoeal illness associated with coccidian-like organism among travellers and foreign residents in Nepal, *Lancet* **341**:1175–1179.

Hoge, C. W., Echeverria, P., Rajah, R., Jacobs, J., Malthouse, S., Chapman, E., Jimenez, L. M., and Shlim, D. R., 1995, Prevalence of *Cyclospora* species and other enteric pathogens among children less than 5 years of age in Nepal, *J. Clin. Microbiol.* **33**:3058–3060.

Huang, P., Weber, J. T., Sosin, D. M., Griffin, P. M., Long, E. G., Murphy, J. J., Kocka, F., Peters, C., and Kallick, C., 1995, The first reported outbreak of diarrheal illness associated with *Cyclospora* in the United States, *Ann. Intern. Med.* **123**:409–414.

Jelinek, T., Lotze, M., Eichenlaub, S., Loscher, T., and Nothdurft, H. D., 1997, Prevalence of infection with *Cryptosporidium parvum* and *Cyclospora cayetanensis* among international travellers, *Gut* **41**:801–804.

Jinneman, K. C., Wetherington, J. H., Hill, W. E., Adams, A. M., Johnson, J. M., Tenge, B. J., Dang, N. L., Manger, R. L., and Wekell, M. M., 1998, Template preparation for PCR and RFLP of amplification products for the detection and identification of *Cyclospora* sp. and *Eimeria* spp. Oocysts directly from raspberries, *J. Food Prot.* **61**:1497–1503.

Jinneman, K. C., Wetherington, J. H., Hill, W. E., Omiescinski, C. J., Adams, A. M., Johnson, J. M., Tenge, B. J., Dang, N. L., and Wekell, M. M., 1999, An oligonucleotide-ligation assay for the differentiation between *Cyclospora* and Eimeria spp. polymerase chain reaction amplification products, *J. Food Prot.* **62**:682–685.

Jonas, C., Van de Perre, P., Reding, P., Burette, A., Deprez, C., Clumeck, N., and Deltenre, M., 1984, Severe digestive complications of AIDS in a group of patients from Zaire, *Acta Gastroenterol. Belg.* **47**:396–402.

Joshi, M., Chowdhary, A. S., Dalal, P. J., and Maniar, J. K., 2002, Parasitic diarrhoea in patients with AIDS, *Natl. Med. J. India* **15**:72–74.

Kan, S. P., and Pathmanathan, R., 1991, Review of sarcocystosis in Malaysia, *Southeast Asian J. Trop. Med. Public Health* **22** (Suppl):129–134.

Katz, D., Kumar, S., Malecki, J., Lowdermilk, M., Koumans, E. H., and Hopkins, R., 1999, Cyclosporiasis associated with imported raspberries, Florida, 1996, *Public Health Rep.* **114**:427–438.

Khan, R. A., and Fong, D., 1991, *Sarcocystis* in caribou (Rangifer tarandus terraenorae) in Newfoundland, *Southeast Asian J. Trop. Med. Public Health* **22**(Suppl):142–143.

Kimmig, P., Piekarski, G., and Heydorn, A. O., 1979, Sarcosporidiosis (*Sarcocystis suihominis*) in man (author's transl), *Immun. Infekt.* **7**:170–177.

Kobayashi, L. M., Kort, M. P., Berlin, O. G., and Bruckner, D. A., 1985, *Isospora* infection in a homosexual man, *Diagn. Microbiol. Infect. Dis.* **3**:363–366.

Kutty, M. K., and Dissanaike, A. S., 1975, A case of human *Sarcocystis* infection in west Malaysia, *Trans. R. Soc. Trop. Med. Hyg.* **69**:503–504.

Lainson, R., and da Silva, B. A., 1999, Intestinal parasites of some diarrhoeic HIV-seropositive individuals in North Brazil, with particular reference to *Isospora belli* Wenyon, 1923 and Dientamoeba fragilis Jepps & Dobell, 1918, *Mem. Inst. Oswaldo Cruz* **94**:611–613.

Lebbad, M., Norrgren, H., Naucler, A., Dias, F., Andersson, S., and Linder, E., 2001, Intestinal parasites in HIV-2 associated AIDS cases with chronic diarrhoea in Guinea-Bissau, *Acta. Trop.* **80**:45–49.

Li, Q. Q., Yang, Z. Q., Zuo, Y. X., Attwood, S. W., Chen, X. W., and Zhang, Y. P., 2002, A PCR-based RFLP analysis of *Sarcocystis cruzi* (Protozoa: Sarcocystidae) in Yunnan Province, PR China, reveals the water buffalo (Bubalus bubalis) as a natural intermediate host, *J. Parasitol.* **88**:1259–1261.

Lian, Z., Ma, J., Wang, Z., Fu, L., Zhou, Z., Li, W., and Wang, X., 1990, Studies on man-cattle-man infection cycle of *Sarcocystis hominis* in Yunnan, *Zhongguo Ji Sheng Chong Xue Yu Ji Sheng Chong Bing Za Zhi* **8**:50–53.

Lindquist, H. D., Bennett, J. W., Hester, J. D., Ware, M. W., Dubey, J. P., and Everson, W. V., 2003, Autofluorescence of *Toxoplasma* gondii and related coccidian oocysts, *J. Parasitol.* **89**:865–867.

Lindsay, S. S., Dubey, J. P., and Blagburn, B. L., 1997, Biology of *Isospora* spp. from humans, nonhuman primates and domestic animals, *Clin. Microb. Rev.* **10**:19–34.

Long, E. G., White, E. H., Carmichael, W. W., Quinlisk, P. M., Raja, R., Swisher, B. L., Daugharty, H., and Cohen, M. T., 1991, Morphologic and staining characteristics of a cyanobacterium-like organism associated with diarrhea, *J. Infect. Dis.* **164**:199–202.

Lopez, A. S., Dodson, D. R., Arrowood, M. J., Orlandi Jr, P. A., da Silva, A. J., Bier, J. W., Hanauer, S. D., Kuster, R. L., Oltman, S., Baldwin, M. S., Won, K. Y., Nace, E. M., Eberhard, M. L., and Herwaldt, B. L., 2001, Outbreak of cyclosporiasis associated with basil in Missouri in 1999, *Clin. Infect. Dis.* **32**:1010–1017.

Lopez, A. S., Bendik, J. M., Alliance, J. Y., Roberts, J. M., da Silva, A. J., Moura, I. N., Arrowood, M. J., Eberhard, M. L., and Herwaldt, B. L., 2003, Epidemiology of *Cyclospora cayetanensis* and other intestinal parasites in a community in Haiti, *J. Clin. Microbiol.* **41**:2047–2054.

Ma, P., Kaufman, D., and Montana, J., 1983, *Isospora belli* diarrheal infection in homosexual men, *AIDS Res.* **1**:327–338.

Madico, G., Gilman, R. H., Miranda, E., Cabrera, L., and Sterling, C. R., 1993, Treatment of *Cyclospora* infections with co-trimoxazole, *Lancet* **342**:122–123.

Madico, G., McDonald, J., Gilman, R. H., Cabrera, L., and Sterling, C. R., 1997, Epidemiology and treatment of *Cyclospora cayetanensis* infection in Peruvian children, *Clin. Infect. Dis.* **24**:977–981.

Maiga, M. Y., Dembele, M. Y., Traore, H. A., Kouyate, M., Traore, A. K., Maiga, I., Bougoudogo, F., Doumbo, O., and Guindo, A., 2002, Gastrointestinal manifestations of AIDS in adults in Mali, *Bull. Soc. Pathol. Exot.* **95**:253–256.

Makni, F., Cheikrouhou, F., and Ayadi, A., 2000, Parasitoses and immunodepression, *Arch. Inst. Pasteur. Tunis.* **77**:51–54.

Mansfield, L. S., and Gajadhar, A. A., 2004, *Cyclospora cayetanensis*, a food- and waterborne coccidian parasite, *Vet. Parasitol.* **126**:73–90.

Markus, M. B., and Frean, J. A., 1993, Occurrence of human *Cyclospora* infection in sub-Saharan Africa, *S. Afr. Med. J.* **83**:862–863.

McLeod, R., Hirabayashi, R. N., Rothman, W., and Remington, J. S., 1980, Necrotizing vasculitis and Sarcocystis: A cause-and-effect relationship? *South. Med. J.* **73**:1380–1383.

Mehrotra, R., Bisht, D., Singh, P. A., Gupta, S. C., and Gupta, R. K., 1996, Diagnosis of human *sarcocystis* infection from biopsies of the skeletal muscle, *Pathology* **28**:281–282.

Meyohas, M. C., Capella, F., Poirot, J. L., Lecomte, I., Binet, D., Eliaszewicz, M., and Frottier, J., 1990, Treatment with doxycycline and nifuroxazide of *Isospora belli* infection in AIDS, *Pathol. Biol. (Paris)* **38**:589–591.

Modigliani, R., Bories, C., Le Charpentier, Y., Salmeron, M., Messing, B., Galian, A., Rambaud, J. C., Lavergne, A., Cochand-Priollet, B., and Desportes, I., 1985, Diarrhoea and malabsorption in acquired immune deficiency syndrome: A study of four cases with special emphasis on opportunistic protozoan infestations, *Gut* **26**:179–187.

Mohandas, S., R., Sud, A., and Malla, N., 2002, Prevalence of intestinal parasitic pathogens in HIV-seropositive individuals in Northern India, *Jpn J. Infect. Dis.* **55**:83–84.

Muller, A., Bialek, R., Fatkenheuer, G., Salzberger, B., Diehl, V., and Franzen, C., 2000, Detection of *Isospora belli* by polymerase chain reaction using primers based on small-subunit ribosomal RNA sequences, *Eur. J. Clin. Microbiol. Infect. Dis.* **19**:631–634.

Ng, E., Markell, E. K., Fleming, R. L., and Fried, M., 1984, Demonstration of *Isospora belli* by acid-fast stain in a patient with acquired immune deficiency syndrome, *J. Clin. Microbiol.* **20**:384–386.

Obana, M., Sagara, H., Aoki, T., Kim, R., Takizawa, Y., Tsunoda, T., Irimajiri, S., and Yamashita, K., 2002, The current status of infectious enteritis in Japan—reports of the "Research Group for Infectious Enteric Diseases, Japan" in the last 5 years (1996–2000), *Kansenshogaku Zasshi* **76**:355–368.

Ono, M., and Ohsumi, T., 1999, Prevalence of *Sarcocystis* spp. cysts in Japanese and imported beef (Loin: Musculus longissimus), *Parasitol. Int.* **48**:91–94.

Orlandi, P. A., and Lampel, K. A., 2000, Extraction-free, filter-based template preparation for rapid and sensitive PCR detection of pathogenic parasitic protozoa, *J. Clin. Microbiol.* **38**:2271–2277.

Ortega, Y. R., Sterling, C. R., Gilman, R. H., Cama, V. A., and Diaz, F., 1993, *Cyclospora* species–a new protozoan pathogen of humans, *N. Engl. J. Med.* **328**:1308–1312.

Ortega, Y. R., Gilman, R. H., and Sterling, C. R., 1994, A new coccidian parasite (Apicomplexa: Eimeriidae) from humans, *J. Parasitol.* **80**:625–629.

Ortega, Y. R., Nagle, R., Gilman, R. H., Watanabe, J., Miyagui, J., Quispe, H., Kanagusuku, P., Roxas, C., and Sterling, C. R., 1997a, Pathologic and clinical findings in patients with cyclosporiasis and a description of intracellular parasite life-cycle stages, *J. Infect. Dis.* **176**:1584–1589.

Ortega, Y. R., Roxas, C. R., Gilman, R. H., Miller, N. J., Cabrera, L., Taquiri, C., and Sterling, C. R., 1997b, Isolation of *Cryptosporidium parvum* and *Cyclospora cayetanensis* from vegetables collected in markets of an endemic region in Peru, *Am. J. Trop. Med. Hyg.* **57**:683–686.

Oryan, A., Moghaddar, N., and Gaur, S. N., 1996, The distribution pattern of *Sarcocystis* species, their transmission and pathogenesis in sheep in Fars Province of Iran, *Vet. Res. Commun.* **20**:243–253.

Pamphlett, R., and O'Donoghue, P., 1990, *Sarcocystis* infection of human muscle, *Aust. N. Z. J. Med.* **20**:705–707.

Pamphlett, R., and O'Donoghue, P., 1992, Antibodies against *Sarcocystis* and *Toxoplasma* in humans with the chronic fatigue syndrome, *Aust. N. Z. J. Med.* **22**:307–308.

Panosian, C. B., 1988, Parasitic diarrhea, *Infect. Dis. Clin. North. Am.* **2**:685–703.

Pape, J. W., Verdier, R. I., Boncy, M., Boncy, J., and Johnson, W. D., Jr., 1994, *Cyclospora* infection in adults infected with HIV. Clinical manifestations, treatment, and prophylaxis, *Ann. Intern. Med.* **121**:654–657.

Pathmanathan, R., and Kan, S. P., 1992, Three cases of human *Sarcocystis* infection with a review of human muscular sarcocystosis in Malaysia, *Trop. Geogr. Med.* **44**:102–108.

Pena, H. F., Ogassawara, S., and Sinhorini, I. L., 2001, Occurrence of cattle *Sarcocystis* species in raw kibbe from Arabian food establishments in the city of Sao Paulo, Brazil, and experimental transmission to humans, *J. Parasitol.* **87**:1459–1465.

Permin, A., Yelifari, L., Bloch, P., Steenhard, N., Hansen, N. P., and Nansen, P., 1999, Parasites in cross-bred pigs in the Upper East region of Ghana, *Vet. Parasitol.* **87**:63–71.

Piekarski, G., Heydorn, A. O., Aryeetey, M. E., Hartlapp, J. H., and Kimmig, P., 1978, Clinical, parasitological and serological investigations in sarcosporidiosis (*Sarcocystis suihominis*) of man (author's transl), *Immun. Infekt.* **6**:153–159.

Pozio, E., 1991, Current status of food-borne parasitic zoonoses in Mediterranean and African regions, *Southeast Asian J. Trop. Med. Public Health* **22**(Suppl):85–87.

Pratdesaba, R. A., Gonzalez, M., Piedrasanta, E., Merida, C., Contreras, K., Vela, C., Culajay, F., Flores, L., and Torres, O., 2001, *Cyclospora cayetanensis* in three populations at risk in Guatemala, *J. Clin. Microbiol.* **39**:2951–2953.

Rabold, J. G., Hoge, C. W., Shlim, D. R., Kefford, C., Rajah, R., and Echeverria, P., 1994, *Cyclospora* outbreak associated with chlorinated drinking water, *Lancet* **344**:1360–1361.

Relman, D. A., Schmidt, T. M., Gajadhar, A., Sogin, M., Cross, J., Yoder, K., Sethabutr, O., and Echeverria, P., 1996, Molecular phylogenetic analysis of *Cyclospora*, the human intestinal pathogen, suggests that it is closely related to Eimeria species, *J. Infect. Dis.* **173**:440–445.

Resiere, D., Vantelon, J. M., Bouree, P., Chachaty, E., Nitenberg, G., and Blot, F., 2003, *Isospora belli* infection in a patient with non-Hodgkin's lymphoma, *Clin. Microbiol. Infect.* **9**:1065–1067.

Ribes, J. A., Seabolt, J. P., and Overman, S. B., 2004, Point prevalence of *Cryptosporidium*, *Cyclospora*, and *Isospora* infections in patients being evaluated for diarrhea, *Am. J. Clin. Pathol.* **122**:28–32.

Rizk, H., and Soliman, M., 2001, Coccidiosis among malnourished children in Mansoura, Dakahlia Governorate, Egypt, *J. Egypt. Soc. Parasitol.* **31**:877–886.

Rommel, M., and Heydorn, A. O., 1972, Contributions to the life cycle of Sarcosporidia. 3. *Isospora hominis* (Railliet and Lucet, 1891) Wenyon, 1923, the sporocyst of the Sarcosporidia of cattle and swine, *Berl Munch Tierarztl Wochenschr* **85**:143–145.

Ros, E., Fueyo, J., Llach, J., Moreno, A., and Latorre, X., 1987, *Isospora belli* infection in patients with AIDS in Catalunya, Spain, *N. Engl. J. Med.* **317**:246–247.

Sadaka, H. A., and Zoheir, M. A., 2001, Experimental studies on cyclosporiosis, *J. Egypt. Soc. Parasitol.* **31**:65–77.

Saito, M., Shibata, Y., Ohno, A., Kubo, M., Shimura, K., and Itagaki, H., 1998, *Sarcocystis suihominis* detected for the first time from pigs in Japan, *J. Vet. Med. Sci.* **60**:307–309.

Saleque, A., and Bhatia, B. B., 1991, Prevalence of *Sarcocystis* in domestic pigs in India, *Vet. Parasitol.* **40**:151–153.

Serra, C. M., Uchoa, C. M., and Coimbra, R. A., 2003, Parasitological study with faecal samples of stray and domiciliated cats (Felis catus domesticus) from the Metropolitan Area of Rio de Janeiro, Brazil, *Rev. Soc. Bras. Med. Trop.* **36**:331–334.

Shein, R., and Gelb, A., 1984, *Isospora belli* in a patient with acquired immunodeficiency syndrome, *J. Clin. Gastroenterol.* **6**:525–528.

Sherchand, J. B., and Cross, J. H., 2001, Emerging pathogen *Cyclospora cayetanensis* infection in Nepal, *Southeast Asian J. Trop. Med. Public Health* **32**(Suppl 2):143–150.

Shields, J. M., and Olson, B. H., 2003, PCR-restriction fragment length polymorphism method for detection of *Cyclospora cayetanensis* in environmental waters without microscopic confirmation, *Appl. Environ. Microbiol.* **69**:4662–4669.

Shlim, D. R., Hoge, C. W., Rajah, R., Scott, R. M., Pandy, P., and Echeverria, P., 1999, Persistent high risk of diarrhea among foreigners in Nepal during the first 2 years of residence, *Clin. Infect. Dis.* **29**:613–616.

Sivapalasingam, S., Friedman, C. R., Cohen, L., and Tauxe, R. V., 2004, Fresh produce: A growing cause of outbreaks of foodborne illness in the United States, 1973 through 1997, *J. Food Prot.* **67**:2342–2353.

Soave, R., 1988, Cryptosporidiosis and isosporiasis in patients with AIDS, *Infect. Dis. Clin. North Am.* **2**:485–493.

Soave, R., Herwaldt, B. L., and Relman, D. A., 1998, Cyclospora, *Infect. Dis. Clin. North Am.* **12**:1–12.

Straka, S., Skracikova, J., Konvit, I., Szilagyiova, M., and Michal, L., 1991, *Sarcocystis* species in Vietnamese workers, *Cesk. Epidemiol. Mikrobiol. Imunol.* **40**:204–208.

Tavarez, L. A., Pena, F., Placencia, F., Mendoza, H. R., and Polanco, D., 1991, Prevalence of protozoans in children with acute diarrheal disease, *Arch. Domin. Pediatr.* **27**:43–47.

Teschareon, S., Jariya, P., and Tipayadarapanich, C., 1983, *Isospora belli* infection as a cause of diarrhoea, *Southeast Asian J. Trop. Med. Public Health* **14**:528–530.

Uga, S., Kimura, D., Kimura, K., and Margono, S. S., 2002, Intestinal parasitic infections in Bekasi district, West Java, Indonesia and a comparison of the infection rates determined by different techniques for fecal examination, *Southeast Asian J. Trop. Med. Public Health* **33**:462–467.

Varma, M., Hester, J. D., Schaefer, F. W. 3rd, Ware, M. W., and Lindquist, H. D., 2003, Detection of *Cyclospora cayetanensis* using a quantitative real-time PCR assay, *J. Microbiol. Methods* **53**:27–36.

Vercruysse, J., Fransen, J., and van Goubergen, M., 1989, The prevalence and identity of *Sarcocystis* cysts in cattle in Belgium, *Zentralbl Veterinarmed. B* **36**:148–153.

Verdier, R. I., Fitzgerald, D. W., Johnson, W. D., Jr., and Pape, J. W., 2000, Trimethoprim-sulfamethoxazole compared with ciprofloxacin for treatment and prophylaxis of *Isospora belli* and *Cyclospora cayetanensis* infection in HIV-infected patients. A randomized, controlled trial, *Ann. Intern. Med.* **132**:885–888.

Visvesvara, G. S., Moura, H., Kovacs-Nace, E., Wallace, S., and Eberhard, M. L., 1997, Uniform staining of *Cyclospora* oocysts in fecal smears by a modified safranin technique with microwave heating, *J. Clin. Microbiol.* **35**:730–733.

Weiss, L. M., Perlman, D. C., Sherman, J., Tanowitz, H., and Wittner, M., 1988, *Isospora belli* infection: Treatment with pyrimethamine, *Ann. Intern. Med.* **109**:474–475.

Whiteside, M. E., Barkin, J. S., May, R. G., Weiss, S. D., Fischl, M. A., and MacLeod, C. L., 1984, Enteric coccidiosis among patients with the acquired immunodeficiency syndrome, *Am. J. Trop. Med. Hyg.* **33**:1065–1072.

Wilairatana, P., Radomyos, P., Radomyos, B., Phraevanich, R., Plooksawasdi, W., Chanthavanich, P., Viravan, C., and Looareesuwan, S., 1996, Intestinal sarcocystosis in Thai laborers, *Southeast Asian J. Trop. Med. Public Health* **27**:43–46.

Wiwanitkit, V., 2001, Intestinal parasitic infections in Thai HIV-infected patients with different immunity status, *BMC Gastroenterol.* **1**:3.

Woldemeskel, M., and Gebreab, F., 1996, Prevalence of sarcocysts in livestock of northwest Ethiopia, *Zentralbl Veterinarmed. B* **43**:55–58.

Wong, K. T., and Pathmanathan, R., 1992, High prevalence of human skeletal muscle sarcocystosis in south-east Asia, *Trans. R. Soc. Trop. Med. Hyg.* **86**:631–632.

Wright, M. S., and Collins, P. A., 1997, Waterborne transmission of *Cryptosporidium*, *Cyclospora* and *Giardia*, *Clin. Lab. Sci.* **10**:287–290.

Yai, L. E., Bauab, A. R., Hirschfeld, M. P., de Oliveira, M. L., and Damaceno, J. T., 1997, The first two cases of *Cyclospora* in dogs, Sao Paulo, Brazil, *Rev. Inst. Med. Trop. Sao Paulo* **39**:177–179.

Yang, Z. Q., Li, Q. Q., Zuo, Y. X., Chen, X. W., Chen, Y. J., Nie, L., Wei, C. G., Zen, J. S., Attwood, S. W., Zhang, X. Z., and Zhang, Y. P., 2002, Characterization of *Sarcocystis* species in domestic animals using a PCR-RFLP analysis of variation in the 18S rRNA gene: A cost-effective and simple technique for routine species identification, *Exp. Parasitol.* **102**:212–217.

Yu, S., 1991, Field survey of *sarcocystis* infection in the Tibet autonomous region, *Zhongguo Yi Xue Ke Xue Yuan Xue Bao* **13**:29–32.

Zerpa, R., Uchima, N., and Huicho, L., 1995, *Cyclospora cayetanensis* associated with watery diarrhoea in Peruvian patients, *J. Trop. Med. Hyg.* **98**:325–329.

Zhang, B. X., Yu, H., Zhang, L. L., Tao, H., Li, Y. Z., Li, Y., Cao, Z. K., Bai, Z. M., and He, Y. Q., 2002, Prevalence survey on *Cyclospora cayetanensis* and *Cryptosporidium* ssp. in diarrhea cases in Yunnan Province, *Zhongguo Ji Sheng Chong Xue Yu Ji Sheng Chong Bing Za Zhi* **20**:106–108.

CHAPTER 4

Cryptosporidium and Cryptosporidiosis

Lihua Xiao and Vitaliano Cama

4.1 PREFACE

Cryptosporidium spp. are apicomplexan parasites that inhabit the brush-borders of the gastrointestinal epithelium (Bird and Smith, 1980). Initially thought to be only a pathogen of young animals such as calves, lambs, piglets, and foals, cryptosporidiosis is now known to be an important cause of enterocolitis, diarrhea, and cholangiopathy in humans (Current *et al.*, 1983). Several *Cryptosporidium* spp. are now recognized to infect humans and more to infect other vertebrates (Xiao *et al.*, 2004a). Healthy children and adults and young animals with cryptosporidiosis usually have a short-term illness accompanied by watery diarrhea, vomiting, malabsorption, and weight loss. In humans and animals with immunodeficiencies, and snakes, however, the infection can be protracted and life-threatening (Hunter and Nichols, 2002).

Cryptosporidium oocysts are environmentally resistant, retain their infectious potential for considerable time in moist environments, such as water, soil, fresh seafood and produce (Rose, 1997), and survive most water disinfection treatments as well (Korich *et al.*, 1990). Two important fecal-oral transmission routes include direct contact with infected persons (person-to-person or anthroponotic transmission) or animals (zoonotic transmission), and consumption of contaminated water (waterborne transmission) or food (foodborne transmission). Thus, *Cryptosporidium* spp. are well recognized water and food-borne pathogens, having caused many outbreaks of human diarrheal disease in the United States and other developed countries (Anonymous, 1984; Current *et al.*, 1983; D'Antonio *et al.*, 1985; Joce *et al.*, 1991; MacKenzie *et al.*, 1994b; Millard *et al.*, 1994). Water and food probably also play an important role in the transmission of cryptosporidiosis in endemic areas, even though the disease burden attributable to them is not fully clear.

4.2 TAXONOMY

Cryptosporidium spp. belong to the family Cryptosporidiidae, which is a member of the phylum Apicomplexa. The exact placement of Cryptosporidiidae in Apicomplexa is uncertain. It was long considered a member of the class Coccidea, in the order of Eimeriida or Eucoccidiorida (Corlis, 1994). Recent phylogenetic studies, however, indicate that *Cryptosporidium* spp. are more related to gregarines than to coccidia (Carreno *et al.*, 1999). Extra-cellular gregarine-like reproductive stages have been described in *Cryptosporidium andersoni* and *Cryptosporidium parvum* (Hijjawi *et al.*, 2002). Thus, *Cryptosporidium* spp. are no longer considered coccidian parasites.

Cryptosporidium spp. were first recognized by Tyzzer in 1907, who described *Cryptosporidium muris* in the stomach of laboratory mice (Tyzzer, 1907, 1910). Later in 1912, Tyzzer described a second species in laboratory mice, *C. parvum* (Tyzzer, 1912). This new species differed from *C. muris* not only by infecting the small intestine instead of the stomach, but also by having smaller oocysts, the environmentally robust stage of the parasite (Upton and Current, 1985).

Over the next 50 years following the initial description of *Cryptosporidium*, these parasites were commonly confused with sporocysts of *Sarcocystis*. Several new *Cryptosporidium* species were described during the period, mostly based on sporocysts of *Sarcocystis* spp. Subsequently, it was thought that because *Cryptosporidium* was closely related to *Eimeria*, *Cryptosporidium* spp. also could not normally be transmitted from one species of animals to another (Levine, 1980). This erroneous concept of strict host specificity led to the description and report of multiple new species during the 1960–1980s, which are no longer considered valid, such as *Cryptosporidium anserinum* in geese (Proctor and Kemp, 1974), *Cryptosporidium agni* in sheep (Barker and Carbonell, 1974), *Cryptosporidium bovis* in neonatal calves (Barker and Carbonell, 1974), *Cryptosporidium rhesi* in monkeys (Levine, 1980), and *Cryptosporidium cuniculus* in rabbits (Inman and Takeuchi, 1979).

Infection and cross-transmission studies conducted in the 1970s and 1980s demonstrated that *Cryptosporidium* isolates could indeed frequently be transmitted from one host species to another (Tzipori *et al.*, 1981a, 1981b, 1982). These findings led to the synonymization of many species into *C. parvum*, and were the basis for proposing the monospecific structure of the genus *Cryptosporidium*. As a result, *C. parvum* was used extensively for the description of *Cryptosporidium* spp. from most mammals including humans (Tzipori *et al.*, 1980; Upton and Current, 1985).

The recent use of molecular methods in the characterization of *Cryptosporidium* has helped to resolve existing confusions in the taxonomy of this genus (Fayer *et al.*, 2000a; Morgan *et al.*, 1999b; Xiao *et al.*, 2000b, 2004a). These molecular tools have been very valuable when used in conjunction with morphological, biological, or host specificity studies. This has resulted in the validation of several *Cryptosporidium* described earlier, such as *Cryptosporidium meleagridis* in birds, *Cryptosporidium wrairi* in guinea pigs, and *Cryptosporidium felis* in cats. It is now well known that various *Cryptosporidium* isolates do have differences in host specificity, but one *Cryptosporidium* sp. usually infect a limited spectrum of animals, especially if the host animals are related. This new *Cryptosporidium* taxonomic paradigm has also led to the establishment of several new *Cryptosporidium* species, such as *Cryptosporidium hominis* (previously known as *C. parvum* genotype 1 or the human genotype) in humans, *C. andersoni* (previously known as *C. muris*-like or *C. muris* bovine genotype) and *C. bovis* (previously known as *Cryptosporidium* bovine genotype B) in weanling calves and adult cattle, *Cryptosporidium canis* (previously known as *C. parvum* dog genotype) in dogs, and *Cryptosporidium suis* (previously known as *Cryptosporidium* pig genotype I) in pigs. Now, there are 15 established *Cryptosporidium* species in fish, reptiles, birds, and mammals (Table 4.1). There are also many host-adapted *Cryptosporidium* genotypes that do not yet have designed species names because of the lack of morphologic and biologic characterizations,

Table 4.1. Currently recognized *Cryptosporidium* species.

Species	Major host	Minor host	Infection site	Reference
C. andersoni	Cattle, bactrian camels	Sheep	Stomach	(Lindsay et al., 2000)
C. baileyi	Chicken, turkeys	Cockatiels, ducks, ostriches, quails	Intestine, respiratory track, bursa	(Current et al., 1986)
C. bovis	Cattle, yaks	Sheep	Intestine	(Fayer et al., 2005)
C. canis	Dogs, foxes, wolves	Humans	Intestine	(Fayer et al., 2001)
C. felis	Cats	Humans, cattle	Intestine	(Iseki, 1979)
C. galli	Chickens, finches, capercalles, grosbeaks		Proventriculus	(Ryan et al., 2003)
C. hominis	Humans, monkeys	Sheep, dugongs	Intestine	(Morgan-Ryan et al., 2002)
C. meleagridis	Turkeys, humans	Parrots	Intestine	(Slavin, 1955)
C. molnari	Fish		Stomach	(Alvarez-Pellitero and Sitja-Bobadilla, 2002)
C. muris	Rodents, bactrian camels	Humans, rock hyrax, mountain goats	Stomach	(Tyzzer, 1910)
C. parvum	Cattle, sheep, goats, deer, humans	Mice, pigs, horses	Intestine	(Upton and Current, 1985)
C. saurophilum	Lizards	Snakes	Intestine	(Koudela and Modry, 1998)
C. serpentis	Snakes, lizards		Stomach	(Tilley et al., 1990)
C. suis	Pigs	Humans	Intestine	(Ryan et al., 2004b)
C. wrairi	Guinea pigs		Intestine	(Vetterling et al., 1971)

such as *Cryptosporidium* horse, rabbit, mouse, ferret, deer mouse, skunk, squirrel, bear, deer, deer-like, cervine, fox, mongoose, wildebeest, duck, woodcock, snake, tortoise, goose I and II, muskrat I and II, opossum I and II, marsupial I and II, and pig II genotypes (Xiao et al., 2004a).

Currently, eight *Cryptosporidium* spp. have been reported in humans: *C. hominis, C. parvum, C. meleagridis, C. felis, C. canis, C. muris, C. suis,* and *Cryptosporidium* cervine genotype. Humans are most frequently infected with *C. hominis* and *C. parvum*. The former almost exclusively infects humans, thus is considered an anthroponotic parasite, whereas the latter infects both humans and domestic and wild ruminants, thus is considered a zoonotic pathogen. The contribution of the two species to human cryptosporidiosis differs among geographic areas, with *C. parvum* responsible for more infection than *C. hominis* in Europe and Kuwait, and *C. hominis* responsible for most human infections in the rest of the world. Other species, such as *C. meleagridis*, *C. felis*, and *C. canis*, are less common. In contrast, *C. muris*, *C. suis*, and *Cryptosporidium* cervine genotypes have been found only in a few human cases (Xiao et al., 2003, 2004a; Xiao and Ryan, 2004). Despite earlier suggestion that unusual zoonotic species usually infect immunocompromised persons, a recent study in Peru suggests that there is no significant difference in the distribution of *Cryptosporidium* species between AIDS patients and children living in the same geographic area (Cama et al., 2003).

4.3 LIFE CYCLE AND DEVELOPMENTAL BIOLOGY

Cryptosporidium spp. are intracellular parasites that primarily infect epithelial cells of the stomach or intestine. The infection site varies according to species, but almost the entire development of *Cryptosporidium* spp. occur between the two lipoprotein layers of the membrane of the epithelial cells (Bird and Smith, 1980), with the exception of *Cryptosporidium molnari*, for which oogonial and sporogonial stages are located deeply within the epithelial cells (Alvarez-Pellitero and Sitja-Bobadilla, 2002; Ryan et al., 2004a). *Cryptosporidium* infections in humans or other susceptible hosts start with the ingestion of viable oocysts, the infectious stage that is environmentally resistant. Upon gastric and duodenal digestion, four sporozoites are liberated from each excysted oocyst, invade the epithelial cells, and develop into trophozoites surrounded by a parasitophorous vacuole. Within the epithelial cells, trophozoites undergo several generations of asexual amplification called merogony, leading to the formation of different types of meronts. The types of meronts depend on *Cryptosporidium* species. For *C. parvum*, there are two types of meronts. The type 1 meront develops six to eight nuclei, giving rise to six to eight merozoites. These stages are morphologically similar to sporozoites and can infect neighboring epithelial cells, forming more type 1 meronts or the new type 2 meronts. The latter develop four nuclei, forming four merozoites. As with type 1 merozoites, these merozoites are released and infect new cells to generate more type 2 meronts, or can differentiate into sexually distinct stages called macro- and micro-gametocytes in a process called gametogony. New oocysts, formed in the epithelial cells from the fusion of macro-gametocytes and micro-gametes, are sporulated *in situ* in a process called sporogony, and contain four sporozoites. It is believed by some that about 20% are "thin walled" and may excyst within the digestive tract of the host, leading to the infection of new cells (autoinfection). The remaining 80% of oocysts are excreted into the environment, are resistant to low temperature, high salinity, and most disinfectants, and can initiate infection in a new host upon ingestion. Thus, the only extracellular stages in the *Cryptosporidium* life cycle are released sporozoites, merozoites, and microgametes, which are briefly in the lumen of the digestive tract (Fayer et al., 1997). However, recently, a gregarine-like stage has been described in

C. andersoni and *C. parvum*, which undergo multiplication through syzgy, a sexual reproduction process involving the end-to-end fusion of two or more parasites (Hijjawi *et al.*, 2002). If verified by others, this would have major implications in our understanding of the *Cryptosporidium* biology, genetics, and transmission.

Like other members of the Apicomplexa, sporozoites, and merozoites of *Cryptosporidium* use the apicomplex for invasion. Unlike other apicomplexan parasites, *Cryptosporidium* spp. have no polar rings and the conoid as part of the apicomplex, with only a relict mitochondrion, no sporocysts and plastids, and no flagelles in micro-gametes. At the contact site between host cells and *Cryptosporidium* developmental stages, there is also a unique electron-dense attachment or feeder organelle, which is supposedly involved in selective transport of nutrients from host cells into developing parasites. The prepatent period (time from ingestion of infective oocysts to the completion of endogenous development and excretion of new oocysts) varies with species, hosts, and infection doses. This is usually between 4 and 14 days.

4.4 EPIDEMIOLOGY AND TRANSMISSION

4.4.1 Cryptosporidiosis in Immunocompetent Persons

In developing countries, human *Cryptosporidium* infection occurs mostly in children younger than five-years old, with peak occurrence of infections and diarrhea in children less than 2 years of age (Bern *et al.*, 2000, 2002; Bhattacharya *et al.*, 1997; Mata, 1986; Newman *et al.*, 1999). Frequent symptoms include diarrhea, abdominal cramps, vomiting, headache, fatigue, and low-grade fever (Nimri and Hijazi, 1994). The diarrhea can be voluminous and watery, but usually resolves within one to two weeks without treatment. Not all infected children have diarrhea or other gastrointestinal symptoms, and the occurrence of diarrhea in children with cryptosporidiosis can be as low as 30% in community-based studies (Bern *et al.*, 2002; Xiao *et al.*, 2001a). Even subclinical cryptosporidiosis exerts a significant adverse effect on child growth, as infected children with no clinical symptoms experience growth faltering, both in weight and in height (Checkley *et al.*, 1997, 1998). *Cryptosporidium*-infected children may never have enough catch-up growth covered for the growth retardation (Checkley *et al.*, 1998; Molbak *et al.*, 1997). Children can have multiple episodes of cryptosporidiosis, implying that the anti-*Cryptosporidium* immunity in children acquired is short-lived or incomplete (Bern *et al.*, 2000, 2002; Newman *et al.*, 1999; Xiao *et al.*, 2001a;). Cryptosporidiosis has been associated with increased child mortality in developing countries (Tumwine *et al.*, 2003).

In developed countries, *Cryptosporidium* infection occurs later in life of children than in developing countries, probably due to later exposures to contaminated environments as a result of better hygiene. In a study conducted in Kuwait, the median age of children with cryptosporidiosis was 4.5 years (Sulaiman *et al.*, 2005). Children in these countries frequently acquire *Cryptosporidium* infection from another infected child attending the same daycare or school, probably via person-to-person transmissions (Alpert *et al.*, 1984; Lacroix *et al.*, 1987; Tangermann *et al.*, 1991; Taylor *et al.*, 1985). Cryptosporidiosis is also common in the elderly in nursing homes, where person-to-person transmission probably also plays a major role in the spread of *Cryptosporidium* infections (Neill *et al.*, 1996). In rural areas, zoonotic infections via direct contact with farm animals have been reported many times, but the relative importance of direct zoonotic transmission of cryptosporidiosis is not entirely clear (Current *et al.*, 1983; Miron *et al.*, 1991). In the general population, a substantial number of adults are probably susceptible to *Cryptosporidium* infection, as sporadic infections occur in all age groups in the United States and United Kingdom, and traveling to developing countries and consumption of contaminated food or water can frequently lead to infection (Dietz and Roberts, 2000; Dietz *et al.*, 2000; Goh *et al.*, 2004; Roy *et al.*, 2004). Hemodialysis patients with chronic renal failure are also frequently infected with *Cryptosporidium* (Chieffi *et al.*, 1998; Turkcapar *et al.*, 2002).

Unlike in developing countries, immunocompetent persons with sporadic cryptosporidiosis in industrialized nations usually have diarrhea (Anonymous, 1990; Assadamongkol *et al.*, 1992; Chmelik *et al.*, 1998; Daoud *et al.*, 1990; Goh *et al.*, 2004; Robertson *et al.*, 2002a; Thomson *et al.*, 1987). The median number of stools per day during the worst period of the infection is 7–9.5 in Australia (Robertson *et al.*, 2002a). The durations of illness are a mean of 12 days in Finland, and a median of 9 days in the United Kingdom and 15–21 days in Australia (Goh *et al.*, 2004; Jokipii and Jokipii, 1986; Robertson *et al.*, 2002a), with a median of 5 days off work or study (Robertson *et al.*, 2002a). Other common symptoms include abdominal pain (in 72.4–91.7% patients), vomiting (in 55.2–70.9% patients), and low-grade fever (in 38.1–48.5% patients) (Goh *et al.*, 2004; Jokipii and Jokipii, 1986; Robertson *et al.*, 2002a). In the United States, United Kingdom, and Australia, 14.4–17.4%, 8.5–22.1%, and 7–11.9% patients with sporadic cryptosporidiosis require hospitalization, respectively (Dietz *et al.*, 2000; Goh *et al.*, 2004; Robertson *et al.*, 2002a).

4.4.2 Cryptosporidiosis in Immunocompromised Persons

Cryptosporidiosis is common in immunocompromised persons, such as AIDS patients, persons with primary immunodeficiency, and cancer and transplant patients undergoing immunosuppressive therapy (Heyworth, 1996; Hunter and Nichols, 2002; McLauchlin *et al.*, 2003). It is frequently associated with chronic, life-threatening diarrhea (Flanigan *et al.*, 1992; Heyworth, 1996; Hunter and Nichols, 2002). In HIV+ persons, the occurrence of cryptosporidiosis increases as the CD4+ lymphocyte cell counts fall, especially below 200 cells/:l (Flanigan *et al.*, 1992; Navin *et al.*, 1999; Pozio *et al.*, 1997). Manabe *et al.* (1998) described four clinical syndromes of cryptosporidiosis in the United States: chronic diarrhea (36% of patients), cholera-like disease (33%), transient diarrhea (15%), and relapsing illness (15%). Sclerosing cholangitis and other biliary involvements, however, are also very common in AIDS patients with cryptosporidiosis (Chen and LaRusso, 2002; French *et al.*, 1995; Hashmey *et al.*, 1997; McGowan *et al.*, 1993; Teixidor *et al.*, 1991; Vakil *et al.*, 1996). Symptoms of cryptosporidiosis in AIDS patients vary in severity, duration, and responses to drug treatment (Flanigan and Graham, 1990; Goodgame *et al.*, 1993; Manabe *et al.*, 1998; McGowan *et al.*, 1993). Much of this variation can be explained by the degree of immunosuppression (Flanigan *et al.*, 1992; McGowan *et al.*, 1993). In addition, variation in the infection site (gastric infection, proximal small intestine infection, ileo-colonic infection, versus pan-enteric infection) has been seen in AIDS patients with cryptosporidiosis (Clayton et al., 1994; Kelly et al., 1998; Lumadue et al., 1998; Ventura et al., 1997), and this anatomic variation may also contribute to differences in disease severity and survival (Clayton et al., 1994; Lumadue et al., 1998). Cryptosporidiosis in AIDS patients is associated with increased mortality and shortened survival (Colford et al., 1996; Manabe et al., 1998)

4.4.3 Transmission Routes and Infection Sources: Anthroponotic Versus Zoonotic Transmission

Cryptosporidium infections normally start with the ingestion of infectious oocysts. This parasite has a worldwide distribution and is ubiquitously present in the environment. Humans can acquire *Cryptosporidium* infections through several transmission routes (Clark, 1999; Griffiths, 1998), such as direct contact with infected persons or animals, and consumption of contaminated water (drinking or recreational) or food. However, the relative role of each in the occurrence of *Cryptosporidium* infection in humans is unclear. Several studies in the United States and Europe have shown that cryptosporidiosis was more common in homosexual men than persons with other HIV-transmission categories (Hashmey et al., 1997; Hellard et al., 2003; Soave et al., 1984), indicating that direct person-to-person or anthroponotic transmission of cryptosporidiosis is common. Contact with persons with diarrhea has been identified as a major risk factor in sporadic *Cryptosporidium* infections in the United States, United Kingdom, and Australia (Hunter et al., 2004b; Robertson et al., 2002a; Roy et al., 2004).

Shortly after the discovery of cryptosporidiosis in humans it has been found that humans can acquire *Cryptosporidium* infection via contact with infected farm animals (Current et al., 1983). However, only a few case control studies assessed the role of zoonotic transmission in the acquisition of cryptosporidiosis in humans. In the United States, United Kingdom, and Australia, contact with farm animals is a major risk factor in the sporadic cases of human cryptosporidiosis (Goh et al., 2004; Hunter et al., 2004b; Robertson et al., 2002a; Roy et al., 2004). Contact with pigs, dogs, or cats is also a risk factor for cryptosporidiosis in children in Guinea-Bissau and Indonesia, (Katsumata et al., 1998; Molbak et al., 1994), but this is actually a protective factor in Australia (Robertson et al., 2002a). A weak association was observed between the occurrence of cryptosporidiosis in HIV+ persons and contact with dogs, but not other animals (Glaser et al., 1998). In other studies, no increased risk in the acquisition of cryptosporidiosis was associated with contact with animals (Nchito et al., 1998; Pereira et al., 2002a).

The distribution of *C. parvum* and *C. hominis* in humans is probably a good indicator of the transmission routes. Thus far, studies conducted in tropical countries such as Peru, Thailand, Malawi, Uganda, Kenya, and South Africa showed a dominance of *C. hominis* in children or HIV+ adults (Gatei et al., 2003; Leav et al., 2002; Peng et al., 2003a; Tiangtip and Jongwutiwes, 2002; Tumwine et al., 2003; Xiao et al., 2001a). In Europe, however, several studies have shown a slightly higher prevalence of *C. parvum* than *C. hominis* in both immunocompetent and immunocompromised persons (Alves et al., 2003b; Chalmers et al., 2002; Guyot *et al.*, 2001; McLauchlin *et al.*, 2000). In contrast, Kuwaiti children were almost exclusively infected with *C. parvum* (Sulaiman *et al.*, 2005). The differences in the distribution of *Cryptosporidium* genotypes in humans is considered an indication of differences in infection sources (Learmonth *et al.*, 2001, 2004; McLauchlin *et al.*, 2000); the occurrence of *C. hominis* in humans is most likely due to anthroponotic transmission, whereas the predominance of *C. parvum* in a population has been considered the result of zoonotic transmission. Thus, in most tropical countries, it is possible that anthroponotic transmission of *Cryptosporidium* play a major role in human cryptosporidiosis; whereas in Europe, both anthroponotic and zoonotic transmissions are important. Indeed, in areas with a high percentage of infections due to *C. parvum*, massive slaughtering of farm animals during foot and mouth disease outbreaks can result in a reduction in the proportion of human infections due to *C. parvum* (Hunter *et al.*, 2003; Smerdon *et al.*, 2003).

Nevertheless, recent subtyping studies have shown that not all *C. parvum* infections in humans are results of zoonotic transmission (Alves *et al.*, 2003b; Mallon *et al.*, 2003b; Xiao *et al.*, 2003). Among the *C. parvum* GP60 subtype families identified, alleles IIa and IIc (previously known as Ic) are the two most common ones. The former has been identified in both humans and ruminants, thus serving as a zoonotic pathogen, whereas the latter has only been seen in humans (Alves *et al.*, 2003b; Peng *et al.*, 2003b; Xiao *et al.*, 2003), thus serving as an anthroponotic pathogen. In Lima, Peru, all *C. parvum* infection in children and HIV+ persons are due to the subtype family IIc, indicating that anthroponotic transmission of *C. parvum* is common in certain areas (Xiao *et al.*, 2004a). Even in the United Kingdom where zoonotic transmission is known to play a significant role in the transmission of human cryptosporidiosis, anthroponotic transmission of *C. parvum* is also common (Mallon *et al.*, 2003a).

4.4.4 Waterborne Transmission

Epidemiologic studies have frequently identified water as a major route of *Cryptosporidium* transmission in disease-endemic areas (Gallaher *et al.*, 1989; Nimri and Hijazi, 1994; Weinstein *et al.*, 1993). In most tropical countries, *Cryptosporidium* transmission in children is usually associated with the rainy season, and waterborne transmission is considered a major route in epidemiology of cryptosporidiosis in these areas (Bern *et al.*, 2000; Bhattacharya *et al.*, 1997; Javier Enriquez *et al.*, 1997; Katsumata *et al.*, 1998; Moodley *et al.*, 1991; Nath *et al.*, 1999; Newman *et al.*, 1999; Peng *et al.*, 2003a; Perch *et al.*, 2001; Tumwine *et al.*, 2003). However, some studies have failed to show a direct linkage between seasonal incidence of cryptosporidiosis and rainfall (Bern *et al.*, 2002).

Seasonal variations in the incidence of human *Cryptosporidium* infection in industrialized nations have also been attributed to waterborne transmission (Brandonisio *et al.*, 1999; Dietz and Roberts, 2000; Dietz *et al.*, 2000; McLauchlin *et al.*, 2000; Roy *et al.*, 2004). In the United States, there are two annual peaks in the number of cryptosporidiosis cases in HIV+ persons: one in spring and one in late summer (Inungu *et al.*, 2000; Sorvillo *et al.*, 1998). In the general population, there is also an annual late summer peak in sporadic cases of cryptosporidiosis (Dietz and Roberts, 2000; Roy *et al.*, 2004). It is generally accepted that the late summer peak of cryptosporidiosis cases is due to recreational activities such as swimming and water sports, suggesting that waterborne transmission may be important in cryptosporidiosis epidemiology. Nevertheless, seasonal transmission of cryptosporidiosis in HIV+ persons is not always associated with rainfall (Sorvillo et al., 1998).

The role of drinking water in sporadic *Cryptosporidium* infection is not clear. In Mexican children living near the United States border, cryptosporidiosis is associated with consumption of municipal water instead of bottled water (Leach et al., 2000). In England, the number of glasses of tap water drunk at home each day is associated with sporadic cases of cryptosporidiosis (Hunter et al., 2004b). In the United States, drinking untreated surface water was identified as a risk factor for the acquisition of *Cryptosporidium* in a small case control study (Gallaher et al., 1989). Residents living in cities with surface-derived drinking water generally have higher blood antibody levels against *Cryptosporidium* antigens than those living in cities with ground water as drinking water, indicating drinking water plays a role in the transmission of human cryptosporidiosis (Frost et al., 2001, 2002, 2003). An earlier study in South Australia also showed an association between consumption of spring water or main water rather than rain water, and the occurrence of cryptosporidiosis (Weinstein et al., 1993). A more recent study in the same area, however, suggested that waterborne transmission in the area was mainly due to swimming in public pools and consumption of unboiled rural water rather than consumption of tap water (Robertson et al., 2002a). Case control studies conducted in both immunocompetent persons and AIDS patients in the United States also have failed to show a direct linkage of *Cryptosporidium* infection to drinking water (Khalakdina et al., 2003; Sorvillo et al., 1994).

Numerous waterborne outbreaks of cryptosporidiosis have occurred in the United States, Canada, United Kingdom, France, Australia, Japan, and other industrialized nations (Dalle et al., 2003; Lemmon et al., 1996; MacKenzie et al., 1994b, 1995; Ong et al., 1999; Smith et al., 1988; Yamamoto et al., 2000). These include outbreaks associated with both drinking water and recreational water (swimming pools and water parks). With the adoption of more stringent treatments of source water by the water industry after the massive cryptosporidiosis outbreak in Milwaukee in 1993, the number of drinking water-associated outbreaks is in decline in the United States and United Kingdom in recent years. Even though five *Cryptosporidium* spp. are commonly found in humans, thus far only *C. parvum* and *C. hominis* are associated with cryptosporidiosis outbreaks, with *C. hominis* responsible for more outbreaks than *C. parvum* (McLauchlin et al., 2000; Peng et al., 1997; Xiao et al., 2003). This is even the case for the United Kingdom, where *C. parvum* is more common than *C. hominis* in the general population. In outbreak settings, immunocompetent adults may have voluminous but self-limiting diarrhea, with or without abdominal cramps, fatigue, vomiting, fever, and other symptoms (MacKenzie et al., 1994a; Yamamoto et al., 2000). Attack rates and incidence of specific clinical symptoms (diarrhea, vomiting, abdominal cramps, headache, fever, etc.) differ among outbreaks, though the reason for these variations is not known (Quiroz et al., 2000).

Surveys conducted in various regions of the United States have demonstrated the presence of *Cryptosporidium* oocysts in 67–100% wastewaters, 24–100% of surface waters, and 3.8–40% drinking waters (LeChevallier *et al.*, 1991a, 1991b; Madore *et al.*, 1987; Rose, 1997). The identity and human infective potential of these waterborne oocysts are not known, although it is likely that not all oocysts are from human-infecting *Cryptosporidium* species. Likewise, the source of the oocyst contamination is also not fully clear. Farm animals and human sewage discharge are generally considered to be major sources of surface water contamination with *C. parvum* (Meinhardt *et al.*, 1996). Because *Cryptosporidium* infection is common in wildlife, it is conceivable that wildlife can also be a source for *Cryptosporidium* oocysts in waters (Rose, 1997). The source for contamination (i.e., with oocysts of human or animal origin) involved in individual outbreaks, however, is frequently not known, largely due to the lack of investigations using suitable strain-specific diagnostic tools.

4.4.5 Foodborne Transmission

The role of food in the transmission of cryptosporidiosis is much less clear. *Cryptosporidium* oocysts have been isolated from several foodstuffs and these have mainly been associated with fruits, vegetables, and shellfish (Table 4.2). A survey of produce sold in Lima, Peru where *Cryptosporidium* is prevalent in humans demonstrated that 14.5% of samples were *Cryptosporidium* positive. In Norway, where sporadic *Cryptosporidium* infection rates are presumably lower, *Cryptosporidium* oocysts were found in 4% of fresh produce (Robertson and Gjerde, 2001b). Oysters, clams, mussels, and cockles in many countries have been shown to be contaminated with *Cryptosporidium* oocysts (Table 4.2). The association of oocyst contamination with these produce is particularly important from a public health viewpoint, as these products are frequently consumed raw without any thermal processing to inactivate oocysts. Mollusc filter feeders such as oysters, mussels, and clams pose a risk because they can concentrate pathogens from large volumes of potentially contaminated water, and *Cryptosporidium* oocysts found in them are frequently viable for extended periods of time (Fayer *et al.*, 1998, 1999, 2002; Freire-Santos *et al.*, 2001; Gomez-Bautista *et al.*, 2000; Gomez-Couso *et al.*, 2003a, 2003b; Tamburrini and Pozio, 1999).

Direct contamination of food by fecal materials from animals or food-handlers has been implicated in several foodborne outbreaks of cryptosporidiosis in industrialized nations (Millard *et al.*, 1994; Quiroz *et al.*, 2000). This is also likely a major source of contamination of fresh produce in endemic areas. Because *Cryptosporidium* oocysts are commonly found in surface water, contamination of fresh produce through irrigation or washing is probably also common (Armon *et al.*, 2002; Robertson and Gjerde, 2001b; Thurston-Enriquez *et al.*, 2002). In addition, marine water may also be contaminated with *Cryptosporidium* oocysts due to sewage discharge and agricultural runoff, which can in turn contaminate shellfish (Fayer *et al.*, 1998; Graczyk *et al.*, 2000). Studies conducted in various countries have found *C. parvum*, *C. hominis*, and *C. meleagridis* in shellfish, but in most areas, *C. parvum* is responsible for more than 80% of the contamination (Table 4.2), indicating agricultural runoff is probably the most important source for *Cryptosporidium* contamination in shellfish.

Table 4.2. Prevalence of *Cryptosporidium* in raw fruits, vegetables, and shellfish.

Food type	Country	Prevalence	Species	Reference
		Vegetables		
Vegetables	Costa Rica	Cilantro leaves: 4/80; Cilantro roots: 7/80; Lettuce: 2/80; Radish: 1/80; Carrot: 1/80; Tomato: 1/80; Cucumber: 1/80; Cabbage: 0/80		(Monge and Arias, 1996; Monge *et al.*, 1996)
Vegetables	Peru	Vegetables (cabbage, celery, cilantro, green onion, ground green chili, Leek, lettuce, parsley, yerba Buena, huacatay): 28/172		(Ortega *et al.*, 1997)
Fruits and vegetables	Norway	Alfalfa: 0/16; Dill: 0/7; Lettuce: 5/125 Mung bean sprouts: 14/149; Mushrooms: 0/55 Parsley: 0/7 Precut salad: 0/38 Radish sprouts: 0/6; Raspberries: 0/10; Strawberries: 0/62		(Robertson and Gjerde, 2001b; Robertson *et al.*, 2002b)
Sprout	Norway			
		Shellfish		
Clams	Spain and Italy	*Dosinia exoleta, Ruditapes philippinarum, Venerupis pullastra, Venerupis rhomboideus, Venus verrucosa*: 1 0/17		(Freire-Santos *et al.*, 2000)
	Spain	*Dosinia exoleta, Venerupis pullastra, Venerupis rhomboideus, Venus verrocosa*: 10/18	*C. parvum* and *C. hominis*	(Gomez-Couso *et al.*, 2004)
	Spain and EU countries	*Dosinia exoleta, Venerupis pullastra, Venerupis rhomboideus, Venus verrocosa*: 20/68		(Gomez-Couso *et al.*, 2003a)
	Italy	*Chamelea gallina*: 2 of 16 pooled clams (30 clams/pool)	*C. parvum*	(Traversa *et al.*, 2004)
	Eastern USA and Canada	Clams: 3/375 (0.8)		(Fayer *et al.*, 2003)

(*continued*)

Table 4.2. (*continued*)

Food type	Country	Prevalence	Species	Reference
Cockles	Spain	*Cerastoderma edule*: postive/6	*C. parvum*	(Gomez-Bautista *et al.*, 2000)
	Spain and EU countries	*Cerastoderma edule*: 5/24		(Gomez-Couso *et al.*, 2003a)
Mussels	Spain	*Mytilus galloprovincialis*: positive/180	*C. parvum*	(Gomez-Bautista *et al.*, 2000)
	Spain	*Mytilus galloprovincialis*: 12/22	*C. parvum*	(Gomez-Couso *et al.*, 2004)
	Spain	*Mytilus galloprovincialis*: 6/15		(Freire-Santos *et al.*, 2000)
	Spain and EU countries	*Mytilus galloprovincialis*: 35/107		(Gomez-Couso *et al.*, 2003a)
	Northern Ireland	*Mytilus edulis*: 2/16	*C. hominis*	(Lowery *et al.*, 2001b)
	Canada	Zebra mussel (*Dreissena ploymorpha* 32/32 pools (514 mussels total)	*C. hominis*	(Graczyk *et al.*, 2001)
	USA	Bent mussel (*Ischadium recurvum*): 14/16		(Graczyk *et al.*, 1999)
	Ireland	*Mytilus edulis*: 3/26 pools (10 mussels/pool)		(Chalmers *et al.*, 1997)
Oysters	Chesapeake Bay, USA	*Crassostrea virginica*: 142/360	*C. parvum* and *C. hominis*	(Fayer *et al.*, 1998)
	Chesapeake Bay, USA	Commercial *Crassostrea virginica*: 182/510	*C. parvum* and *C. hominis*	(Fayer *et al.*, 1999)
	Chesapeake Bay, USA	*Crassostrea virginica*: 331/1590	*C. parvum* and *C. hominis*	(Fayer *et al.*, 2002)
	Eastern USA and Canada	*Crassostrea virginica*: 32/550 (5.8%)	*C. parvum*, *C. hominis*, *C. meleagridis*	(Fayer *et al.*, 2003)
	Spain	*Ostrea edulis*: 5/6		(Freire-Santos *et al.*, 2000)
	Spain	*Ostrea edulis*: 6/9	*C. parvum* and *C. hominis*	(Gomez-Couso *et al.*, 2004)
	Spain and EU countries	*Ostrea edulis*: 23/42		(Gomez-Couso *et al.*, 2003a)

Very few case control studies have examined the role of potentially contaminated food as a risk factor in the acquisition of *Cryptosporidium* infection in endemic areas. A study conducted on children in Brazil failed to show any association between *Cryptosporidium* infection and diet or type of food hygiene (Pereira *et al.*, 2002a). Case control studies conducted in the United States, United Kingdom, and Australia have actually shown that eating raw vegetables has a protective role against *Cryptosporidium* infection in immunocompetent persons (Hunter *et al.*, 2004b; Robertson *et al.*, 2002a; Roy *et al.*, 2004). Nevertheless, foodborne outbreaks of cryptosporidiosis occurs frequently in the United States, United Kingdom, and other industrialized nations, usually due to consumption of contaminated fresh produce, apple cider, or milk (Anonymous, 1996, 1997, 1998; Gelletlie *et al.*, 1997; Millard *et al.*, 1994; Quiroz *et al.*, 2000). It is estimated that about 10% *Cryptosporidium* infections in the United States are foodborne (Mead *et al.*, 1999).

4.5 DETECTION AND DIAGNOSIS

4.5.1 Serologic Methods

Humans and animals infected with *Cryptosporidium* spp. develop antibodies against *Cryptosporidium* antigens (Mead *et al.*, 1988). Electrophoretic and Western blot analysis showed that specific antibody response appeared between day 4 and 15 post inoculation. The two main target antigens had apparent molecular weights of 15–17 and 23 kDa (Reperant *et al.*, 1994). These two antigens, Cp17 (also called gp15) and Cp23 (also called the 27 kDa antigen), have been used by many researchers in the detection of *Cryptosporidium* antibodies by enzyme-linked immunosorbent assays (ELISA) or Western blot (Caputo *et al.*, 1999; Frost *et al.*, 1998; Priest *et al.*, 1999, 2001). Usually, native Cp17 extracted by Triton from oocysts of *C. parvum* and recombinant Cp23 expressed in *E. coli* are used in these assays (Priest *et al.*, 1999; Wang *et al.*, 2003). ELISA methods using these two antigens generally have higher sensitivity and specificity than earlier methods (Leach *et al.*, 2000; Okhuysen *et al.*, 1998; Zu *et al.*, 1994) that use crude oocyst antigens (Priest *et al.*, 1999). Most researchers use both antigens in serologic studies. ELISA based on Cp17 and Cp23 have been used in many studies of *Cryptosporidium* transmission in immunocompromised persons (Eisenberg *et al.*, 2001), children (Steinberg *et al.*, 2004), the general community (Frost *et al.*, 2001, 2002, 2003, 2004), and in investigations of cryptosporidiosis outbreaks (McDonald *et al.*, 2001). Recently, a multiplex bead assay based on these two antigens has been developed for the detection of *Cryptosporidium* antibodies in sera and oral fluids (Moss *et al.*, 2004). These serologic assays are not intended for the diagnosis of active *Cryptosporidium* infection, as antibodies to both the 27- and 17-kDa antigens have a half-life of about 12 weeks (Priest *et al.*, 2001).

4.5.2 Methods for Detection of *Cryptosporidium* in Stool Specimens

At the moment, almost all active *Cryptosporidium* infections are diagnosed by analysis of stool specimens. Examination of intestinal or biliary biopsy is sometimes used in the diagnosis of cryptosporidiosis in AIDS patients (Clayton *et al.*, 1994).

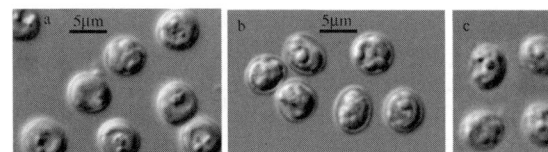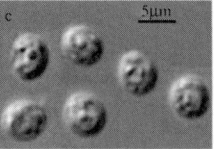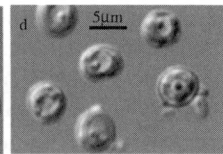

Figure 4.1. Oocysts of *Cryptosporidium parvum* (a) *C. hominis* (b) *C. meleagridis* (c) and *C. suis* (d) under differential interference contrast microscopy.

However, the sensitivity of the diagnosis depends on the location of tissues examined; duodenum is usually infected with *Cryptosporidium* only at high-intensity infection (Genta *et al.*, 1993), and the terminal ileum has significantly higher detection rates than the duodenum (Greenberg *et al.*, 1996). Thus, upper endoscopic biopsies are much less sensitive than lower endoscopic biopsies in diagnosing cryptosporidiosis. However, lower endoscopy is generally considered too invasive and risky for many AIDS patients.

Stool specimens are usually collected fresh or in fixative solutions such as 2.5% potassium dichromate or 10% buffered formalin (Garcia *et al.*, 1983), and are concentrated using either traditional ethyl acetate (Dubey, 1993) or Weber-modified ethyl-acetate concentration (Weber *et al.*, 1992). Sometimes other concentration methods such as sucrose, salt, or cesium chloride floatation are also used (Deng and Cliver, 1999b; Fayer *et al.*, 2000b; Kuczynska and Shelton, 1999; Kuhn *et al.*, 2002; Webster *et al.*, 1996), but they are mostly used in the analysis of fecal specimens from animals, which generally do not have as much lipids as human stool specimens. A variety of methods are used in the detection of *Cryptosporidium* in concentrated stool specimens, including microscopy, immunoassays, and molecular techniques (Arrowood, 1997). If clinical specimens will be analyzed by molecular methods, formalin should not be used as a fixative, as it would interfere with the analysis and reduce the efficiency of PCR amplification.

4.5.2.1 Microscopy

Concentrated stool specimens can be examined by microscopy in several ways. Frequently, when the number of oocysts is high, direct wet mount is made and *Cryptosporidium* oocysts are detected by bright-field microscopy. This allows the observation of oocysts morphology and more accurate measurement of oocysts, which is frequently needed in biologic studies. More often, differential interference contrast (DIC) is used in microscopy, which produces better images and visualization of internal structures of oocysts (Fig. 4.1). Morphology and morphometrics measurements, however, are generally not enough for *Cryptosporidium* species differentiation (Fall *et al.*, 2003; Xiao *et al.*, 2004a), as many species of *Cryptosporidium* look similar under microscopes and have similar morphometrics measurements (Table 4.3, Fig. 4.1). In general, oocysts of gastric *Cryptosporidium* species are bigger and more ovoid and those of intestinal species are smaller and more spherical (Table 4.3).

More often, *Cryptosporidium* oocysts in concentrated stool specimens are detected by microscopy after staining of the fecal smears. Many special stains have

Table 4.3. Morphometric measurements of established *Cryptosporidium* species[a].

Species	No. of oocysts measured	Length in μm (mean)	Width in μm (mean)	Length/width (mean)	Reference
Gastric					
C. andersoni	50	6.0–8.1 (7.4)	5.0–6.5 (5.5)	1.07–1.50 (1.35)	(Lindsay et al., 2000)
C. galli	50	8.0–8.5 (8.25)	6.2–6.4 (6.30)	1.30	(Ryan et al., 2003)
C. molnari[b]	22	3.23–5.45 (4.72)	3.02–5.04 (4.47)	1.00–1.17 (1.05)	(Alvarez-Pellitero and Sitja-Bobadilla, 2002)
C. muris	25	8.0–9.0 (8.4)	5.6–6.4 (6.1)	1.25–1.61 (1.38)	(Palmer et al., 2003)
C. serpentis	37	5.82–6.06 (5.94)	4.35–5.19 (5.11)	1.14–1.20 (1.17)	(Xiao et al., 2004c)
Intestinal					
C. baileyi[c]	25	5.6–6.3 (6.2)	4.5–4.8 (4.6)	1.2–1.4 (1.4)	(Current et al., 1986)
C. bovis	50	4.76–5.35 (4.89)	4.17–4.76 (4.63)	1.06	(Fayer et al., 2005)
C. canis	200	3.68–5.88 (4.95)	3.68–5.88 (4.71)	1.04–1.06 (1.05)	(Fayer et al., 2001)
C. felis	40	3.2–5.1 (4.6)	3.0–4.0 (4.0)	1.15	(Sargent et al., 1998)
C. hominis	100	4.4–5.9 (5.20)	4.4–5.4 (4.86)	1.00–1.09 (1.07)	(Morgan-Ryan et al., 2002)
C. meleagridis	55	4.93 (CL = 0.06)[d]	4.40 (CL = 0.05)[d]	1.12 (CL = 0.02)[d]	(Xiao et al., 2004a)
C. parvum	100	4.70–6.00 (5.19)	4.41–5.95 (4.90)	1.05–1.06 (1.06)	(Fayer et al., 2001)
C. saurophilum	20	4.81–5.07 (4.94)	4.35–4.63 (4.49)	1.11–1.17 (1.14)	(Xiao et al., 2004c)
C. suis	50	4.4–4.9 (4.6)	4.0–4.3 (4.2)	1.1	(Ryan et al., 2004b)
C. wrairi	30	4.8–5.6 (5.4)	4.0–5.0 (4.6)	1.04–1.33 (1.17)	(Tilley et al., 1991)

[a] Whenever possible, measurements from parasites confirmed by molecular or biologic characterizations are quoted.
[b] Also found in the intestine.
[c] Also found in the respiratory tract.
[d] CL: 95% confidence limit.

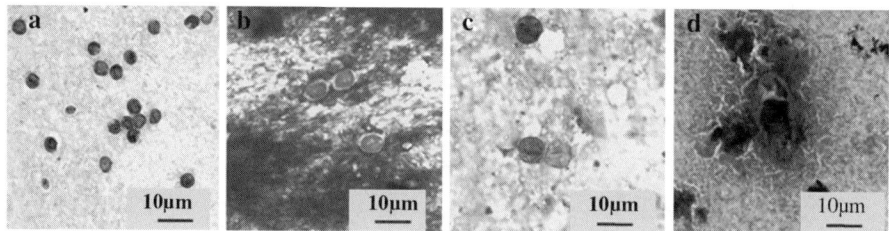

Figure 4.2. Acid-fast stained oocysts of *Cryptosporidium hominis* (a), *C. muris* (b), *Isospora belli* (c), and *Cyclospora cayetanensis* (d) under bright-field microscopy.

been used in the detection of *Cryptosporidium* oocysts, but acid-fast stains are the most often used (Arrowood, 1997). Modified acid-fast staining is very commonly used in developing countries because of its low cost, easy use, no need for special microscopes, and simultaneous detection of several other pathogens such as *Isospora* and *Cyclospora* (Fig. 4.2). Two acid-fast staining widely used in *Cryptosporidium* oocyst detection are the modified Ziehl-Neelsen acid-fast staining and modified Kinyoun's acid-fast staining (Arrowood, 1997).

Recently, immunofluorescence assays (IFA) have been used increasingly in *Cryptosporidium* oocyst detection by microscopy, especially in industrialized nations. Compared to acid-fast staining, IFA has higher sensitivity and specificity (Arrowood and Sterling, 1989; Johnston *et al.*, 2003; Quilez *et al.*, 1996). Many commercial IFA kits are marketed for the diagnosis of *Cryptosporidium*, some of which include reagents allowing simultaneous detection of *Giardia* cysts (Fig. 4.3). These include Merifluor *Cryptosporidium/Giardia* kit from Meridian Bioscience,

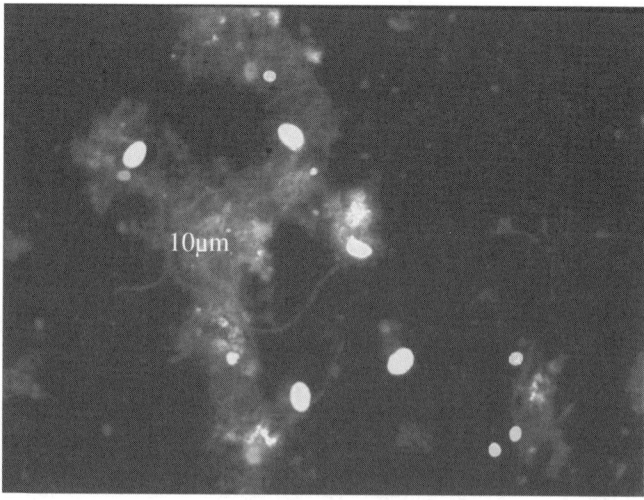

Figure 4.3. *Cryptosporidium parvum* oocysts (small apple green objects) and *Giardia duodenalis* cysts (large apple green objects) under immunofluorescence microscopy.

Giardia/Crypto IF kit from TechLab, Monofluo *Cryptosporidium* kit from Sanofi Diagnostics Pasteur, Crypto/*Giardia* Cel kit from TCS Biosciences, and Aqua-Glo G/C kit from Waterborne, etc. Because of the high sensitivity and specificity, IFA has been used in some studies as the gold standard or as a reference test (Garcia and Shimizu, 1997; Johnston *et al.*, 2003). It has been shown that most antibodies used in immunofluorescence detection of *Cryptosporidium* oocysts recognize carbohydrate epitopes on the oocyst wall (Moore *et al.*, 1998; Yu *et al.*, 2002). As the monoclonal antibodies used in commercial IFA kits react with oocysts of almost all *Cryptosporidium* species, IFA cannot make diagnosis at the species level (Graczyk *et al.*, 1996; Yu *et al.*, 2002).

The sensitivity of most microscopic methods is probably low. The detection limit for the combination of ethyl acetate concentration and IFA was shown to be 10,000 oocysts per gram of liquid stool and 50,000 oocysts per gram of formed stool (Weber *et al.*, 1991; Webster *et al.*, 1996). The sensitivity of acid-fast staining was 10-fold lower (Weber *et al.*, 1991), probably because acid-fast stains do not always consistently stain all oocysts (Garcia *et al.*, 1987). Replacing the ethyl acetate concentration procedure with sucrose, cesium chloride, or sodium chloride floatation can increase the sensitivity to 30–200 oocysts per gram of feces in animal studies (Deng and Cliver, 1999b; Fayer *et al.*, 2000b; Kuczynska and Shelton, 1999). These concentration techniques, however, are rarely used in diagnostic analysis of human stool specimens. For now, it is recommended that whenever possible, multiple specimens from each patient should be examined in the diagnosis of *Cryptosporidium* infection, as carriers with low oocyst shedding are common (Roberts *et al.*, 1989), and examination of individual specimens can lead to the detection of only 53% of infections (Greenberg *et al.*, 1996).

4.5.2.2 Antigen Detection by Immunoassays

Cryptosporidium infection can also be diagnosed by the detection of *Cryptosporidium* antigens in stool specimens by immunoassays. Antigen-capture-based enzyme immunoassays (EIA) have been used in the diagnosis of cryptosporidiosis since 1990 (Anusz *et al.*, 1990; Chapman *et al.*, 1990; Garcia and Shimizu, 1997; Rosenblatt and Sloan, 1993; Ungar, 1990). In recent years, they have gained popularity because of the ability to screen a large number of samples and an experienced microscopists is not required. Several commercial EIA kits are commonly used, such as the Alexon-Trend ProSpecT *Cryptosporidium* Microplate Assay and Meridian Premier *Cryptosporidium* kit. High specificity (99–100%) has been generally reported for these EIA kits (Dagan *et al.*, 1995; Garcia and Shimizu, 1997; Johnston *et al.*, 2003; Parisi and Tierno, 1995; Siddons *et al.*, 1992). Various sensitivities, however, have been reported, ranging from 70 (Johnston *et al.*, 2003) to 94–100% (Bialek *et al.*, 2002; Dagan *et al.*, 1995; Garcia and Shimizu, 1997; Parisi and Tierno, 1995; Rosenblatt and Sloan, 1993; Siddons *et al.*, 1992). Nevertheless, occasional false-positivity of EIA kits is known to occur in the detection of *Cryptosporidium* (Chapman *et al.*, 1990), and at least one manufacturer's recall of EIA kits has occurred because of high nonspecificity (Anonymous, 1999). These kits generally do not perform well when the number of oocysts in specimens is small (Ignatius *et al.*, 1997; Johnston *et al.*, 2003). Almost all EIA kits are for the detection of only *Cryptosporidium*, but a triage parasite panel EIA has also been marketed for simultaneous detection of *Giardia duodenalis, Entamoeba histolytica/E. dispar*, and *Cryptosporidium* antigens in human stool specimens (Garcia *et al.*, 2000). Most of the EIA kits have been evaluated only with human stool specimens. Their usefulness in the detection of *Cryptosporidium* spp. in animals may be compromised by the high specificity of antibodies. For example, the ProSpecT *Cryptosporidium* EIA does not detect many *Cryptosporidium* species that are genetically distant from *C. parvum*, such as *C. muris, C. andersoni, Cryptosporidium Serpentis*, and *Cryptosporidium baileyi* (Graczyk *et al.*, 1996).

In the last few years, at least four lateral flow immunochromatographic assays have been marketed for rapid detection of *Cryptosporidium* in stool specimens: the ImmunoCard STAT! *Cryptosporidium/Giardia* rapid assay (Meridian Bioscience), ColorPAC *Cryptosporidium/Giardia* rapid assay (Becton Dickinson), RIDA Quick *Cryptosporidium/Giardia* Combi (R-Biopharm), and the *Cryptosporidium* Dipstick (Cypress Diagnostics) (Garcia and Shimizu, 2000; Garcia *et al.*, 2003; Johnston *et al.*, 2003). In a few evaluation studies conducted with two of the assays, they have been shown to have high specificities (> 98%) (Garcia and Shimizu, 2000; Garcia *et al.*, 2003; Johnston *et al.*, 2003; Katanik *et al.*, 2001). The sensitivities of these assays were also high (98–100%) in earlier studies (Garcia and Shimizu, 2000; Garcia *et al.*, 2003; Katanik *et al.*, 2001). However, a recent study has shown a sensitivity of 68% for one of the assays (Johnston *et al.*, 2003). These rapid assays have also been plagued by quality problems and have been subjected to several manufacturer's recalls because of false positivity (Anonymous, 2002, 2004).

4.5.2.3 Molecular Methods

Molecular techniques, especially PCR and PCR-related methods, have been developed and used in the detection and differentiation of *Cryptosporidium* spp. for many years. Earlier PCR methods (Chrisp and LeGendre, 1994; Johnson *et al.*, 1995; Laxer *et al.*, 1991; Webster *et al.*, 1993) do not have the ability for species-differentiation or genotyping, and can thus only be used in the determination of the presence or absence of *Cryptosporidium* spp. The primer sequences of these techniques, with the exception of those by Johnson *et al.* (1995), are mostly based on undefined genomic sequences from *C. parvum* bovine isolates. These sequences tend to be more polymorphic than structural and house-keeping genes, therefore the primers based on them are unlikely to efficiently amplify DNA from *Cryptosporidium* spp. (such as *C. muris, C. baileyi, C. serpentis, C. canis*, and *C. felis*) and genotypes (such as the fox, skunk, and opossum genotypes) that are more distant from *C. parvum*.

Several PCR-RFLP based genotyping tools have been developed for the detection and differentiation of *Cryptosporidium* at the species level (Amar *et al.*, 2004; Awad-el-Kariem *et al.*, 1994; Kimbell *et al.*, 1999; Leng *et al.*, 1996; Lowery *et al.*, 2000; Nichols *et al.*, 2003; Sturbaum *et al.*, 2001; Xiao *et al.*, 1999a, 1999b). Most of these techniques are based on the SSU rRNA gene. However, one of the method uses an array of primers (23 primers in a nested PCR) to cover all combinations of sequence heterogeneity in the primer region of the COWP gene (Amar *et al.*, 2004). Unfortunately, primers of some of the SSU rRNA-based techniques (Awad-el-Kariem *et al.*, 1994; Kimbell *et al.*, 1999; Leng *et al.*, 1996) used conserved sequences of eukaryotic organisms. Therefore, these primers also amplify DNA from organisms other than *Cryptosporidium* (Sulaiman *et al.*, 1999). The technique by Sturbaum *et al.* (2001) also amplifies DNA of dinoflagellates (Sturbaum *et al.*, 2002). A PCR-RFLP analysis of the internal transcribed spacers of the rRNA gene can also differentiate *C. felis* from *C. parvum* (Morgan *et al.*, 1999a). Nucleotide sequencing-based approaches have also been developed for the differentiation of various *Cryptosporidium* spp. (Morgan *et al.*, 1998, 1999a; Sulaiman *et al.*, 2000, 2002; Ward *et al.*, 2002). Not all these molecular techniques, however, are diagnostic methods by nature because some of them use long amplicons (Sulaiman *et al.*, 2000, 2002), and some also amplify other apicomplexan parasites and dinoflagellates (Ward *et al.*, 2002).

Other genotyping techniques are mostly for the differentiation of *C. parvum* and *C. hominis* (Bonnin *et al.*, 1996; Carraway *et al.*, 1996, 1997; Morgan *et al.*, 1995, 1996, 1997; Patel *et al.*, 1998, 1999; Peng *et al.*, 1997; Rochelle *et al.*, 1999; Spano *et al.*, 1997, 1998; Sulaiman *et al.*, 1998; Widmer, 1998;). Both parasites have been identified in humans, but *C. hominis* (the anthroponotic genotype) has been almost exclusively found in humans; whereas the *C. parvum* (the zoonotic genotype) infects humans, ruminants, and a few other animals. Many of the genotyping tools used in these studies, however, cannot detect and differentiate other *Cryptosporidium* spp. or genotypes. Their usefulness in the analysis of human stool specimens is compromised by the failure to detect *C. canis* and *C. felis*. Indeed, a recent study has compared the ability of 10 commonly used genotyping tools in detecting seven human-pathogenic *Cryptosporidium* species/genotypes. With the exception of SSU rRNA-based PCR tools, which detected all seven *Cryptosporidium* species/genotypes, most of the genotyping tools examined had only the ability to detect *C. parvum*, *C. hominis*, and *C. meleagridis* (Jiang and Xiao, 2003).

Several subtyping tools have also been developed to characterize the diversity within the *C. parvum* or *C. hominis*. One of the most commonly used techniques is microsatellite analysis. Even though initial characterizations of eight microsatellite loci had identified only limited intragenotypic genetic diversity in *C. parvum* and *C. hominis* (Aiello *et al.*, 1999), more recent studies have identified several microsatellite sequences that seem to be more variable (Alves *et al.*, 2003a; Caccio *et al.*, 2000, 2001; Feng *et al.*, 2000; Mallon *et al.*, 2003a, 2003b; Widmer *et al.*, 2004). Although not a strict microsatellite locus by definition, results of a series of recent studies have shown high sequence polymorphism in the gene of 60 kDa glycoprotein precursor (GP60; also known as gp15/45/60, gp40/15) (Leav *et al.*, 2002; Peng *et al.*, 2001, 2003a, 2003b; Strong *et al.*, 2000; Sturbaum *et al.*, 2003; Sulaiman *et al.*, 2001; Wu *et al.*, 2003; Zhou *et al.*, 2003). Most of the genetic heterogeneity in the gene is present in the number of a tri-nucleotide repeat (TCA, TCG, or TCT), although extensive sequence differences are also present between groups (allele families) of subtypes. Other subtyping tools include sequence analysis of HSP70 (Peng *et al.*, 2003a; Sulaiman *et al.*, 2001), heteroduplex analysis and nucleotide sequencing of the double-stranded RNA (Leoni *et al.*, 2003; Xiao *et al.*, 2001b), and single-strand conformation polymorphism (SSCP)-based analysis of the second internal transcribed spacer (ITS-2) (Gasser *et al.*, 2003, 2004). A multilocus mini- and micro-satellite subtyping tool for *C. parvum* and *C. hominis* have also been developed (Mallon *et al.*, 2003a, 2003b). The usefulness of subtyping tools has been demonstrated by the analysis of samples from foodborne and waterborne outbreaks of cryptosporidiosis (Glaberman *et al.*, 2002; Leoni *et al.*, 2003; Sulaiman *et al.*, 2001; Xiao *et al.*, 2001b, 2003).

A few PCR related techniques have also been used in the quantitation and viability evaluation of *Cryptosporidium* oocysts. An excystation procedure prior to DNA extraction and PCR (excystation-PCR) has been developed to detect viable *C. parvum* oocysts (Filkorn *et al.*, 1994; Wagner-Wiening and Kimmig, 1995). Similarly, others have used a combination of cell culture and PCR (Di Giovanni *et al.*, 1999; Rochelle *et al.*, 1996; LeChevallier *et al.*, 2003) or RT-PCR (Rochelle *et al.*, 1997b) (CC-PCR or CC-RT-PCR) to detect viable *Cryptosporidium* oocysts. Because in theory RNA is less stable than DNA and breaks down quickly by the released RNAse during cell death, several reverse transcription-PCR (RT-PCR) techniques have been described for the detection of viable oocysts (Hallier-Soulier and Guillot, 2003; Jenkins *et al.*, 2000; Kaucner and Stinear, 1998; Stinear *et al.*, 1996; Widmer *et al.*, 1999). However, RNA breakdown is a slow process, which may lead to an overestimate of the viability of oocysts (Fontaine and Guillot, 2003). By nature, most of the techniques do not differentiate *Cryptosporidium* species or genotypes, although one research group used sequence analysis to determine genotypes (Di Giovanni *et al.*, 1999; LeChevallier *et al.*, 2003). More recently, several real-time PCR methods have been developed, which allow quick detection and even quantification of *Cryptosporidium* oocysts (Fontaine and Guillot, 2002, 2003; Higgins *et al.*, 2001; Limor *et al.*, 2002; MacDonald *et al.*, 2002; Tanriverdi *et al.*, 2002;). One of the techniques can differentiate *C. parvum* from *C. hominis* (Tanriverdi *et al.*, 2002), whereas another can differentiate the five common *Cryptosporidium* species in humans (Limor *et al.*, 2002). A new integrated detection assay combining capture of double-stranded RNA with probe-coated beads, RT-PCR, and lateral flow chromatography has also been developed, which should also shorten detection time (Kozwich *et al.*, 2000).

Molecular tools other than PCR have also been developed for the detection and/or differentiation of *Cryptosporidium*. Fluorescence *in situ* hybridization (FISH) or colorimetric *in situ* hybridization of probes to the SSU rRNA has been used in the detection or viability evaluation of *C. parvum* oocysts (Lindquist *et al.*, 2001b; Rochelle *et al.*, 2001; Smith *et al.*, 2004; Vesey *et al.*, 1998). It probably does not have higher sensitivity than microscopy, but with further development, it may be used in the differentiation of the species/genotypes of *Cryptosporidium* oocysts on microscope slides. Nucleic acid sequence-based amplification (NASBA) has been used in the detection of viable *C. parvum* oocysts (Baeumner *et al.*, 2001). More recently, a biosensor technique for the detection of viable *C. parvum* oocysts has also been described (Baeumner *et al.*, 2004), and a microarray technique based on HSP70 sequence polymorphism has been developed to differentiate *Cryptosporidium* genotypes (Straub *et al.*, 2002).

The following are some of the examples of the usages of molecular tools in epidemiologic investigations of human *Cryptosporidium* infections (Xiao and Ryan, 2004; Xiao *et al.*, 2003).

(A) Establishment of the identity of *Cryptosporidium* spp. in humans. We can now identify the species of *Cryptosporidium* that infects humans, the potential for non-*C. parvum Cryptosporidium* spp. to infect humans, the proportion of infections attributable to each species in various socioeconomic and epidemiologic settings, and the heterogeneity within each species causing human infections (Pedraza-Diaz *et al.*, 2000; Pieniazek *et al.*, 1999; Xiao *et al.*, 2001a).

(B) Identification of infection or contamination sources. When used in conjunction with traditional epidemiologic investigations, molecular tools can help identify the source of infection or contamination: Anthroponotic versus zoonotic *Cryptosporidium* infection, farm animal or companion animal origin versus wildlife origin. With a large sample size, molecular tools can help assess the human infective potential of *Cryptosporidium* spp. from various animals that are in frequent contact with humans. With higher resolution tools, molecular techniques can make a direct linkage between human cases of cryptosporidiosis and contamination sources (contaminated food item or water source, human index case, e.g., a foodhandler, animal reservoir) (Alves *et al.*, 2003b; Glaberman *et al.*, 2002; Hunter *et al.*, 2003; Learmonth *et al.*, 2004; McLauchlin *et al.*, 2000).

(C) Characterization of transmission dynamics of cryptosporidiosis in communities. High-resolution molecular tools can help to distinguish cryptosporidiosis point-source outbreaks from endemic but unrelated clusters of cases. These tools may also serve to identify common transmission pathways, distinguish multiple episodes of infections in humans, elucidate mechanisms of immunity against homologous and heterologous *Cryptosporidium* spp., and differentiate new episodes of infection from reactivation of latent infection (Alves *et al.*, 2003b; Cama *et al.*, 2003; Hunter *et al.*, 2004b; Peng *et al.*, 2003a; Xiao *et al.*, 2001a).

(D) Characterization of clinical spectrum and pathobiology of cryptosporidiosis. Molecular tools can improve understanding of the mechanisms underlying the variable clinical presentations and attack rates in outbreaks, variations in disease spectrum in AIDS patients, and differences in infection sites and pathophysiology caused by *Cryptosporidium* spp. In addition to host susceptibility, it is likely that the genetic diversity of *Cryptosporidium* spp. plays an important role in the clinical and pathologic spectrum of human cryptosporidiosis (Hashim *et al.*, 2004; Hunter *et al.*, 2004a; McLauchlin *et al.*, 1999; Pereira *et al.*, 2002b; Xiao *et al.*, 2001a).

4.5.3 Methods for Detection of *Cryptosporidium* Oocysts in Environmental Samples

4.5.3.1 Detection of *Cryptosporidium* Oocysts in Water Samples
Currently, the identification of *Cryptosporidium* oocysts in environmental samples is largely made by the use of IFA after concentration processes (EPA ICR method, EPA method1622/1623, United Kingdom SCA method, and United Kingdom regulatory method) (Lindquist *et al.*, 2001a). This generally requires the filtration of 10–100 L or more water, concentration and isolation of oocysts, staining of oocysts with FITC-labeled *Cryptosporidium* antibodies, and examination and quantitation of oocysts by microscopy. In the ICR or SCA method, nominal 1 μm 10" cartridge filters are used for filtration and floatation (using Percoll, sucrose or tripotassium citrate) is used

in oocyst concentration. In method 1622/1622 and the United Kingdom regulatory method, capsule filters are used in filtration and immounomagnetic separation is used in oocyst concentration. In addition, 4', 6- diamidoino-2-phenylindole (DAPI) vital dye is used in these newer methods for counterstaining. As a result, the sensitivity and accuracy of the newer methods have been improved. The recovery rates of the EPA method 1622/1623 for *Cryptosporidium* oocysts have been reported to be between 10 and 75% for surface water (DiGiorgio *et al.*, 2002; Hsu, 2003; LeChevallier *et al.*, 2003; Simmons *et al.*, 2001; Ware *et al.*, 2003). The EPA methods 1622 and 1623 can be downloaded at http://www.epa.gov/nerlcwww/1622ap01.pdf and http://www.epa.gov/waterscience/methods/1623.pdf, respectively. The United Kingdom regulatory method can be downloaded at http://www.dwi.gov.uk/regs/crypto/pdf/sop%20part%202.pdf. It should be noted that cross-reactivity of the monoclonal antibodies used in the IMS and IFA kits has been reported with dinoflagellates (Sturbaum *et al.*, 2002) and algae (Rodgers *et al.*, 1995), which may interfere with accurate detection and quantitation of *Cryptosporidium* oocysts in water, and requires careful examinations of oocyst internal structure by DAPI staining and DIC microscopy.

Because IFA detects oocysts from all *Cryptosporidium* spp., the species distribution of *Cryptosporidium* oocysts in environmental samples cannot be assessed. Although many surface water samples contain *Cryptosporidium* oocysts, it is unlikely that all of these oocysts are from human-pathogenic species or genotypes, because only five *Cryptosporidium* spp. (*C. parvum*, *C. hominis*, *C. meleagridis*, *C. canis*, and *C. felis*) are responsible for most human *Cryptosporidium* infections. Information on the source of *Cryptosporidium* contamination is necessary for accurate risk assessment, effective evaluation, and selection of management practices for reducing *Cryptosporidium* contamination in surface water and the risk of cryptosporidiosis. Thus, identification of oocysts to the species/genotype level is of significant public health importance.

The performance of many PCR methods in the analysis of environmental samples have been evaluated with *Cryptosporidium* negative samples seeded with known numbers of *C. parvum* oocysts. In early studies, PCR or RT-PCT was performed on DNA extracted directly from water concentrates seeded with *Cryptosporidium* oocysts with no oocyst isolation procedures or mere Percoll-sucrose floatation (Chung *et al.*, 1998, 1999; Kaucner and Stinear, 1998; Mayer and Palmer, 1996; Monis and Saint, 2001; Rochelle *et al.*, 1997a, 1997b; Sluter *et al.*, 1997; Stinear *et al.*, 1996). Variable sensitivities were reported by these studies, ranging from 1 to more than 100 oocysts per sample. Many researchers observed an inhibitory effect of surface water on PCR (Chung *et al.*, 1998; Johnson *et al.*, 1995; Lowery *et al.*, 2000; Rochelle *et al.*, 1997a; Sluter *et al.*, 1997; Xiao *et al.*, 2000a). Thus, almost all recent techniques have used an IMS procedure prior to cell culture and/or DNA extraction to remove PCR inhibitors or contaminants present in water samples (Di Giovanni *et al.*, 1999; Hallier-Soulier and Guillot, 1999, 2000, 2003; Johnson *et al.*, 1995; Jellison *et al.*, 2002; Kostrzynska *et al.*, 1999; Lowery *et al.*, 2000, 2001a, 2001b; Nichols *et al.*, 2003; Rimhanen-Finne *et al.*, 2002; Sturbaum *et al.*, 2002; Ward *et al.*, 2002; Wu *et al.*, 2000; Xiao *et al.*, 2000a, 2001c).

The presence of host-adapted *Cryptosporidium* species and genotypes make it possible to develop genotyping tools to determine whether the *Cryptosporidium* oocysts found in waters are from human-infective species, and to track the source of *Cryptosporidium* oocyst contamination in water. One of such techniques, the SSU rRNA-based nested PCR-RFLP method, has been successfully used in conjunction with IMS in the detection and differentiation of *Cryptosporidium* oocysts present in storm water, raw surface water, and wastewater (Xiao *et al.*, 2000a, 2001c, 2004b). In one study, 29 water samples were collected after storms, from a stream that contributes to the New York City Water Supply system and analyzed. They showed the presence of 12 wildlife genotypes of *Cryptosporidium* in 27 samples. Twelve of the 27 PCR positive samples had multiple genotypes. Four of the genotypes were traced to sources (*C. baileyi* from birds, an unnamed species from snakes, and 2 genotypes from opossums), whereas the rest were presumed to be wildlife genotypes that have never been found in humans or domestic animals, suggesting that wildlife was a major contributor for *Cryptosporidium* oocyst contamination in storm water (runoffs) in the area studied. This finding was consistent with the environmental setting (catchments were forested and isolated from agricultural activities) of the sampling site (Xiao *et al.*, 2000a).

The same technique was used in the analysis of raw surface water samples collected from different locations (Maryland, Wisconsin, Illinois, Texas, Missouri, Kansas, Michigan, Virginia, and Iowa) in the United States. A total of 55 samples were analyzed, 25 of which produced positive PCR amplification. Only 4 *Cryptosporidium* genotypes (*C. parvum*, *C. hominis*, *C. andersoni*, and *C. baileyi*) were found, all of which are parasites commonly found in farm animals and/or humans, indicating that humans and farm animals are major sources of *Cryptosporidium* oocyst contamination in these waters. Similar results were also obtained from 49 raw wastewater samples (10 or 50 ml of grab samples) collected from a treatment plant in Milwaukee, WI, 12 of which were positive for *Cryptosporidium*. Seven *Cryptosporidium* spp. (*C. parvum*, *C. hominis*, *C. andersoni*, *C. muris*, *C. canis*, *C. felis*, and *Cryptosporidium* cervine genotype) were found, with *C. andersoni* as the most common *Cryptosporidium*. As expected, the diversity of *Cryptosporidium* spp. found in source and wastewaters was much lower than that in storm waters (Xiao *et al.*, 2001c).

Two SSU rRNA-based PCR-sequencing tools and one other SSU-based PCR-RFLP tool have also been used successfully in the differentiation of *Cryptosporidium* oocysts in surface and wastewater samples (Jellison *et al.*, 2002; Nichols *et al.*, 2003; Ward *et al.*, 2002). Sequences of *C. muris*, *C. andersoni*, and presumed *C. baileyi* were obtained from seven samples of surface water from a watershed in Massachusetts (Jellison *et al.*, 2002). Analysis of 17 positive surface water samples and 6 wastewater samples from Germany and Switzerland showed the presence of 8 *Cryptosporidium* genotypes, with *C. parvum*, *C. hominis*, *C. muris*, and *C. andersoni* as the most prevalent species, and 4 samples having *C. baileyi* and 3 unidentified wildlife genotypes (Ward *et al.*, 2002). In a recent study conducted in the United Kingdom, all 14 finished water samples examined were positive for *C. hominis* by a new SSU rRNA-based PCR-RFLP tool (Nichols *et al.*, 2003). Results of these recent studies support the conclusion that humans, farm animals, and wildlife all contribute to *Cryptosporidium* oocyst contamination in water.

Promising results in the genotyping of *Cryptosporidium* spp. in water samples have also been generated in recent studies using other techniques. HSP70 sequence analysis of cell culture-PCR amplified products revealed the presence of six sequence types of *C. parvum* in raw surface water samples and filter backwash water samples, all of which were from *C. parvum*, *C. hominis*, and *Cryptosporidium* mouse genotypes (Di Giovanni *et al.*, 1999), suggesting that farm animals, rodents, and humans were responsible for *Cryptosporidium* oocyst contamination in these waters. This was confirmed more recently in a more extensive study, in which infectious *C. parvum* and *C. hominis* oocysts were detected in 22 of 560 surface water samples, with *C. parvum* found in more than 90% of the positive samples (LeChevallier *et al.*, 2003). Analysis of six river water samples by a HSP70-based RT-PCR technique also showed the presence of *C. parvum* and *C. meleagridis* in two samples (Karasudani *et al.*, 2001). Using sequencing analysis of TRAP-C2, *C. parvum* was found in 11 of 214 surface and finished water samples in Northern Ireland in one study and in 2 of 10 river water and sewage effluent samples in another study (Lowery *et al.*, 2001a, 2001b). However, HSP70 and TRAP-C2-based primers are unlikely to amplify DNA of species genetically distant from *C. parvum* (Jiang and Xiao, 2003), and the primers used in the study by Karasudani *et al.* (2001) were previously shown to have poor specificity (Kaucner and Stinear, 1998).

4.5.3.2 Detection of *Cryptosporidium* Oocysts in Food Samples

The detection of parasites in food matrices has been a major challenge to parasitologists and food safety professionals for many years. First, there is a wide range of sample matrices. Second, the volume of materials needing to be analyzed is often huge when compared to the technical abilities of most traditional methods. Third, the load of parasites likely to be present is usually low. As a result, the recovery rate of detection methods for parasites in foodstuff can be very low (Bier, 1991).

The first step in the detection of *Cryptosporidium* oocysts in food after sampling is the elution of parasites from different food matrices. In the case of fruits, leafy greens or fresh produce, parasites can be recovered by washing the produce samples in 0.025M phosphate buffered saline, pH 7.25 (Ortega *et al.*, 1997). Sometimes, detergents (1% sodium dodecyl sulfate and 0.1% Tween 80, or the membrane filter elution buffer from EPA method 1623) and sonication (3–10 minutes) are also used to facilitate the elution of parasites from the food matrices (Bier, 1991; Robertson and Gjerde, 2000). The parasites are then concentrated by centrifugation and examined directly or after immunofluorescence staining (Bier, 1991; Ortega *et al.*, 1997). Sometimes, a sucrose floatation step is included to further purify parasites (Bier, 1991). Initially, this procedure was reported to have a low recovery rate of 1% for *Cryptosporidium* oocysts in cabbage and lettuce, two relatively simple food matrices (Bier, 1991). However, moderate recovery rates of 18.2–25.2% were subsequently reported for a variety of fresh produces (Ortega *et al.*, 1997). More recently, IMS has been used in the recovery of *Cryptosporidium* oocysts and *Giardia* cysts from fruits and vegetables, which has resulted in an improvement in recovery of parasites from lettuce, Chinese leaves, and strawberries to 42% for *Cryptosporidium* and 67% for

Giardia (Robertson and Gjerde, 2000). This new method includes washing procedures, sonication, IMS, immunofluorescence staining, and microscopy (Robertson and Gjerde, 2000, 2001a).

The detection of *Cryptosporidium* oocysts in shellfish is relatively easy compared to detection in vegetables, largely because the amount of materials for analysis is smaller and the number of oocysts potentially present is generally higher. In large molluscs such as oysters, mussels, and large clams, gills are usually removed with scissors, and washed by vortexing and centrifugation. *Cryptosporidium* oocysts present are examined and quantitated by microscopy after immunofluorescence staining. Sometimes, hemolymph is also harvested and *Cryptosporidium* oocysts in hemocytes are examined by immunofluorescence (Fayer *et al.*, 1998). With smaller molluscs such as small mussels and clams, the hemolymph or homogenized whole shellfish or gastrointestinal tract is generally examined individually or in pools (Graczyk *et al.*, 1999, 2001; Tamburrini and Pozio, 1999).

PCR has not been used in the analysis of fresh produce, but in theory, IMS-purified oocysts from fresh produces can be genotyped by molecular techniques using the same procedures developed for the analysis of water samples. Many studies have used PCR to genotype *Cryptosporidium* oocysts found in shellfish (Fayer *et al.*, 1999, 2002, 2003; Gomez-Bautista *et al.*, 2000; Gomez-Couso *et al.*, 2004; Graczyk *et al.*, 2001), which is useful in tracking the sources of contamination.

4.6 TREATMENT

Numerous pharmaceutical compounds have been screened for anti-*Cryptosporidium* activities *in vitro* or in laboratory animals. Some of those showing promise have been used in the experimental treatment of cryptosporidiosis in humans, but few have been shown to be effective in controlled clinical trials (Hunter and Nichols, 2002). Oral or intravenous rehydration is used whenever severe diarrhea is associated with *Cryptosporidium* infection (Hoepelman, 1996). Nitazoxanide (NTZ) is the only FDA approved drug for the treatment of pediatric cryptosporidiosis. It has also been recently approved for the treatment of Giardiasis in adult patients. Clinical trials have demonstrated that NTZ can shorten clinical disease and reduce parasite load (Amadi *et al.*, 2002; Rossignol *et al.*, 2001). This drug, however, is not yet approved for the treatment of *Cryptosporidium* infections in immunodeficient people, even though it is likely to be partially effective (Doumbo *et al.*, 1997; Rossignol *et al.*, 1998). For this population, paramomycin and spiramycin have been used in the treatment of some patients, but their efficacy remains unproven (Hewitt *et al.*, 2000). Thus, rehydration is still the major supportive treatment in AIDS patients (Hoepelman, 1996).

In industrialized nations, the most effective treatment and prophylaxis for cryptosporidiosis in AIDS patients is the use of highly active antiretroviral therapy (HAART) (Carr *et al.*, 1998; Miao *et al.*, 2000). Nonetheless, it is believed that the eradication and prevention of the infection is directly related to the replenishment of CD4+ cells in treated persons, rather than antiparasitic activities of these drugs (Carr *et al.*, 1998), even though some of the protease inhibitors used in HAART, such as indinavir, nelfinavir, and ritonavir, have been shown to have anti-cryptosporidial activities *in vitro* and in laboratory animals (Hommer *et al.*, 2003; Mele *et al.*, 2003). Relapse of cryptosporidiosis is common in AIDS patients who have stopped taking HAART (Carr *et al.*, 1998; Maggi *et al.*, 2000).

4.7 CONTROL OF *CRYPTOSPORIDIUM* CONTAMINATION IN WATER AND FOOD

Cryptosporidium oocysts are very environmentally robust, with the capability for long-term survival in a variety of natural environments and resistance to most disinfectants. Unlike the majority of bacteria and viruses, *Cryptosporidium* spp. have an environmentally resistant resting stage in the form of the oocyst as part of its complex life cycle. The wall of oocysts allows the organism to remain viable for a considerable period, resist various harsh environmental challenges, and await the opportunity to infect a new susceptible host. Table 4.4 summarizes the survival of *Cryptosporidium* oocysts in a variety of matrices under controlled conditions in selected studies. *Cryptosporidium* oocysts can survive for months in soil, fresh water, and seawater. Thus, natural contamination of the environment can accumulate over time and the contaminated environment may be a reservoir of viable oocysts for long periods of time. For example, Tamburrini and Pozio (1999) reported that oocysts remain infective in seawater for up to one year and can be filtered out by benthic mussels, which retain their infectivity.

A combination of filtration and disinfection is required for controlling *Cryptosporidium* oocysts in water, which also helps to reduce the contamination in foods and beverages. Physical removal of oocysts from drinking water through coagulation, sedimentation, and filtration is the primary defense against waterborne cryptosporidiosis (Rose, 1997). Deficiencies in any one of these processes have been shown to directly affect the efficiency of overall oocyst removal (Medema *et al.*, 2003). Properly operated conventional treatment (coagulation/flocculation, sedimentation, filtration, and disinfection) can remove 99% or more of oocysts (Hashimoto *et al.*, 2001; Hijnen *et al.*, 2004; Hsu and Yeh, 2003). One of the critical times when oocysts can breach the filtration barrier is following backwash (Karanis *et al.*, 1996). For this reason, optimization of the backwash procedure, including the addition of coagulants, or filtering of waste can minimize the passage of oocysts.

Chlorination alone has not been successful in eliminating waterborne *Cryptosporidium* oocysts. As much as 80mg/l of free chlorine or monochloramine requires 90 minutes to produce 90% oocyst inactivation (Korich *et al.*, 1990). Chlorine dioxide, on the other hand, seems to be more effective than free chlorine. Peeters *et al.* (1989) reported that 0.43mg/l of chlorine dioxide (ClO_2) reduced infectivity within 15 min, although some oocysts remained viable. Korich *et al.* (1990) reported approximately 90% inactivation of oocysts exposed to 1.3mg/l of chlorine dioxide for 60 min. In contrast, ozone and ultra violet (UV) radiation have shown the most promise as effective inactivation practices. An initial concentration of 1.11mg/l ozone for 6 min was shown to inactivate viable oocysts at a concentration of 10^4 oocysts/ml (Peeters *et al.*, 1989). Korich *et al.* (1990) reported that exposure

Table 4.4. Percentage reduction in *Cryptosporidium* oocyst viability with different treatments.

	Treatment	%Reduction*	Reference
Water	60 days at natural condition (*C. parvum*)	54	(Kato et al., 2001)
	120 days at natural condition (*C. parvum*)	89	(Kato et al., 2001)
Soil	60 days at natural condition (*C. parvum*)	61	(Kato et al., 2001)
	120 days at natural condition (*C. parvum*)	90	(Kato et al., 2001)
Silage	106 days (*C. parvum*)	46–62	(Merry et al., 1997)
Mineral water	4°C for 12 weeks (*C. parvum*)	1–11	(Nichols et al., 2004)
	20°C for 12 weeks (*C. parvum*)	22–59	(Nichols et al., 2004)
	4.5% NaCl at 22°C for 8 days (*C. hominis*)	77	(Dawson et al., 2004)
	4.5% NaCl 9 days at 22°C (*C. parvum*)	57	(Dawson et al., 2004)
	9% Ethanol 7 days at 22°C (*C. hominis*)	77	(Dawson et al., 2004)
	9% Ethanol 8 days at 22°C (*C. hominis*)	66	(Dawson et al., 2004)
	40% Ethanol 8 days at 22°C (*C. hominis*)	72	(Dawson et al., 2004)
Food and beverage treatment	20% Glycerol 7 days at 4°C (*C. hominis*)	57	(Dawson et al., 2004)
	20% Glycerol 13 days at 4°C (*C. parvum*)	85	(Dawson et al., 2004)
	20% Glycerol 13 days at 22°C (*C. parvum*)	87	(Dawson et al., 2004)
	20% Glycerol 14 days at 4°C (*C. parvum*)	53	(Dawson et al., 2004)
	50% Sucrose 7 days at 22°C (*C. hominis*)	100	(Dawson et al., 2004)
	50% Sucrose 8 days at 22°C (*C. hominis*)	86	(Dawson et al., 2004)
	50% Sucrose 9 days at 22°C (*C. parvum*)	90	(Dawson et al., 2004)

(*continued*)

Table 4.4. (*continued*)

	Treatment	%Reduction*	Reference
Water and water treatment	Frozen at −22°C for 297 h (*C. parvum*)	86	(Robertson *et al.*, 1992)
	4°C for 176 days (*C. parvum*)	57–66	(Robertson *et al.*, 1992)
	In seawater at 4°C for 35 days (*C. parvum*)	22–31	(Robertson *et al.*, 1992)
	1.5 ppm aluminum at room temperature for 7 mins (*C. parvum*)	3–4	(Robertson *et al.*, 1992)
	16 ppm ferric sulfate at room temperature for 1 h (*C. parvum*)	37	(Robertson *et al.*, 1992)
	0.2% calcium hydroxide (lime) at room temperature for 1 h (*C. parvum*)	30	(Robertson *et al.*, 1992)
	250–270 nm UV radiation at 2 mJ/cm^2 (*C. parvum*)	1.8–2.3 log	(Linden *et al.*, 2001)

to 1mg/l ozone inactivated between 90 and 99% of oocysts (2.8×10^5/ml) in water at 25°C. Ninety-nine to 99.9% inactivation was achieved when the exposure time was increased to 10 mins. In addition to ozone treatment, UV radiation has now been rapidly adopted by the water industry for inactivation of *Cryptosporidium* oocysts in water (Bukhari *et al.*, 2004; Lorenzo-Lorenzo *et al.*, 1993). UV light between 250 and 270 nm in wavelengths have been shown to reduce *C. parvum* oocyst infectivity at 2 mJ/cm^2 (Linden *et al.*, 2001). Higher doses can lead to higher inactivation rates (Craik *et al.*, 2001). Most chemicals used in floccation during the first step of water treatment have only a limited effect on the viability of *Cryptosporidium* oocysts at the practical concentrations (Robertson *et al.*, 1992).

Cryptosporidium oocysts can contaminate food through many pathways. These include (i) introduction to the foodstuff through contaminated raw ingredients, e.g., unwashed lettuce destined for ready-to-eat salads; (ii) introduction during food processing due to addition of contaminated water, as an important ingredient of the foodstuff, e.g., in soft drinks production; (iii) introduction during food processing as a contaminant of equipment cleaning with non-potable water; (iv) introduction of the parasite through pest infestations, e.g., cockroaches, house flies, mice, and rats; and (v) introduction of the parasite to processed foodstuffs from positive food handlers. The associated risk from each of these potential routes of entry of oocysts into the foodstuff should be controlled through an integrated HACCP (hazard analysis and critical control point) management.

The effect of food processing and storage practices on the viability of potentially contaminated *Cryptosporidium* oocysts in food and beverage depends on the nature of the treatment. Snap freezing is detrimental to the survival of *Cryptosporidium* oocysts (Fayer and Nerad, 1996; Robertson et al., 1992), but if suspended in water, stored at $-20°$ C for 24 h, and then transferred to $-70°$C, *C. muris* oocysts can survive for at least 15 months (Rhee and Park, 1996). Some *C. parvum* oocysts can survive freezing in water at higher temperatures ($-20-22°$C) for 1–32 days (Deng and Cliver, 1999a; Fayer and Nerad, 1996; Robertson et al., 1992). As expected, air drying for 4 h kills almost all *C. parvum* oocysts (Deng and Cliver, 1999a; Robertson et al., 1992). *Cryptosporidium* oocysts can survive high temperatures for only short durations. Oocysts of *C. parvum* lose infectivity at 72.4°C or higher within 1 min, or when the temperature is held at 64.2°C or higher for 2 min (Fayer, 1994). The high-temperature-short-time conditions (71.7°C for 15 s) used in commercial pasteurization are sufficient to destroy infectivity of *C. parvum* oocysts in milk and apple cider (Harp et al., 1996; Deng and Cliver, 2001). Most oocysts of *C. parvum* can survive for at least 10 days during the process of yogurt making and storage, but cannot survive the ice cream making process (Deng and Cliver, 1999a).

Not much is known on the survival of *Cryptosporidium* oocysts in beverage. Although high salinities reduce the survival of *C. parvum* in water (Fayer et al., 1998), oocysts can maintain viability for months in natural mineral water, especially at low temperatures (Nichols et al., 2004). There is some reduction in oocyst viability in acidified and carbonated beverages (Friedman et al., 1997). Oocysts of *C. hominis* stored in 9 or 40% ethanol for 7 or 8 days at 22°C suffer 66–77% reductions in viability (Dawson et al., 2004). However, their ability for long-term survival in ethanol is not clear.

Not all food preservatives have detrimental effects on the viability of *Cryptosporidium* oocysts. Considerable viability is maintained when *C. parvum* oocysts are stored at 4°C or 22°C in media containing citric, acetic, or lactic acid (Dawson et al., 2004). Oocysts of *C. parvum* and *C. hominis* kept in 4.5% sodium chloride at 22°C for 8 or 9 days have 57–77% reduction in viability. Similar losses in viability also occur in oocysts stored in 20% glycerol at 4°C or 22°C for 13 or 14 days. Storage in 50% sucrose at 22°C, however, is detrimental to most *C. parvum* and *C. hominis* oocysts (Dawson et al., 2004).

Presently, there is no solid recommendation regarding the management of *Cryptosporidium*-positive food handlers within the food-processing sector. The mean duration of the illness has previously been reported as 12.2 days; however, the range in duration is 2 to 26 days (Jokipii and Jokipii, 1986). Oocyst excretion times have varied widely from 6.9 days (range 1–15 days) after the cessation of symptoms, to 2 months and greater in a small proportion of patients. Thus, it is impossible to predict the carrier status of persons based on cessation of symptoms. In addition, microbiological screening for carrier status in infected persons is problematic as symptomatic patients may have intermittently negative stool specimens (Jokipii and Jokipii, 1986). Other studies have shown that asymptomatic carriers are found in 0.4% of the general population in Australia (Hellard et al., 2000) and 6.4% of immunocompetent children in the United States (Pettoello-Mantovani et al., 1995). Thus, consumers of ready-to-eat foodstuffs are vulnerable to potential contamination of products by food-handlers with both symptomatic and asympotomatic cryptosporidiosis (Quiroz et al., 2000). Therefore, it is important that other general hygienic practices such as hand washing and glove wearing are also implemented as part of the HACCP management to minimize food poisoning due to cryptosporidiosis and other pathogens.

REFERENCES

Aiello, A. E., Xiao, L. H., Limor, J. R., Liu, C., Abrahamsen, M. S., and Lal, A. A., 1999, Microsatellite analysis of the human and bovine genotypes of *Cryptosporidium parvum*, *J. Eukaryot. Microbiol.* **46**:46S–47S.

Alpert, G., Bell, L. M., Kirkpatrick, C. E., Budnick, L. D., Campos, J. M., Friedman, H. M., and Plotkin, S. A., 1984, Cryptosporidiosis in a day-care center, *N. Engl. J. Med.* **311**:860–861.

Alvarez-Pellitero, P., and Sitja-Bobadilla, A., 2002, *Cryptosporidium molnari* n. sp. (Apicomplexa: Cryptosporidiidae) infecting two marine fish species, Sparus aurata L. and Dicentrarchus labrax L, *Int. J. Parasitol.* **32**:1007–1021.

Alves, M., Matos, O., and Antunes, F., 2003a, Microsatellite analysis of *Cryptosporidium hominis* and *C. parvum* in Portugal: A preliminary study, *J. Eukaryot. Microbiol.* **50**(Suppl):529–530.

Alves, M., Xiao, L., Sulaiman, I., Lal, A. A., Matos, O., and Antunes, F., 2003b, Subgenotype analysis of *Cryptosporidium* isolates from humans, cattle, and zoo ruminants in Portugal, *J. Clin. Microbiol.* **41**:2744–2747.

Amadi, B., Mwiya, M., Musuku, J., Watuka, A., Sianongo, S., Ayoub, A., and Kelly, P., 2002, Effect of nitazoxanide on morbidity and mortality in Zambian children with cryptosporidiosis: A randomised controlled trial, *Lancet* **360**:1375–1380.

Amar, C. F., Dear, P. H., and McLauchlin, J., 2004, Detection and identification by real time PCR/RFLP analyses of *Cryptosporidium* species from human faeces, *Lett. Appl. Microbiol.* **38**:217–222.

Anonymous, 1984, Cryptosporidiosis among children attending day-care centers—Georgia, Pennsylvania, Michigan, California, New Mexico, *Morb. Mortal. Wkly. Rep.* **33**:599–601.

Anonymous, 1990, Cryptosporidiosis in England and Wales: Prevalence and clinical and epidemiological features. Public Health Laboratory Service Study Group, *BMJ* **300**:774–777.

Anonymous, 1996, Foodborne outbreak of diarrheal illness associated with *Cryptosporidium parvum*—Minnesota, 1995, *Morb. Mortal. Wkly. Rep.* **45**:783–784.

Anonymous, 1997, Outbreaks of *Escherichia coli* O157:H7 infection and cryptosporidiosis associated with drinking unpasteurized apple cider—Connecticut and New York, October 1996, *Morb. Mortal. Wkly. Rep.* **46**:4–8.

Anonymous, 1998, Foodborne outbreak of cryptosporidiosis—Spokane, Washington, 1997, *Morb. Mortal. Wkly. Rep.* **47**:565–567.

Anonymous, 1999, False-positive laboratory tests for *Cryptosporidium* involving an enzyme-linked immunosorbent assay—United States, November 1997-March 1998, *Morb. Mortal. Wkly. Rep.* **48**:4–8.

Anonymous, 2002, Manufacturer's recall of rapid assay kits based on false positive *Cryptosporidium* antigen tests—Wisconsin, 2001–2002, *Morb. Mortal. Wkly. Rep.* **51**:189.

Anonymous, 2004, Manufacturer's recall of rapid cartridge assay kits on the basis of false-positive *Cryptosporidium* antigen tests—Colorado, 2004, *Morb. Mortal. Wkly. Rep.* **53**:198.

Anusz, K. Z., Mason, P. H., Riggs, M. W., and Perryman, L. E., 1990, Detection of *Cryptosporidium parvum* oocysts in bovine feces by monoclonal antibody capture enzyme-linked immunosorbent assay, *J. Clin. Microbiol.* **28**:2770–2774.

Armon, R., Gold, D., Brodsky, M., and Oron, G., 2002, Surface and subsurface irrigation with effluents of different qualities and presence of *Cryptosporidium* oocysts in soil and on crops, *Water Sci. Technol.* **46**:115–122.

Arrowood, M. J., 1997, Diagnosis, In Fayer, R. (ed), *Cryptosporidium and Cryptosporidiosis*, CRC Press, Boca Raton, FL, pp. 43–64.

Arrowood, M. J., and Sterling, C. R., 1989, Comparison of conventional staining methods and monoclonal antibody-based methods for *Cryptosporidium* oocyst detection, *J. Clin. Microbiol.* **27**:1490–1495.

Assadamongkol, K., Gracey, M., Forbes, D., and Varavithya, W., 1992, *Cryptosporidium* in 100 Australian children, *Southeast Asian J. Trop. Med. Public Health* **23**:132–137.

Awad-el-Kariem, F. M., Warhurst, D. C., and McDonald, V., 1994, Detection and species identification of *Cryptosporidium* oocysts using a system based on PCR and endonuclease restriction, *Parasitology* **109**:19–22.

Baeumner, A. J., Humiston, M. C., Montagna, R. A., and Durst, R. A., 2001, Detection of viable oocysts of *Cryptosporidium parvum* following nucleic acid sequence based amplification, *Anal. Chem.* **73**:1176–1180.

Baeumner, A. J., Pretz, J., and Fang, S., 2004, A universal nucleic Acid sequence biosensor with nanomolar detection limits, *Anal. Chem.* **76**:888–894.

Barker, I. K., and Carbonell, P. L., 1974, *Cryptosporidium agni* sp.n. from lambs, and *Cryptosporidium bovis* sp.n. from a calf, with observations on the oocyst, *Z. Parasitenkd.* **44**:289–298.

Bern, C., Hernandez, B., Lopez, M. B., Arrowood, M. J., De Merida, A. M., and Klein, R. E., 2000, The contrasting epidemiology of *Cyclospora* and *Cryptosporidium* among outpatients in Guatemala, *Am. J. Trop. Med. Hyg.* **63**:231–235.

Bern, C., Ortega, Y., Checkley, W., Roberts, J. M., Lescano, A. G., Cabrera, L., Verastegui, M., Black, R. E., Sterling, C., and Gilman, R. H., 2002, Epidemiologic differences between cyclosporiasis and cryptosporidiosis in Peruvian children, *Emerg. Infect. Dis.* **8**:581–585.

Bhattacharya, M. K., Teka, T., Faruque, A. S., and Fuchs, G. J., 1997, *Cryptosporidium* infection in children in urban Bangladesh, *J. Trop. Pediatr.* **43**:282–286.

Bialek, R., Binder, N., Dietz, K., Joachim, A., Knobloch, J., and Zelck, U. E., 2002, Comparison of fluorescence, antigen and PCR assays to detect *Cryptosporidium parvum* in fecal specimens, *Diagn. Microbiol. Infect. Dis.* **43**:283–288.

Bier, J. W., 1991, Isolation of parasites on fruits and vegetables, *Southeast Asian J. Trop. Med. Public Health* **22**(Suppl):144–145.

Bird, R. G., and Smith, M. D., 1980, Cryptosporidiosis in man: Parasite life cycle and fine structural pathology, *J. Pathol.* **132**:217–233.

Bonnin, A., Fourmaux, M. N., Dubremetz, J. F., Nelson, R. G., Gobet, P., Harly, G., Buisson, M., Puygauthier-Toubas, D., Gabriel-Pospisil, G., Naciri, M., and Camerlynck, P., 1996, Genotyping human and bovine isolates of *Cryptosporidium parvum* by polymerase chain reaction-restriction fragment length polymorphism analysis of a repetitive DNA sequence, *FEMS Microbiol. Lett.* **137**:207–211.

Brandonisio, O., Maggi, P., Panaro, M. A., Lisi, S., Andriola, A., Acquafredda, A., and Angarano, G., 1999, Intestinal protozoa in HIV-infected patients in Apulia, South Italy, *Epidemiol. Infect.* **123**:457–462.

Bukhari, Z., Abrams, F., and LeChevallier, M., 2004, Using ultraviolet light for disinfection of finished water, *Water Sci. Technol.* **50**:173–178.

Caccio, S., Homan, W., Camilli, R., Traldi, G., Kortbeek, T., and Pozio, E., 2000, A microsatellite marker reveals population heterogeneity within human and animal genotypes of *Cryptosporidium parvum*, *Parasitology* **120**:237–244.

Caccio, S., Spano, F., and Pozio, E., 2001, Large sequence variation at two microsatellite loci among zoonotic (genotype C) isolates of *Cryptosporidium parvum*, *Int. J. Parasitol.* **31**:1082–1086.

Cama, V. A., Bern, C., Sulaiman, I. M., Gilman, R. H., Ticona, E., Vivar, A., Kawai, V., Vargas, D., Zhou, L., and Xiao, L., 2003, *Cryptosporidium* species and genotypes in HIV-positive patients in Lima, Peru, *J. Eukaryot. Microbiol.* **50**(Suppl):531–533.

Caputo, C., Forbes, A., Frost, F., Sinclair, M. I., Kunde, T. R., Hoy, J. F., and Fairley, C. K., 1999, Determinants of antibodies to *Cryptosporidium* infection among gay and bisexual men with HIV infection, *Epidemiol. Infect.* **122**:291–297.

Carr, A., Marriott, D., Field, A., Vasak, E., and Cooper, D. A., 1998, Treatment of HIV-1-associated microsporidiosis and cryptosporidiosis with combination antiretroviral therapy, *Lancet* **351**:256–261.

Carraway, M., Tzipori, S., and Widmer, G., 1996, Identification of genetic heterogeneity in the *Cryptosporidium parvum* ribosomal repeat, *Appl. Environ. Microbiology.* **62**:712–716.

Carraway, M., Tzipori, S., and Widmer, G., 1997, A new restriction fragment length polymorphism from *Cryptosporidium parvum* identifies genetically heterogeneous parasite populations and genotypic changes following transmission from bovine to human hosts, *Infect. Immun.* **65**:3958–3960.

Carreno, R. A., Martin, D. S., and Barta, J. R., 1999, *Cryptosporidium* is more closely related to the gregarines than to coccidia as shown by phylogenetic analysis of apicomplexan parasites inferred using small-subunit ribosomal RNA gene sequences, *Parasitol. Res.* **85**:899–904.

Chalmers, R. M., Sturdee, A. P., Mellors, P., Nicholson, V., Lawlor, F., Kenny, F., and Timpson, P., 1997, *Cryptosporidium parvum* in environmental samples in the Sligo area, Republic of Ireland: A preliminary report, *Lett. Appl. Microbiol.* **25**:380–384.

Chalmers, R. M., Elwin, K., Thomas, A. L., and Joynson, D. H., 2002, Infection with unusual types of *Cryptosporidium* is not restricted to immunocompromised patients, *J. Infect. Dis.* **185**:270–271.

Chapman, P. A., Rush, B. A., and McLauchlin, J., 1990, An enzyme immunoassay for detecting *Cryptosporidium* in faecal and environmental samples, *J. Med. Microbiol.* **32**:233–237.

Checkley, W., Epstein, L. D., Gilman, R. H., Black, R. E., Cabrera, L., and Sterling, C. R., 1998, Effects of *Cryptosporidium parvum* infection in Peruvian children: Growth faltering and subsequent catch-up growth, *Am. J. Epidemiol.* **148**:497–506.

Checkley, W., Gilman, R. H., Epstein, L. D., Suarez, M., Diaz, J. F., Cabrera, L., Black, R. E., and Sterling, C. R., 1997, Asymptomatic and symptomatic cryptosporidiosis: Their acute effect on weight gain in Peruvian children, *Am. J. Epidemiol.* **145**:156–163.

Chen, X. M., and LaRusso, N. F., 2002, Cryptosporidiosis and the Pathogenesis of AIDS-Cholangiopathy, *Semin. Liver Dis.* **22**:277–290.

Chieffi, P. P., Sens, Y. A., Paschoalotti, M. A., Miorin, L. A., Silva, H. G., and Jabur, P., 1998, Infection by *Cryptosporidium parvum* in renal patients submitted to renal transplant or hemodialysis, *Rev. Soc. Bras. Med. Trop.* **31**:333–337.

Chmelik, V., Ditrich, O., Trnovcova, R., and Gutvirth, J., 1998, Clinical features of diarrhoea in children caused by *Cryptosporidium parvum*, *Folia Parasitol.* **45**:170–172.

Chrisp, C. E., and LeGendre, M., 1994, Similarities and differences between DNA of *Cryptosporidium parvum* and C. wrairi detected by the polymerase chain reaction, *Folia Parasitol.* **41**:97–100.

Chung, E., Aldom, J. E., Carreno, R. A., Chagla, A. H., Kostrzynska, M., Lee, H., Palmateer, G., Trevors, J. T., Unger, S., Xu, R., and De Grandis, S. A., 1999, PCR-based quantitation of *Cryptosporidium parvum* in municipal water samples, *J. Microbiol. Methods* **38**:119–130.

Chung, E., Aldom, J. E., Chagla, A. H., Kostrzynska, M., Lee, H., Palmateer, G., Trevors, J. T., Unger, S., and Degrandis, S., 1998, Detection of *Cryptosporidium parvum* oocysts in municipal water samples by the polymerase chain reaction, *J. Microbiol. Methods* **33**:171–180.

Clark, D. P., 1999, New insights into human cryptosporidiosis, *Clin. Microbiol. Rev.* **12**:554–563.

Clayton, F., Heller, T., and Kotler, D. P., 1994, Variation in the enteric distribution of cryptosporidia in acquired immunodeficiency syndrome, *Am. J. Clin. Pathol.* **102**:420–425.

Colford, J. M., Jr., Tager, I. B., Hirozawa, A. M., Lemp, G. F., Aragon, T., and Petersen, C., 1996, Cryptosporidiosis among patients infected with human immunodeficiency virus. Factors related to symptomatic infection and survival, *Am. J. Epidemiol.* **144**:807–816.

Corlis, J. O., 1994, An interim utilitarian ('user-friendly') hierarchical classification and characterization of the protists, *Acta Protozool.* **33**:1–51.

Craik, S. A., Weldon, D., Finch, G. R., Bolton, J. R., and Belosevic, M., 2001, Inactivation of *Cryptosporidium parvum* oocysts using medium- and low-pressure ultraviolet radiation, *Water Res.* **35**:1387–1398.

Current, W. L., Reese, N. C., Ernst, J. V., Bailey, W. S., Heyman, M. B., and Weinstein, W. M., 1983, Human cryptosporidiosis in immunocompetent and immunodeficient persons. Studies of an outbreak and experimental transmission, *N. Engl. J. Med.* **308**:1252–1257.

Current, W. L., Upton, S. J., and Haynes, T. B., 1986, The life cycle of *Cryptosporidium baileyi* n. sp. (Apicomplexa, Cryptosporidiidae) infecting chickens, *J. Protozool.* **33**:289–296.

Dagan, R., Fraser, D., El-On, J., Kassis, I., Deckelbaum, R., and Turner, S., 1995, Evaluation of an enzyme immunoassay for the detection of *Cryptosporidium* spp. in stool specimens from infants and young children in field studies, *Am. J. Trop. Med. Hyg.* **52**:134–138.

Dalle, F., Roz, P., Dautin, G., Di-Palma, M., Kohli, E., Sire-Bidault, C., Fleischmann, M. G., Gallay, A., Carbonel, S., Bon, F., Tillier, C., Beaudeau, P., and Bonnin, A., 2003, Molecular characterization of isolates of waterborne *Cryptosporidium* spp. collected during an outbreak of gastroenteritis in South Burgundy, France, *J. Clin. Microbiol.* **41**:2690–2693.

D'Antonio, R. G., Winn, R. E., Taylor, J. P., Gustafson, T. L., Current, W. L., Rhodes, M. M., Gary, G. W., Jr., and Zajac, R. A., 1985, A waterborne outbreak of cryptosporidiosis in normal hosts, *Ann. Intern. Med.* **103**:886–888.

Daoud, A. S., Zaki, M., Pugh, R. N., al-Mutairi, G., al-Ali, F., and el-Saleh, Q., 1990, *Cryptosporidium* gastroenteritis in immunocompetent children from Kuwait, *Trop. Geogr. Med.* **42**:113–118.

Dawson, D. J., Samuel, C. M., Scrannage, V., and Atherton, C. J., 2004, Survival of *Cryptosporidium* species in environments relevant to foods and beverages, *J. Appl. Microbiol.* **96**:1222–1229.

Deng, M. Q., and Cliver, D. O., 1999a, *Cryptosporidium parvum* studies with dairy products, *Int. J. Food Microbiol.* **46**:113–121.

Deng, M. Q., and Cliver, D. O., 1999b, Improved immunofluorescence assay for detection of *Giardia* and *Cryptosporidium* from asymptomatic adult cervine animals, *Parasitol. Res.* **85**:733–736.

Deng, M. Q., and Cliver, D. O., 2001, Inactivation of *Cryptosporidium parvum* oocysts in cider by flash pasteurization, *J. Food Prot.* **64**:523–527.

Di Giovanni, G. D., Hashemi, F. H., Shaw, N. J., Abrams, F. A., LeChevallier, M. W., and Abbaszadegan, M., 1999, Detection of infectious *Cryptosporidium parvum* oocysts in surface and filter backwash water samples by immunomagnetic separation and integrated cell culture-PCR, *Appl. Environ. Microbiol.* **65**:3427–3432.

Dietz, V. J., and Roberts, J. M., 2000, National surveillance for infection with *Cryptosporidium parvum*, 1995–1998: What have we learned?, *Public Health Rep.* **115**:358–363.

Dietz, V., Vugia, D., Nelson, R., Wicklund, J., Nadle, J., McCombs, K. G., and Reddy, S., 2000, Active, multisite, laboratory-based surveillance for *Cryptosporidium parvum*, *Am. J. Trop. Med. Hyg.* **62**:368–372.

DiGiorgio, C. L., Gonzalez, D. A., and Huitt, C. C., 2002, *Cryptosporidium* and *Giardia* recoveries in natural waters by using environmental protection agency method 1623, *Appl. Environ. Microbiol.* **68**:5952–5955.

Doumbo, O., Rossignol, J. F., Pichard, E., Traore, H. A., Dembele, T. M., Diakite, M., Traore, F., and Diallo, D. A., 1997, Nitazoxanide in the treatment of cryptosporidial diarrhea and other intestinal parasitic infections associated with acquired immunodeficiency syndrome in tropical Africa, *Am. J. Trop. Med. Hyg.* **56**:637–639.

Dubey, J. P., 1993, Intestinal protozoa infections, *Vet. Clin. North Am. Small Anim. Pract.* **23**:37–55.

Eisenberg, J. N., Priest, J. W., Lammie, P. J., and Colford, J. M., Jr., 2001, The Serologic response to *Cryptosporidium* in HIV-infected persons: Implications for epidemiologic research, *Emerg. Infect. Dis.* **7**:1004–1009.

Fall, A., Thompson, R. C., Hobbs, R. P., and Morgan-Ryan, U., 2003, Morphology is not a reliable tool for delineating species within *Cryptosporidium*, *J. Parasitol.* **89**:399–402.

Fayer, R., 1994, Effect of high temperature on infectivity of *Cryptosporidium parvum* oocysts in water, *Appl. Environ. Microbiol.* **60**:2732–2735.

Fayer, R., and Nerad, T., 1996, Effects of low temperatures on viability of *Cryptosporidium parvum* oocysts, *Appl. Environ. Microbiol.* **62**:1431–1433.

Fayer, R., Speer, C. A., and Dubey, J. P., 1997, The general biology of *Cryptosporidium*, In Fayer, R. (ed), *Cryptosporidium and Cryptosporidiosis*, CRC Press, Boca Raton, FL, pp. 1–41.

Fayer, R., Graczyk, T. K., Lewis, E. J., Trout, J. M., and Farley, C. A., 1998, Survival of infectious *Cryptosporidium parvum* oocysts in seawater and eastern oysters (*Crassostrea virginica*) in the Chesapeake Bay, *App. Environ. Microbiol.* **64**:1070–1074.

Fayer, R., Lewis, E. J., Trout, J. M., Graczyk, T. K., Jenkins, M. C., Higgins, J., Xiao, L., and Lal, A. A., 1999, *Cryptosporidium parvum* in oysters from commercial harvesting sites in the Chesapeake Bay, *Emerg. Infect. Dis.* **5**:706–710.

Fayer, R., Morgan, U., and Upton, S. J., 2000a, Epidemiology of *Cryptosporidium*: Transmission, detection and identification, *Int. J. Parasitol.* **30**:1305–1322.

Fayer, R., Trout, J. M., Graczyk, T. K., and Lewis, E. J., 2000b, Prevalence of *Cryptosporidium*, *Giardia* and *Eimeria* infections in post-weaned and adult cattle on three Maryland farms, *Vet. Parasitol.* **93**:103–112.

Fayer, R., Trout, J. M., Xiao, L., Morgan, U. M., Lai, A. A., and Dubey, J. P., 2001, *Cryptosporidium canis* n. sp. from domestic dogs, *J. Parasitol.* **87**:1415–1422.

Fayer, R., Trout, J. M., Lewis, E. J., Xiao, L., Lal, A., Jenkins, M. C., and Graczyk, T. K., 2002, Temporal variability of *Cryptosporidium* in the Chesapeake Bay, *Parasitol. Res.* **88**:998–1003.

Fayer, R., Trout, J. M., Lewis, E. J., Santin, M., Zhou, L., Lal, A. A., and Xiao, L., 2003, Contamination of Atlantic coast commercial shellfish with *Cryptosporidium*, *Parasitol. Res.* **89**:141–145.

Fayer, R., Santin, M., and Xiao, L., 2005, *Cryptosporidium bovis* n. sp. (Apicomplexa: Cryptosporidiidae) in cattle (*Bos taurus*), *J. Parasitol.* **91**:624–629.

Feng, X., Rich, S. M., Akiyoshi, D., Tumwine, J. K., Kekitiinwa, A., Nabukeera, N., Tzipori, S., and Widmer, G., 2000, Extensive polymorphism in *Cryptosporidium parvum* identified by multilocus microsatellite analysis, *Appl. Environ. Microbiol.* **66**:3344–3349.

Filkorn, R., Wiedenmann, A., and Botzenhart, K., 1994, Selective detection of viable *Cryptosporidium* oocysts by PCR, *Zentralbl. Hyg. Umweltmed.* **195**:489–494.

Flanigan, T. P., and Graham, R., 1990, Extended spectrum of symptoms in cryptosporidiosis, *Am. J. Med.* **89**:252.

Flanigan, T., Whalen, C., Turner, J., Soave, R., Toerner, J., Havlir, D., and Kotler, D., 1992, *Cryptosporidium* infection and CD4 counts, *Ann. Intern. Med.* **116**:840–842.

Fontaine, M., and Guillot, E., 2002, Development of a TaqMan quantitative PCR assay specific for *Cryptosporidium parvum*, *FEMS Microbiol. Lett.* **214**:13.

Fontaine, M., and Guillot, E., 2003, Study of 18S rRNA and rDNA stability by real-time RT-PCR in heat-inactivated *Cryptosporidium parvum* oocysts, *FEMS Microbiol. Lett.* **226**:237–243.

Freire-Santos, F., Oteiza-Lopez, A. M., Vergara-Castiblanco, C. A., Ares-Mazas, E., Alvarez-Suarez, E., and Garcia-Martin, O., 2000, Detection of *Cryptosporidium* oocysts in bivalve molluscs destined for human consumption, *J. Parasitol.* **86**:853–854.

Freire-Santos, F., Oteiza-Lopez, A. M., Castro-Hermida, J. A., Garcia-Martin, O., and Ares-Mazas, M. E., 2001, Viability and infectivity of oocysts recovered from clams, Ruditapes philippinarum, experimentally contaminated with *Cryptosporidium parvum*, *Parasitol. Res.* **87**:428–430.

French, A. L., Beaudet, L. M., Benator, D. A., Levy, C. S., Kass, M., and Orenstein, J. M., 1995, Cholecystectomy in patients with AIDS: Clinicopathologic correlations in 107 cases, *Clin. Infect. Dis.* **21**:852–858.

Friedman, D. E., Pattern, K. A., Rose, J. B., and Marney, M. C., 1997, The potential for *Cryptosporidium parvum* oocyst survival in beverages associated with contaminated tap water, *J. Food Safety* **17**:125–132.

Frost, F. J., de la Cruz, A. A., Moss, D. M., Curry, M., and Calderon, R. L., 1998, Comparisons of ELISA and Western blot assays for detection of *Cryptosporidium* antibody, *Epidemiol. Infect.* **121**:205–211.

Frost, F. J., Muller, T., Craun, G. F., Calderon, R. L., and Roefer, P. A., 2001, Paired city *Cryptosporidium* serosurvey in the southwest USA, *Epidemiol. Infect.* **126**:301–307.

Frost, F. J., Muller, T., Craun, G. F., Lockwood, W. B., and Calderon, R. L., 2002, Serological evidence of endemic waterborne *Cryptosporidium* infections, *Ann. Epidemiol.* **12**:222–227.

Frost, F. J., Kunde, T. R., Muller, T. B., Craun, G. F., Katz, L. M., Hibbard, A. J., and Calderon, R. L., 2003, Serological responses to *Cryptosporidium* antigens among users of surface- vs. ground-water sources, *Epidemiol. Infect.* **131**:1131–1138.

Frost, F. J., Muller, T. B., Calderon, R. L., and Craun, G. F., 2004, Analysis of serological responses to *Cryptosporidium* antigen among NHANES III participants, *Ann. Epidemiol.* **14**:473–478.

Gallaher, M. M., Herndon, J. L., Nims, L. J., Sterling, C. R., Grabowski, D. J., and Hull, H. F., 1989, Cryptosporidiosis and surface water, *Am. J. Public Health* **79**:39–42.

Garcia, L. S., Bruckner, D. A., Brewer, T. C., and Shimizu, R. Y., 1983, Techniques for the recovery and identification of *Cryptosporidium* oocysts from stool specimens, *J. Clin. Microbiol.* **18**:185–190.

Garcia, L. S., Brewer, T. C., and Bruckner, D. A., 1987, Fluorescence detection of *Cryptosporidium* oocysts in human fecal specimens by using monoclonal antibodies, *J. Clin. Microbiol.* **25**:119–121.

Garcia, L. S., and Shimizu, R. Y., 1997, Evaluation of nine immunoassay kits (enzyme immunoassay and direct fluorescence) for detection of *Giardia* lamblia and *Cryptosporidium parvum* in human fecal specimens, *J. Clin. Microbiol.* **35**:1526–1529.

Garcia, L. S., and Shimizu, R. Y., 2000, Detection of *Giardia* lamblia and *Cryptosporidium parvum* antigens in human fecal specimens using the ColorPAC combination rapid solid-phase qualitative immunochromatographic assay, *J. Clin. Microbiol.* **38**:1267–1268.

Garcia, L. S., Shimizu, R. Y., and Bernard, C. N., 2000, Detection of *Giardia lamblia, Entamoeba histolytica/Entamoeba dispar*, and *Cryptosporidium parvum* antigens in human fecal specimens using the triage parasite panel enzyme immunoassay, *J. Clin. Microbiol.* **38**:3337–3340.

Garcia, L. S., Shimizu, R. Y., Novak, S., Carroll, M., and Chan, F., 2003, Commercial assay for detection of *Giardia lamblia* and *Cryptosporidium parvum* antigens in human fecal specimens by rapid solid-phase qualitative immunochromatography, *J. Clin. Microbiol.* **41**:209–212.

Gasser, R. B., Abs, E. L. O. Y. G., Prepens, S., and Chalmers, R. M., 2004, An improved 'cold SSCP' method for the genotypic and subgenotypic characterization of *Cryptosporidium*, *Mol. Cell Probes* **18**:329–332.

Gasser, R. B., El-Osta, Y. G., and Chalmers, R. M., 2003, Electrophoretic analysis of genetic variability within *Cryptosporidium parvum* from imported and autochthonous cases of human cryptosporidiosis in the United Kingdom, *Appl. Environ. Microbiol.* **69**:2719–2730.

Gatei, W., Greensill, J., Ashford, R. W., Cuevas, L. E., Parry, C. M., Cunliffe, N. A., Beeching, N. J., and Hart, C. A., 2003, Molecular analysis of the 18S rRNA gene of *Cryptosporidium* parasites from patients with or without human immunodeficiency virus infections living in Kenya, Malawi, Brazil, the United Kingdom, and Vietnam, *J. Clin. Microbiol.* **41**:1458–1462.

Gelletlie, R., Stuart, J., Soltanpoor, N., Armstrong, R., and Nichols, G., 1997, Cryptosporidiosis associated with school milk, *Lancet* **350**:1005–1006.

Genta, R. M., Chappell, C. L., White, A. C., Jr., Kimball, K. T., and Goodgame, R. W., 1993, Duodenal morphology and intensity of infection in AIDS-related intestinal cryptosporidiosis, *Gastroenterology* **105**:1769–1775.

Glaberman, S., Moore, J. E., Lowery, C. J., Chalmers, R. M., Sulaiman, I., Elwin, K., Rooney, P. J., Millar, B. C., Dooley, J. S., Lal, A. A., and Xiao, L., 2002, Three drinking-water-associated cryptosporidiosis outbreaks, Northern Ireland, *Emerg. Infect. Dis.* **8**:631–633.

Glaser, C. A., Safrin, S., Reingold, A., and Newman, T. B., 1998, Association between *Cryptosporidium* infection and animal exposure in HIV-infected individuals, *J. Acquir. Immune Defic. Syndr. Hum. Retrovirol.* **17**:79–82.

Goh, S., Reacher, M., Casemore, D. P., Verlander, N. Q., Chalmers, R., Knowles, M., Williams, J., Osborn, K., and Richards, S., 2004, Sporadic cryptosporidiosis, North Cumbria, England, 1996–2000, *Emerg. Infect. Dis.* **10**:1007–1015.

Gomez-Bautista, M., Ortega-Mora, L. M., Tabares, E., Lopez-Rodas, V., and Costas, E., 2000, Detection of infectious *Cryptosporidium parvum* oocysts in mussels (*Mytilus galloprovincialis*) and cockles (*Cerastoderma edule*), *Appl. Environ. Microbiol.* **66**:1866–1870.

Gomez-Couso, H., Freire-Santos, F., Martinez-Urtaza, J., Garcia-Martin, O., and Ares-Mazas, M. E., 2003a, Contamination of bivalve molluscs by *Cryptosporidium* oocysts: The need for new quality control standards, *Int. J. Food Microbiol.* **87**:97–105.

Gomez-Couso, H., Freire-Santos, F., Ortega-Inarrea, M. R., Castro-Hermida, J. A., and Ares-Mazas, M. E., 2003b, Environmental dispersal of *Cryptosporidium parvum* oocysts and cross transmission in cultured bivalve molluscs, *Parasitol. Res.* **90**:140–142.

Gomez-Couso, H., Freire-Santos, F., Amar, C. F., Grant, K. A., Williamson, K., Ares-Mazas, M. E., and McLauchlin, J., 2004, Detection of *Cryptosporidium* and *Giardia* in molluscan shellfish by multiplexed nested-PCR, *Int. J. Food Microbiol.* **91**:279–288.

Goodgame, R. W., Genta, R. M., White, A. C., and Chappell, C. L., 1993, Intensity of infection in AIDS-associated cryptosporidiosis, *J. Infect. Dis.* **167**:704–709.

Graczyk, T. K., Cranfield, M. R., and Fayer, R., 1996, Evaluation of commercial enzyme immunoassay (EIA) and immunofluorescent antibody (FA) test kits for detection of *Cryptosporidium* oocysts of species other than *Cryptosporidium parvum*, *Am. J. Trop. Med. Hyg.* **54**:274–279.

Graczyk, T. K., Fayer, R., Lewis, E. J., Trout, J. M., and Farley, C. A., 1999, *Cryptosporidium* oocysts in Bent mussels (*Ischadium recurvum*) in the Chesapeake Bay, *Parasitol. Res.* **85**:518–521.

Graczyk, T. K., Fayer, R., Trout, J. M., Jenkins, M. C., Higgins, J., Lewis, E. J., and Farley, C. A., 2000, Susceptibility of the Chesapeake Bay to environmental contamination with *Cryptosporidium parvum*, *Environ. Res.* **82**:106–112.

Graczyk, T. K., Marcogliese, D. J., de Lafontaine, Y., Da Silva, A. J., Mhangami-Ruwende, B., and Pieniazek, N. J., 2001, *Cryptosporidium parvum* oocysts in zebra mussels (*Dreissena polymorpha*): Evidence from the St Lawrence River, *Parasitol. Res.* **87**:231–234.

Greenberg, P. D., Koch, J., and Cello, J. P., 1996, Diagnosis of *Cryptosporidium parvum* in patients with severe diarrhea and AIDS, *Dig. Dis. Sci.* **41**:2286–2290.

Griffiths, J. K., 1998, Human cryptosporidiosis: Epidemiology, transmission, clinical disease, treatment, and diagnosis, *Adv. Parasitol.* **40**:37–85.

Guyot, K., Follet-Dumoulin, A., Lelievre, E., Sarfati, C., Rabodonirina, M., Nevez, G., Cailliez, J. C., Camus, D., and Dei-Cas, E., 2001, Molecular characterization of *Cryptosporidium* isolates obtained from humans in France, *J. Clin. Microbiol.* **39**:3472–3480.

Hallier-Soulier, S., and Guillot, E., 1999, An immunomagnetic separation polymerase chain reaction assay for rapid and ultra-sensitive detection of *Cryptosporidium parvum* in drinking water, *FEMS Microbiol. Lett.* **176**:285–289.

Hallier-Soulier, S., and Guillot, E., 2000, Detection of cryptosporidia and *Cryptosporidium parvum* oocysts in environmental water samples by immunomagnetic separation-polymerase chain reaction, *J. Appl. Microbiol.* **89**:5–10.

Hallier-Soulier, S., and Guillot, E., 2003, An immunomagnetic separation-reverse transcription polymerase chain reaction (IMS-RT-PCR) test for sensitive and rapid detection of viable waterborne *Cryptosporidium parvum*, *Environ. Microbiol.* **5**:592–598.

Harp, J. A., Fayer, R., Pesch, B. A., and Jackson, G. J., 1996, Effect of pasteurization on infectivity of *Cryptosporidium parvum* oocysts in water and milk, *Appl. Environ. Microbiol.* **62**:2866–2868.

Hashim, A., Clyne, M., Mulcahy, G., Akiyoshi, D., Chalmers, R., and Bourke, B., 2004, Host cell tropism underlies species restriction of human and bovine *Cryptosporidium parvum* genotypes, *Infect. Immun.* **72**:6125–6131.

Hashimoto, A., Hirata, T., and Kunikane, S., 2001, Occurrence of *Cryptosporidium* oocysts and *Giardia* cysts in a conventional water purification plant, *Water Sci. Technol.* **43**:89–92.

Hashmey, R., Smith, N. H., Cron, S., Graviss, E. A., Chappell, C. L., and White, A. C., Jr., 1997, Cryptosporidiosis in Houston, Texas. A report of 95 cases, *Medicine (Baltimore)* **76**:118–139.

Hellard, M. E., Sinclair, M. I., Fairley, C. K., Andrews, R. M., Bailey, M., Black, J., Dharmage, S. C., and Kirk, M. D., 2000, An outbreak of cryptosporidiosis in an urban swimming pool: Why are such outbreaks difficult to detect? *Aust. N. Z. J. Public Health* **24**:272–275.

Hellard, M., Hocking, J., Willis, J., Dore, G., and Fairley, C., 2003, Risk factors leading to *Cryptosporidium* infection in men who have sex with men, *Sex Transm. Infect.* **79**:412–414.

Hewitt, R. G., Yiannoutsos, C. T., Higgs, E. S., Carey, J. T., Geiseler, P. J., Soave, R., Rosenberg, R., Vazquez, G. J., Wheat, L. J., Fass, R. J., Antoninievic, Z., Walawander, A. L., Flanigan, T. P., and Bender, J. F., 2000, Paromomycin: No more effective than placebo for treatment of cryptosporidiosis in patients with advanced human immunodeficiency virus infection. AIDS Clinical Trial Group, *Clin. Infect. Dis.* **31**:1084–1092.

Heyworth, M. F., 1996, Parasitic diseases in immunocompromised hosts. Cryptosporidiosis, isosporiasis, and strongyloidiasis, *Gastroenterol. Clin. North Am.* **25**:691–707.

Higgins, J. A., Fayer, R., Trout, J. M., Xiao, L., Lal, A. A., Kerby, S., and Jenkins, M. C., 2001, Real-time PCR for the detection of *Cryptosporidium parvum*, *J. Microbiol. Methods* **47**:323–337.

Hijjawi, N. S., Meloni, B. P., Ryan, U. M., Olson, M. E., and Thompson, R. C., 2002, Successful *in vitro* cultivation of *Cryptosporidium* andersoni: Evidence for the existence of novel extracellular stages in the life cycle and implications for the classification of *Cryptosporidium*, *Int. J. Parasitol.* **32**:1719–1726.

Hijnen, W. A., Schijven, J. F., Bonne, P., Visser, A., and Medema, G. J., 2004, Elimination of viruses, bacteria and protozoan oocysts by slow sand filtration, *Water Sci. Technol.* **50**:147–154.

Hoepelman, A. I., 1996, Current therapeutic approaches to cryptosporidiosis in immunocompromised patients, *J. Antimicrob. Chemother.* **37**:871–880.

Hommer, V., Eichholz, J., and Petry, F., 2003, Effect of antiretroviral protease inhibitors alone, and in combination with paromomycin, on the excystation, invasion and *in vitro* development of *Cryptosporidium parvum*, *J. Antimicrob. Chemother.* **52**:359–364.

Hsu, B. M., 2003, Evaluation of analyzing methods for *Giardia* and *Cryptosporidium* in a Taiwan water treatment plant, *J. Parasitol.* **89**:369–371.

Hsu, B. M., and Yeh, H. H., 2003, Removal of *Giardia* and *Cryptosporidium* in drinking water treatment: A pilot-scale study, *Water Res.* **37**:1111–1117.

Hunter, P. R., and Nichols, G., 2002, Epidemiology and clinical features of *Cryptosporidium* infection in immunocompromised patients, *Clin. Microbiol. Rev.* **15**:145–154.

Hunter, P. R., Chalmers, R. M., Syed, Q., Hughes, L. S., Woodhouse, S., and Swift, L., 2003, Foot and mouth disease and cryptosporidiosis: Possible interaction between two emerging infectious diseases, *Emerg. Infect. Dis.* **9**:109–112.

Hunter, P. R., Hughes, S., Woodhouse, S., Raj, N., Syed, Q., Chalmers, R. M., Verlander, N. Q., and Goodacre, J., 2004a, Health sequelae of human cryptosporidiosis in immunocompetent patients, *Clin. Infect. Dis.* **39**:504–510.

Hunter, P. R., Hughes, S., Woodhouse, S., Syed, Q., Verlander, N. Q., Chalmers, R. M., Morgan, K., Nichols, G., Beeching, N., and Osborn, K., 2004b, Sporadic cryptosporidiosis case-control study with genotyping, *Emerg. Infect. Dis.* **10**:1241–1249.

Ignatius, R., Eisenblatter, M., Regnath, T., Mansmann, U., Futh, U., Hahn, H., and Wagner, J., 1997, Efficacy of different methods for detection of low *Cryptosporidium parvum* oocyst numbers or antigen concentrations in stool specimens, *Eur. J. Clin. Microbiol. Infect. Dis.* **16**:732–736.

Inman, L. R., and Takeuchi, A., 1979, Spontaneous cryptosporidiosis in an adult female rabbit, *Vet. Pathol.* **16**:89–95.

Inungu, J. N., Morse, A. A., and Gordon, C., 2000, Risk factors, seasonality, and trends of cryptosporidiosis among patients infected with human immunodeficiency virus, *Am. J. Trop. Med. Hyg.* **62**:384–387.

Iseki, M., 1979, *Cryptosporidium felis* sp. n. (Protozoa: Eimeriorina) from the domestic cat, *Jap. J. Parasitol.* **28**:285–307.

Javier Enriquez, F., Avila, C. R., Ignacio Santos, J., Tanaka-Kido, J., Vallejo, O., and Sterling, C. R., 1997, *Cryptosporidium* infections in Mexican children: Clinical, nutritional, enteropathogenic, and diagnostic evaluations, *Am. J. Trop. Med. Hyg.* **56**:254–257.

Jellison, K. L., Hemond, H. F., and Schauer, D. B., 2002, Sources and species of *Cryptosporidium* oocysts in the Wachusett reservoir watershed, *Appl. Environ. Microbiol.* **68**:569–575.

Jenkins, M. C., Trout, J., Abrahamsen, M. S., Lancto, C. A., Higgins, J., and Fayer, R., 2000, Estimating viability of *Cryptosporidium parvum* oocysts using reverse

transcriptase-polymerase chain reaction (RT-PCR) directed at mRNA encoding amyloglucosidase, *J. Microbiol. Methods* **43**:97–106.
Jiang, J., and Xiao, L., 2003, An evaluation of molecular diagnostic tools for the detection and differentiation of human-pathogenic *Cryptosporidium* spp, *J. Eukaryot Microbiol.* **50** (Suppl):542–547.
Joce, R. E., Bruce, J., Kiely, D., Noah, N. D., Dempster, W. B., Stalker, R., Gumsley, P., Chapman, P. A., Norman, P., Watkins, J., *et al.*, 1991, An outbreak of cryptosporidiosis associated with a swimming pool, *Epidemiol. Infect.* **107**:497–508.
Johnson, D. W., Pieniazek, N. J., Griffin, D. W., Misener, L., and Rose, J. B., 1995, Development of a PCR protocol for sensitive detection of *Cryptosporidium* oocysts in water samples, *Appl. Environ. Microbiol.* **61**:3849–3855.
Johnston, S. P., Ballard, M. M., Beach, M. J., Causer, L., and Wilkins, P. P., 2003, Evaluation of three commercial assays for detection of *Giardia* and *Cryptosporidium* organisms in fecal specimens, *J. Clin. Microbiol.* **41**:623–626.
Jokipii, L., and Jokipii, A. M., 1986, Timing of symptoms and oocyst excretion in human cryptosporidiosis, *N. Engl. J. Med.* **315**:1643–1647.
Karanis, P., Schoenen, D., and Seitz, H. M., 1996, *Giardia* and *Cryptosporidium* in backwash water from rapid sand filters used for drinking water production, *Zentralbl. Bakteriol.* **284**:107–114.
Karasudani, T., Aoki, S., Takeuchi, J., Okuyama, M., Oseto, M., Matsuura, S., Asai, T., and Inouye, H., 2001, Sensitive detection of *Cryptosporidium* oocysts in environmental water samples by reverse transcription-PCR, *Jpn. J. Infect. Dis.* **54**:122–124.
Katanik, M. T., Schneider, S. K., Rosenblatt, J. E., Hall, G. S., and Procop, G. W., 2001, Evaluation of ColorPAC *Giardia/Cryptosporidium* rapid assay and ProSpecT *Giardia/Cryptosporidium* microplate assay for detection of *Giardia* and *Cryptosporidium* in fecal specimens, *J. Clin. Microbiol.* **39**:4523–4525.
Kato, S., Jenkins, M. B., Ghiorse, W. C., Fogarty, E. A., and Bowman, D. D., 2001, Inactivation of *Cryptosporidium parvum* oocysts in field soil, *Southeast Asian J. Trop. Med. Public Health* **32**(Suppl 2):183–189.
Katsumata, T., Hosea, D., Wasito, E. B., Kohno, S., Hara, K., Soeparto, P., and Ranuh, I. G., 1998, Cryptosporidiosis in Indonesia: A hospital-based study and a community-based survey, *Am. J. Trop. Med. Hyg.* **59**:628–632.
Kaucner, C., and Stinear, T., 1998, Sensitive and rapid detection of viable *Giardia* cysts and *Cryptosporidium parvum* oocysts in large-volume water samples with wound fiberglass cartridge filters and reverse transcription-PCR, *Appl. Environ. Microbiol.* **64**:1743–1749.
Kelly, P., Makumbi, F. A., Carnaby, S., Simjee, A. E., and Farthing, M. J., 1998, Variable distribution of *Cryptosporidium parvum* in the intestine of AIDS patients revealed by polymerase chain reaction, *Europ. J. Gastroenterol. Hepatol.* **10**:855–858.
Khalakdina, A., Vugia, D. J., Nadle, J., Rothrock, G. A., and Colford, J. M., Jr., 2003, Is drinking water a risk factor for endemic cryptosporidiosis? A case-control study in the immunocompetent general population of the San Francisco Bay Area, *BMC Public Health* **3**:11.
Kimbell, L. M., Miller, D. L., Chavez, W., and Altman, N., 1999, Molecular analysis of the 18S rRNA gene of *Cryptosporidium* serpentis in a wild-caught corn snake (*Elaphe guttata guttata*) and a five-species restriction fragment length polymorphism-based assay that can additionally discern *C. parvum* from *C. wrairi*, *Appl. Environ. Microbiol.* **65**:5345–5349.
Korich, D. G., Mead, J. R., Madore, M. S., Sinclair, N. A., and Sterling, C. R., 1990, Effects of ozone, chlorine dioxide, chlorine, and monochloramine on *Cryptosporidium parvum* oocyst viability, *Appl. Environ. Microbiol.* **56**:1423–1428.

Kostrzynska, M., Sankey, M., Haack, E., Power, C., Aldom, J. E., Chagla, A. H., Unger, S., Palmateer, G., Lee, H., Trevors, J. T., and De Grandis, S. A., 1999, Three sample preparation protocols for polymerase chain reaction based detection of *Cryptosporidium parvum* in environmental samples, *J. Microbiol. Methods.* **35**:65–71.

Koudela, B., and Modry, D., 1998, New species of *Cryptosporidium* (Apicomplexa, Cryptosporidiidae) from lizards, *Folia Parasitol.* **45**:93–100.

Kozwich, D., Johansen, K. A., Landau, K., Roehl, C. A., Woronoff, S., and Roehl, P. A., 2000, Development of a novel, rapid integrated *Cryptosporidium parvum* detection assay, *Appl. Environ. Microbiol.* **66**:2711–2717.

Kuczynska, E., and Shelton, D. R., 1999, Method for detection and enumeration of *Cryptosporidium parvum* oocysts in feces, manures, and soils, *Appl. Environ. Microbiol.* **65**:2820–2826.

Kuhn, R. C., Rock, C. M., and Oshima, K. H., 2002, Occurrence of *Cryptosporidium* and *Giardia* in wild ducks along the Rio Grande River Valley in Southern New Mexico, *Appl. Environ. Microbiol.* **68**:161–165.

Lacroix, C., Berthier, M., Agius, G., Bonneau, D., Pallu, B., and Jacquemin, J. L., 1987, *Cryptosporidium* oocysts in immunocompetent children: Epidemiologic investigations in the day-care centers of Poitiers, France, *Eur. J. Epidemiol.* **3**:381–385.

Laxer, M. A., Timblin, B. K., and Patel, R. J., 1991, DNA sequences for the specific detection of *Cryptosporidium parvum* by the polymerase chain reaction, *Am. J. Trop. Med. Hyg.* **45**:688–694.

Leach, C. T., Koo, F. C., Kuhls, T. L., Hilsenbeck, S. G., and Jenson, H. B., 2000, Prevalence of *Cryptosporidium parvum* infection in children along the Texas-Mexico border and associated risk factors, *Am. J. Trop. Med. Hyg.* **62**:656–661.

Learmonth, J., Ionas, G., Pita, A., and Cowie, R., 2001, Seasonal shift in *Cryptosporidium parvum* transmission cycles in New Zealand, *J. Eukaryot. Microbiol.* 34S–35S.

Learmonth, J. J., Ionas, G., Ebbett, K. A., and Kwan, E. S., 2004, Genetic characterization and transmission cycles of *Cryptosporidium* species isolated from humans in new zealand, *Appl. Environ. Microbiol.* **70**:3973–3978.

Leav, B. A., Mackay, M. R., Anyanwu, A., RM, O. C., Cevallos, A. M., Kindra, G., Rollins, N. C., Bennish, M. L., Nelson, R. G., and Ward, H. D., 2002, Analysis of sequence diversity at the highly polymorphic Cpgp40/15 locus among *Cryptosporidium* isolates from human immunodeficiency virus-infected children in South Africa, *Infect. Immun.* **70**:3881–3890.

LeChevallier, M. W., Norton, W. D., and Lee, R. G., 1991a, *Giardia* and *Cryptosporidium* spp. in filtered drinking water supplies, *Appl. Environ. Microbiol.* **57**:2617–2621.

LeChevallier, M. W., Norton, W. D., and Lee, R. G., 1991b, Occurrence of *Giardia* and *Cryptosporidium* spp. in surface water supplies, *Appl. Environ. Microbiol.* **57**:2610–2616.

LeChevallier, M. W., Di Giovanni, G. D., Clancy, J. L., Bukhari, Z., Bukhari, S., Rosen, J. S., Sobrinho, J., and Frey, M. M., 2003, Comparison of method 1623 and cell culture-PCR for detection of *Cryptosporidium* spp. in source waters, *Appl. Environ. Microbiol.* **69**:971–979.

Lemmon, J. M., McAnulty, J. M., and Bawden-Smith, J., 1996, Outbreak of cryptosporidiosis linked to an indoor swimming pool, *Med. J. Aust.* **165**:613–616.

Leng, X., Mosier, D. A., and Oberst, R. D., 1996, Differentiation of *Cryptosporidium parvum*, *C. muris*, and *C. baileyi* by PCR-RFLP analysis of the 18S rRNA gene, *Vet. Parasitol.* **62**:1–7.

Leoni, F., Gallimore, C. I., Green, J., and McLauchlin, J., 2003, Molecular epidemiological analysis of *Cryptosporidium* isolates from humans and animals by using a heteroduplex mobility assay and nucleic acid sequencing based on a small double-stranded RNA element, *J. Clin. Microbiol.* **41**:981–992.

Levine, N. D., 1980, Some corrections of coccidian (Apicomplexa: Protozoa) nomenclature, *J. Parasitol.* **66**:830–834.
Limor, J. R., Lal, A. A., and Xiao, L., 2002, Detection and differentiation of *Cryptosporidium* parasites that are pathogenic for humans by real-time PCR, *J. Clin. Microbiol.* **40**:2335–2338.
Linden, K. G., Shin, G., and Sobsey, M. D., 2001, Comparative effectiveness of UV wavelengths for the inactivation of *Cryptosporidium parvum* oocysts in water, *Water Sci. Technol.* **43**:171–174.
Lindquist, H. D., Bennett, J. W., Ware, M., Stetler, R. E., Gauci, M., and Schaefer, F. W., 2001a, Testing methods for detection of *Cryptosporidium* spp. in water samples, *Southeast Asian J. Trop. Med. Public Health* **32**:190–194.
Lindquist, H. D., Ware, M., Stetler, R. E., Wymer, L., and Schaefer, F. W., 3rd, 2001b, A comparison of four fluorescent antibody-based methods for purifying, detecting, and confirming *Cryptosporidium parvum* in surface waters, *J. Parasitol.* **87**:1124–1131.
Lindsay, D. S., Upton, S. J., Owens, D. S., Morgan, U. M., Mead, J. R., and Blagburn, B. L., 2000, *Cryptosporidium andersoni* n. sp. (Apicomplexa: Cryptosporiidae) from cattle, *Bos taurus*, *J. Eukaryot. Microbiol.* **47**:91–95.
Lorenzo-Lorenzo, M. J., Ares-Mazas, M. E., Villacorta-Martinez de Maturana, I., and Duran-Oreiro, D., 1993, Effect of ultraviolet disinfection of drinking water on the viability of *Cryptosporidium parvum* oocysts, *J. Parasitol.* **79**:67–70.
Lowery, C. J., Moore, J. E., Millar, B. C., Burke, D. P., McCorry, K. A., Crothers, E., and Dooley, J. S., 2000, Detection and speciation of *Cryptosporidium* spp. in environmental water samples by immunomagnetic separation, PCR and endonuclease restriction, *J. Med. Microbiol.* **49**:779–785.
Lowery, C. J., Moore, J. E., Millar, B. C., McCorry, K. A., Xu, J., Rooney, P. J., and Dooley, J. S., 2001a, Occurrence and molecular genotyping of *Cryptosporidium* spp. in surface waters in Northern Ireland, *J. Appl. Microbiol.* **91**:774–779.
Lowery, C. J., Nugent, P., Moore, J. E., Millar, B. C., Xiru, X., and Dooley, J. S., 2001b, PCR-IMS detection and molecular typing of *Cryptosporidium parvum* recovered from a recreational river source and an associated mussel (*Mytilus edulis*) bed in Northern Ireland, *Epidemiol. Infect.* **127**:545–553.
Lumadue, J. A., Manabe, Y. C., Moore, R. D., Belitsos, P. C., Sears, C. L., and Clark, D. P., 1998, A clinicopathologic analysis of AIDS-related cryptosporidiosis, *AIDS* **12**:2459–2466.
MacDonald, L. M., Sargent, K., Armson, A., Thompson, R. C., and Reynoldson, J. A., 2002, The development of a real-time quantitative-PCR method for characterisation of a *Cryptosporidium parvum in vitro* culturing system and assessment of drug efficacy, *Mol. Biochem. Parasitol.* **121**:279–282.
MacKenzie, W. R., Hoxie, N. J., Proctor, M. E., Gradus, M. S., Blair, K. A., Peterson, D. E., Kazmierczak, J. J., Addiss, D. G., Fox, K. R., and Rose, J. B., 1994a, A massive outbreak in Milwaukee of *Cryptosporidium* infection transmitted through the public water supply, *N. Engl. J. Med.* **331**:161–167.
MacKenzie, W. R., Hoxie, N. J., Proctor, M. E., Gradus, M. S., Blair, K. A., Peterson, D. E., Kazmierczak, J. J., Addiss, D. G., Fox, K. R., Rose, J. B. *et al.*, 1994b, A massive outbreak in Milwaukee of *Cryptosporidium* infection transmitted through the public water supply, *N. Engl. J. Med.* **331**:161–167.
MacKenzie, W. R., Kazmierczak, J. J., and Davis, J. P., 1995, An outbreak of cryptosporidiosis associated with a resort swimming pool, *Epidemiol. Infect.* **115**:545–553.

Madore, M. S., Rose, J. B., Gerba, C. P., Arrowood, M. J., and Sterling, C. R., 1987, Occurrence of *Cryptosporidium* oocysts in sewage effluents and selected surface waters, *J. Parasitol.* **73**:702–705.

Maggi, P., Larocca, A. M., Quarto, M., Serio, G., Brandonisio, O., Angarano, G., and Pastore, G., 2000, Effect of antiretroviral therapy on cryptosporidiosis and microsporidiosis in patients infected with human immunodeficiency virus type 1, *Eur. J. Clin. Microbiol. Infect. Dis.* **19**:213–217.

Mallon, M., MacLeod, A., Wastling, J., Smith, H., Reilly, B., and Tait, A., 2003a, Population structures and the role of genetic exchange in the zoonotic pathogen *Cryptosporidium parvum*, *J. Mol. Evol.* **56**:407–417.

Mallon, M. E., MacLeod, A., Wastling, J. M., Smith, H., and Tait, A., 2003b, Multilocus genotyping of *Cryptosporidium parvum* Type 2: Population genetics and sub-structuring, *Infect. Genet. Evol.* **3**:207–218.

Manabe, Y. C., Clark, D. P., Moore, R. D., Lumadue, J. A., Dahlman, H. R., Belitsos, P. C., Chaisson, R. E., and Sears, C. L., 1998, Cryptosporidiosis in patients with AIDS—correlates of disease and survival, *Clin. Infect. Dis.* **27**:536–542.

Mata, L., 1986, *Cryptosporidium* and other protozoa in diarrheal disease in less developed countries, *Pediatr. Infect. Dis.* **5**:S117–S130.

Mayer, C. L., and Palmer, C. J., 1996, Evaluation of PCR, nested PCR, and fluorescent antibodies for detection of *Giardia* and *Cryptosporidium* species in wastewater, *Appl. Environ. Microbiol.* **62**:2081–2085.

McDonald, A. C., Mac Kenzie, W. R., Addiss, D. G., Gradus, M. S., Linke, G., Zembrowski, E., Hurd, M. R., Arrowood, M. J., Lammie, P. J., and Priest, J. W., 2001, *Cryptosporidium parvum*-specific antibody responses among children residing in Milwaukee during the 1993 waterborne outbreak, *J. Infect. Dis.* **183**:1373–1379.

McGowan, I., Hawkins, A. S., and Weller, I. V., 1993, The natural history of cryptosporidial diarrhoea in HIV-infected patients, *AIDS* **7**:349–354.

McLauchlin, J., Pedraza-Diaz, S., Amar-Hoetzeneder, C., and Nichols, G. L., 1999, Genetic characterization of *Cryptosporidium* strains from 218 patients with diarrhea diagnosed as having sporadic cryptosporidiosis, *J. Clin. Microbiol.* **37**:3153–3158.

McLauchlin, J., Amar, C., Pedraza-Diaz, S., and Nichols, G. L., 2000, Molecular epidemiological analysis of *Cryptosporidium* spp. in the United Kingdom: Results of genotyping *Cryptosporidium* spp. in 1705 fecal samples from humans and 105 fecal samples from livestock animals, *J. Clin. Microbiol.* **38**:3984–3990.

McLauchlin, J., Amar, C. F., Pedraza-Diaz, S., Mieli-Vergani, G., Hadzic, N., and Davies, E. G., 2003, Polymerase chain reaction-based diagnosis of infection with *Cryptosporidium* in children with primary immunodeficiencies, *Pediatr. Infect. Dis. J.* **22**:329–335.

Mead, J. R., Arrowood, M. J., and Sterling, C. R., 1988, Antigens of *Cryptosporidium* sporozoites recognized by immune sera of infected animals and humans, *J. Parasitol.* **74**:135–143.

Mead, P. S., Slutsker, L., Dietz, V., McCaig, L. F., Bresee, J. S., Shapiro, C., Griffin, P. M., and Tauxe, R. V., 1999, Food-related illness and death in the United States, *Emerg. Infect. Dis.* **5**:607–625.

Medema, G. J., Hoogenboezem, W., van der Veer, A. J., Ketelaars, H. A., Hijnen, W. A., and Nobel, P. J., 2003, Quantitative risk assessment of *Cryptosporidium* in surface water treatment, *Water Sci. Technol.* **47**:241–247.

Meinhardt, P. L., Casemore, D. P., and Miller, K. B., 1996, Epidemiologic aspects of human cryptosporidiosis and the role of waterborne transmission, *Epidemiol. Rev.* **18**:118–136.

Mele, R., Morales, M. A., Tosini, F., and Pozio, E., 2003, Indinavir reduces *Cryptosporidium parvum* infection in both *in vitro* and *in vivo* models, *Int. J. Parasitol.* **33**:757–764.

Merry, R. J., Mawdsley, J. L., Brooks, A. E., and Davies, D. R., 1997, Viability of *Cryptosporidium parvum* during ensilage of perennial ryegrass, *J. Appl. Microbiol.* **82**:115–120.
Miao, Y. M., Awad-El-Kariem, F. M., Franzen, C., Ellis, D. S., Muller, A., Counihan, H. M., Hayes, P. J., and Gazzard, B. G., 2000, Eradication of cryptosporidia and microsporidia following successful antiretroviral therapy, *J. Acquir. Immune Defic. Syndr.* **25**:124–129.
Millard, P. S., Gensheimer, K. F., Addiss, D. G., Sosin, D. M., Beckett, G. A., Houck-Jankoski, A., and Hudson, A., 1994, An outbreak of cryptosporidiosis from fresh-pressed apple cider, *JAMA* **272**:1592–1596.
Miron, D., Kenes, J., and Dagan, R., 1991, Calves as a source of an outbreak of cryptosporidiosis among young children in an agricultural closed community, *Pediat. Infect. Dis. J.* **10**:438–441.
Molbak, K., Aaby, P., Hojlyng, N., and da Silva, A. P., 1994, Risk factors for *Cryptosporidium* diarrhea in early childhood: A case-control study from Guinea-Bissau, West Africa, *Am. J. Epidemiol.* **139**:734–740.
Molbak, K., Andersen, M., Aaby, P., Hojlyng, N., Jakobsen, M., Sodemann, M., and da Silva, A. P., 1997, *Cryptosporidium* infection in infancy as a cause of malnutrition: A community study from Guinea-Bissau, west Africa, *Am. J. Clin. Nutr.* **65**:149–152.
Monge, R., and Arias, M. L., 1996, Presence of various pathogenic microorganisms in fresh vegetables in Costa Rica, *Arch. Latinoam. Nutr.* **46**:292–294.
Monge, R., Chinchilla, M., and Reyes, L., 1996, Seasonality of parasites and intestinal bacteria in vegetables that are consumed raw in Costa Rica, *Rev. Biol. Trop.* **44**:369–375.
Monis, P. T., and Saint, C. P., 2001, Development of a nested-PCR assay for the detection of *Cryptosporidium parvum* in finished water, *Water Res.* **35**:1641–1648.
Moodley, D., Jackson, T. F., Gathiram, V., and van den Ende, J., 1991, *Cryptosporidium* infections in children in Durban. Seasonal variation, age distribution and disease status, *South Afr. Med. J.* **79**:295–297.
Moore, A. G., Vesey, G., Champion, A., Scandizzo, P., Deere, D., Veal, D., and Williams, K. L., 1998, Viable *Cryptosporidium parvum* oocysts exposed to chlorine or other oxidising conditions may lack identifying epitopes, *Int. J. Parasitol.* **28**:1205–1212.
Morgan, U. M., Constantine, C. C., P, O. D., Meloni, B. P., PA, O. B., and Thompson, R. C., 1995, Molecular characterization of *Cryptosporidium* isolates from humans and other animals using random amplified polymorphic DNA analysis, *Am. J. Trop. Med. Hyg.* **52**:559–564.
Morgan, U. M., Pa, O. B., and Thompson, R. C., 1996, The development of diagnostic PCR primers for *Cryptosporidium* using RAPD-PCR, *Mol. Biochem. Parasitol.* **77**:103–108.
Morgan, U. M., Constantine, C. C., Forbes, D. A., and Thompson, R. C., 1997, Differentiation between human and animal isolates of *Cryptosporidium parvum* using rDNA sequencing and direct PCR analysis, *J. Parasitol.* **83**:825–830.
Morgan, U. M., Deplazes, P., Forbes, D. A., Spano, F., Hertzberg, H., Sargent, K. D., Elliot, A., and Thompson, R. C., 1999a, Sequence and PCR-RFLP analysis of the internal transcribed spacers of the rDNA repeat unit in isolates of *Cryptosporidium* from different hosts, *Parasitology* **118**:49–58.
Morgan, U. M., Xiao, L., Fayer, R., Lal, A. A., and Thompson, R. C., 1999b, Variation in *Cryptosporidium*: Towards a taxonomic revision of the genus, *Int. J. Parasitol.* **29**:1733–1751.
Morgan, U. M., Sargent, K. D., Deplazes, P., Forbes, D. A., Spano, F., Hertzberg, H., Elliot, A., and Thompson, R. C., 1998, Molecular characterization of *Cryptosporidium* from various hosts, *Parasitology* **117**:31–37.

Morgan-Ryan, U. M., Fall, A., Ward, L. A., Hijjawi, N., Sulaiman, I., Fayer, R., Thompson, R. C., Olson, M., Lal, A., and Xiao, L., 2002, *Cryptosporidium hominis* n. sp. (Apicomplexa: Cryptosporidiidae) from *Homo sapiens*, *J. Eukaryot. Microbiol.* **49**:433–440.

Moss, D. M., Montgomery, J. M., Newland, S. V., Priest, J. W., and Lammie, P. J., 2004, Detection of *Cryptosporidium* antibodies in sera and oral fluids using multiplex bead assay, *J. Parasitol.* **90**:397–404.

Nath, G., Choudhury, A., Shukla, B. N., Singh, T. B., and Reddy, D. C. S., 1999, Significance of *Cryptosporidium* in acute diarrhoea in North-Eastern India, *J. Med. Microbiol.* **48**:523–526.

Navin, T. R., Weber, R., Vugia, D. J., Rimland, D., Roberts, J. M., Addiss, D. G., Visvesvara, G. S., Wahlquist, S. P., Hogan, S. E., Gallagher, L. E., Juranek, D. D., Schwartz, D. A., Wilcox, C. M., Stewart, J. M., Thompson, S. E. R., and Bryan, R. T., 1999, Declining CD4+ T-lymphocyte counts are associated with increased risk of enteric parasitosis and chronic diarrhea: Results of a 3-year longitudinal study, *J. AIDS* **20**:154–159.

Nchito, M., Kelly, P., Sianongo, S., Luo, N. P., Feldman, R., Farthing, M., and Baboo, K. S., 1998, Cryptosporidiosis in urban Zambian children: An analysis of risk factors, *Am. J. Trop. Med. Hyg.* **59**:435–437.

Neill, M. A., Rice, S. K., Ahmad, N. V., and Flanigan, T. P., 1996, Cryptosporidiosis: An unrecognized cause of diarrhea in elderly hospitalized patients, *Clin. Infect. Dis.* **22**:168–170.

Newman, R. D., Sears, C. L., Moore, S. R., Nataro, J. P., Wuhib, T., Agnew, D. A., Guerrant, R. L., and Lima, A. A. M., 1999, Longitudinal study of *Cryptosporidium* infection in children in northeastern Brazil, *J. Infect. Dis.* **180**:167–175.

Nichols, R. A., Campbell, B. M., and Smith, H. V., 2003, Identification of *Cryptosporidium* spp. oocysts in United Kingdom noncarbonated natural mineral waters and drinking waters by using a modified nested PCR-restriction fragment length polymorphism assay, *Appl. Environ. Microbiol.* **69**:4183–4189.

Nichols, R. A., Paton, C. A., and Smith, H. V., 2004, Survival of *Cryptosporidium parvum* oocysts after prolonged exposure to still natural mineral waters, *J. Food Prot.* **67**:517–523.

Nimri, L. F., and Hijazi, S. S., 1994, *Cryptosporidium*. A cause of gastroenteritis in preschool children in Jordan, *J. Clin. Gastroenterol.* **19**:288–291.

Okhuysen, P. C., Chappell, C. L., Sterling, C. R., Jakubowski, W., and DuPont, H. L., 1998, Susceptibility and serologic response of healthy adults to reinfection with *Cryptosporidium parvum*, *Infect. Immun.* **66**:441–443.

Ong, C. S. L., Eisler, D. L., Goh, S. H., Tomblin, J., Awad-El-Kariem, F. M., Beard, C. B., Xiao, L. H., Sulaiman, I., Lal, A., Fyfe, M., King, A., Bowie, W. R., and Isaac-Renton, J. L., 1999, Molecular epidemiology of cryptosporidiosis outbreaks and transmission in British Columbia, Canada, *Am. J. Trop. Med. Hyg.* **61**:63–69.

Ortega, Y. R., Roxas, C. R., Gilman, R. H., Miller, N. J., Cabrera, L., Taquiri, C., and Sterling, C. R., 1997, Isolation of *Cryptosporidium parvum* and *Cyclospora cayetanensis* from vegetables collected in markets of an endemic region in Peru, *Am. J. Trop. Med. Hyg.* **57**:683–686.

Palmer, C. J., Xiao, L., Terashima, A., Guerra, H., Gotuzzo, E., Saldias, G., Bonilla, J. A., Zhou, L., Lindquist, A., and Upton, S. J., 2003, *Cryptosporidium muris*, a rodent pathogen, recovered from a human in Peru, *Emerg. Infect. Dis.* **9**:1174–1176.

Parisi, M. T., and Tierno, P. M., Jr., 1995, Evaluation of new rapid commercial enzyme immunoassay for detection of *Cryptosporidium* oocysts in untreated stool specimens, *J. Clin. Microbiol.* **33**:1963–1965.

Patel, S., Pedraza-Diaz, S., and McLauchlin, J., 1999, The identification of *Cryptosporidium* species and *Cryptosporidium parvum* directly from whole faeces by analysis of a multiplex

PCR of the 18S rRNA gene and by PCR/RFLP of the *Cryptosporidium* outer wall protein (COWP) gene, *Int. J. Parasitol.* **29**:1241–1247.

Patel, S., Pedraza-Diaz, S., McLauchlin, J., and Casemore, D. P., 1998, Molecular characterisation of *Cryptosporidium parvum* from two large suspected waterborne outbreaks. Outbreak Control Team South and West Devon 1995, Incident Management Team and Further Epidemiological and Microbiological Studies Subgroup North Thames 1997, *Commun. Dis. Pub. Health* **1**:231–233.

Pedraza-Diaz, S., Amar, C., and McLauchlin, J., 2000, The identification and characterisation of an unusual genotype of *Cryptosporidium* from human faeces as *Cryptosporidium meleagridis*, *FEMS Microbiol. Lett.* **189**:189–194.

Peeters, J. E., Mazas, E. A., Masschelein, W. J., Villacorta Martiez de Maturana, I., and Debacker, E., 1989, Effect of disinfection of drinking water with ozone or chlorine dioxide on survival of *Cryptosporidium parvum* oocysts, *Appl. Environ. Microbiol.* **55**:1519–1522.

Peng, M. M., Xiao, L., Freeman, A. R., Arrowood, M. J., Escalante, A. A., Weltman, A. C., Ong, C. S., Mac Kenzie, W. R., Lal, A. A., and Beard, C. B., 1997, Genetic polymorphism among *Cryptosporidium parvum* isolates: Evidence of two distinct human transmission cycles, *Emerg. Infect. Dis.* **3**:567–573.

Peng, M. M., Matos, O., Gatei, W., Das, P., Stantic-Pavlinic, M., Bern, C., Sulaiman, I. M., Glaberman, S., Lal, A. A., and Xiao, L., 2001, A comparison of *Cryptosporidium* subgenotypes from several geographic regions, *J. Eukaryot. Microbiol.* 28S–31S.

Peng, M. M., Meshnick, S. R., Cunliffe, N. A., Thindwa, B. D., Hart, C. A., Broadhead, R. L., and Xiao, L., 2003a, Molecular epidemiology of cryptosporidiosis in children in Malawi, *J. Eukaryot. Microbiol.* **50**(Suppl):557–559.

Peng, M. M., Wilson, M. L., Holland, R. E., Meshnick, S. R., Lal, A. A., and Xiao, L., 2003b, Genetic diversity of *Cryptosporidium* spp. in cattle in Michigan: Implications for understanding the transmission dynamics, *Parasitol. Res.* **90**:175–180.

Perch, M., Sodemann, M., Jakobsen, M. S., Valentiner-Branth, P., Steinsland, H., Fischer, T. K., Lopes, D. D., Aaby, P., and Molbak, K., 2001, Seven years' experience with *Cryptosporidium parvum* in Guinea-Bissau, West Africa, *Ann. Trop. Paediatr.* **21**:313–318.

Pereira, M. D., Atwill, E. R., Barbosa, A. P., Silva, S. A., and Garcia-Zapata, M. T., 2002a, Intra-familial and extra-familial risk factors associated with *Cryptosporidium parvum* infection among children hospitalized for diarrhea in Goiania, Goias, Brazil, *Am. J. Trop. Med. Hyg.* **66**:787–793.

Pereira, S. J., Ramirez, N. E., Xiao, L., and Ward, L. A., 2002b, Pathogenesis of Human and Bovine *Cryptosporidium parvum* in Gnotobiotic Pigs, *J. Infect. Dis.* **186**:715–718.

Pettoello-Mantovani, M., Di Martino, L., Dettori, G., Vajro, P., Scotti, S., Ditullio, M. T., and Guandalini, S., 1995, Asymptomatic carriage of intestinal *Cryptosporidium* in immunocompetent and immunodeficient children: A prospective study, *Pediatr. Infect. Dis. J.* **14**:1042–1047.

Pieniazek, N. J., Bornay-Llinares, F. J., Slemenda, S. B., da Silva, A. J., Moura, I. N., Arrowood, M. J., Ditrich, O., and Addiss, D. G., 1999, New *Cryptosporidium* genotypes in HIV-infected persons, *Emerg. Infect. Dis.* **5**:444–449.

Pozio, E., Rezza, G., Boschini, A., Pezzotti, P., Tamburrini, A., Rossi, P., Di Fine, M., Smacchia, C., Schiesari, A., Gattei, E., Zucconi, R., and Ballarini, P., 1997, Clinical cryptosporidiosis and human immunodeficiency virus (HIV)-induced immunosuppression: Findings from a longitudinal study of HIV-positive and HIV-negative former injection drug users, *J. Infect. Dis.* **176**:969–975.

Priest, J. W., Kwon, J. P., Moss, D. M., Roberts, J. M., Arrowood, M. J., Dworkin, M. S., Juranek, D. D., and Lammie, P. J., 1999, Detection by enzyme immunoassay of serum immunoglobulin G antibodies that recognize specific *Cryptosporidium parvum* antigens, *J. Clin. Microbiol.* **37**:1385–1392.

Priest, J. W., Li, A., Khan, M., Arrowood, M. J., Lammie, P. J., Ong, C. S., Roberts, J. M., and Isaac-Renton, J., 2001, Enzyme immunoassay detection of antigen-specific immunoglobulin g antibodies in longitudinal serum samples from patients with cryptosporidiosis, *Clin. Diagn. Lab. Immunol.* **8**:415–423.

Proctor, S. J., and Kemp, R. L., 1974, *Cryptosporidium anserinum* sp. N. (Sporozoa) in a domestic goose Anser anser L., from Iowa, *J. Protozool.* **21**:664–666.

Quilez, J., Sanchez-Acedo, C., Clavel, A., del Cacho, E., and Lopez-Bernad, F., 1996, Comparison of an acid-fast stain and a monoclonal antibody-based immunofluorescence reagent for the detection of *Cryptosporidium* oocysts in faecal specimens from cattle and pigs, *Vet. Parasitol.* **67**:75–81.

Quiroz, E. S., Bern, C., MacArthur, J. R., Xiao, L., Fletcher, M., Arrowood, M. J., Shay, D. K., Levy, M. E., Glass, R. I., and Lal, A., 2000, An outbreak of cryptosporidiosis linked to a foodhandler, *J. Infect. Dis.* **181**:695–700.

Reperant, J. M., Naciri, M., Iochmann, S., Tilley, M., and Bout, D. T., 1994, Major antigens of *Cryptosporidium parvum* recognised by serum antibodies from different infected animal species and man, *Vet. Parasitol.* **55**:1–13.

Rhee, J. K., and Park, B. K., 1996, Survival of *Cryptosporidium muris* (strain MCR) oocysts under cryopreservation, *Korean J. Parasitol.* **34**:155–157.

Rimhanen-Finne, R., Horman, A., Ronkainen, P., and Hanninen, M. L., 2002, An IC-PCR method for detection of *Cryptosporidium* and *Giardia* in natural surface waters in Finland, *J. Microbiol. Methods* **50**:299–303.

Roberts, W. G., Green, P. H., Ma, J., Carr, M., and Ginsberg, A. M., 1989, Prevalence of cryptosporidiosis in patients undergoing endoscopy: Evidence for an asymptomatic carrier state, *Am. J. Med.* **87**:537–539.

Robertson, L. J., Campbell, A. T., and Smith, H. V., 1992, Survival of *Cryptosporidium parvum* oocysts under various environmental pressures, *Appl. Environ. Microbiol.* **58**:3494–3500.

Robertson, L. J., and Gjerde, B., 2000, Isolation and enumeration of *Giardia* cysts, *Cryptosporidium* oocysts, and Ascaris eggs from fruits and vegetables, *J. Food Prot.* **63**:775–778.

Robertson, L. J., and Gjerde, B., 2001a, Factors affecting recovery efficiency in isolation of *Cryptosporidium* oocysts and *Giardia* cysts from vegetables for standard method development, *J. Food Prot.* **64**:1799–1805.

Robertson, B., Sinclair, M. I., Forbes, A. B., Veitch, M., Kirk, M., Cunliffe, D., Willis, J., and Fairley, C. K., 2002a, Case-control studies of sporadic cryptosporidiosis in Melbourne and Adelaide, Australia, *Epidemiol. Infect.* **128**:419–431.

Robertson, L. J., and Gjerde, B., 2001b, Occurrence of parasites on fruits and vegetables in Norway, *J. Food Prot.* **64**:1793–1798.

Robertson, L. J., Johannessen, G. S., Gjerde, B. K., and Loncarevi, S., 2002b, Microbiological analysis of seed sprouts in Norway, *Int. J. Food Microbiol.* **75**:119–126.

Rochelle, P. A., De Leon, R., Stewart, M. H., and Wolfe, R. L., 1997a, Comparison of primers and optimization of PCR conditions for detection of *Cryptosporidium parvum* and *Giardia* lamblia in water, *Appl. Environ. Microbiol.* **63**:106–114.

Rochelle, P. A., Ferguson, D. M., Handojo, T. J., De Leon, R., Stewart, M. H., and Wolfe, R. L., 1996, Development of a rapid detection procedure for *Cryptosporidium*, using *in vitro* cell culture combined with PCR, *J. Eukaryot. Microbiol.* **43**:72S.

Rochelle, P. A., Ferguson, D. M., Handojo, T. J., De Leon, R., Stewart, M. H., and Wolfe, R. L., 1997b, An assay combining cell culture with reverse transcriptase PCR to detect and determine the infectivity of waterborne *Cryptosporidium parvum*, *Appl. Environ. Microbiol.* **63**:2029–2037.

Rochelle, P. A., Jutras, E. M., Atwill, E. R., De Leon, R., and Stewart, M. H., 1999, Polymorphisms in the beta-tubulin gene of *Cryptosporidium parvum* differentiate between isolates based on animal host but not geographic origin, *J. Parasitol.* **85**:986–989.

Rochelle, P. A., Ferguson, D. M., Johnson, A. M., and De Leon, R., 2001, Quantitation of *Cryptosporidium parvum* infection in cell culture using a colorimetric *in situ* hybridization assay, *J. Eukaryot. Microbiol.* **48**:565–574.

Rodgers, M. R., Flanigan, D. J., and Jakubowski, W., 1995, Identification of algae which interfere with the detection of *Giardia* cysts and *Cryptosporidium* oocysts and a method for alleviating this interference, *Appl. Environ. Microbiol.* **61**:3759–3763.

Rose, J. B., 1997, Environmental ecology of *Cryptosporidium* and public health implications, *Ann. Rev. Public Health* **18**:135–161.

Rosenblatt, J. E., and Sloan, L. M., 1993, Evaluation of an enzyme-linked immunosorbent assay for detection of *Cryptosporidium* spp. in stool specimens, *J. Clin. Microbiol.* **31**:1468–1471.

Rossignol, J. F., Hidalgo, H., Feregrino, M., Higuera, F., Gomez, W. H., Romero, J. L., Padierna, J., Geyne, A., and Ayers, M. S., 1998, A double-'blind' placebo—controlled study of nitazoxanide in the treatment of cryptosporidial diarrhoea in AIDS patients in Mexico, *Trans. Roy. Soc. Trop. Med. Hyg.* **92**:663–666.

Rossignol, J. F., Ayoub, A., and Ayers, M. S., 2001, Treatment of diarrhea caused by *Cryptosporidium parvum*: A prospective randomized, double-blind, placebo-controlled study of Nitazoxanide, *J. Infect. Dis.* **184**:103–106.

Roy, S. L., DeLong, S. M., Stenzel, S. A., Shiferaw, B., Roberts, J. M., Khalakdina, A., Marcus, R., Segler, S. D., Shah, D. D., Thomas, S., Vugia, D. J., Zansky, S. M., Dietz, V., and Beach, M. J., 2004, Risk factors for sporadic cryptosporidiosis among immunocompetent persons in the United States from 1999 to 2001, *J. Clin. Microbiol.* **42**:2944–2951.

Ryan, U. M., Xiao, L., Read, C., Sulaiman, I. M., Monis, P., Lal, A. A., Fayer, R., and Pavlasek, I., 2003, A redescription of *Cryptosporidium galli* Pavlasek, 1999 (Apicomplexa: Cryptosporidiidae) from birds, *J. Parasitol.* **89**:809–813.

Ryan, U., O'Hara, A., and Xiao, L., 2004a, Molecular and biological characterization of a *Cryptosporidium* molnari-like isolate from a guppy (*Poecilia reticulata*), *Appl. Environ. Microbiol.* **70**:3761–3765.

Ryan, U. M., Monis, P., Enemark, H. L., Sulaiman, I., Samarasinghe, B., Read, C., Buddle, R., Robertson, I., Zhou, L., Thompson, R. C., and Xiao, L., 2004b, *Cryptosporidium suis* n. sp. (Apicomplexa: Cryptosporidiidae) in pigs (*Sus scrofa*), *J. Parasitol.* **90**:769–773.

Sargent, K. D., Morgan, U. M., Elliot, A., and Thompson, R. C., 1998, Morphological and genetic characterisation of *Cryptosporidium* oocysts from domestic cats, *Vet. Parasitol.* **77**:221–227.

Siddons, C. A., Chapman, P. A., and Rush, B. A., 1992, Evaluation of an enzyme immunoassay kit for detecting *Cryptosporidium* in faeces and environmental samples, *J. Clin. Pathol.* **45**:479–482.

Simmons, O. D., 3rd, Sobsey, M. D., Heaney, C. D., Schaefer, F. W., 3rd, and Francy, D. S., 2001, Concentration and detection of *Cryptosporidium* oocysts in surface water samples by method 1622 using ultrafiltration and capsule filtration, *Appl. Environ. Microbiol.* **67**:1123–1127.

Slavin, D., 1955, *Cryptosporidium meleagridis* (sp. nov.), *J. Comp. Path* **65**:262–270.

Sluter, S. D., Tzipori, S., and Widmer, G., 1997, Parameters affecting polymerase chain reaction detection of waterborne *Cryptosporidium parvum* oocysts, *Appl. Microbiol. Biotechnol.* **48**:325–330.

Smerdon, W. J., Nichols, T., Chalmers, R. M., Heine, H., and Reacher, M. H., 2003, Foot and mouth disease in livestock and reduced cryptosporidiosis in humans, England and Wales, *Emerg. Infect. Dis.* **9**:22–28.

Smith, H. V., Girdwood, R. W., Patterson, W. J., Hardie, R., Green, L. A., Benton, C., Tulloch, W., Sharp, J. C., and Forbes, G. I., 1988, Waterborne outbreak of cryptosporidiosis, *Lancet* **2**:1484.

Smith, J. J., Gunasekera, T. S., Barardi, C. R., Veal, D., and Vesey, G., 2004, Determination of *Cryptosporidium parvum* oocyst viability by fluorescence *in situ* hybridization using a ribosomal RNA-directed probe, *J. Appl. Microbiol.* **96**:409–417.

Soave, R., Danner, R. L., Honig, C. L., Ma, P., Hart, C. C., Nash, T., and Roberts, R. B., 1984, Cryptosporidiosis in homosexual men, *Ann. Intern. Med.* **100**:504–511.

Sorvillo, F., Lieb, L. E., Nahlen, B., Miller, J., Mascola, L., and Ash, L. R., 1994, Municipal drinking water and cryptosporidiosis among persons with AIDS in Los Angeles County, *Epidemiol. Infect.* **113**:313–320.

Sorvillo, F., Beall, G., Turner, P. A., Beer, V. L., Kovacs, A. A., Kraus, P., Masters, D., and Kerndt, P. R., 1998, Seasonality and factors associated with cryptosporidiosis among individuals with HIV infection, *Epidemiol. Infect.* **121**:197–204.

Spano, F., Putignani, L., McLauchlin, J., Casemore, D. P., and Crisanti, A., 1997, PCR-RFLP analysis of the *Cryptosporidium* oocyst wall protein (COWP) gene discriminates between *C. wrairi* and *C. parvum*, and between *C. parvum* isolates of human and animal origin, *FEMS Microbiol. Lett.* **150**:209–217.

Spano, F., Putignani, L., Guida, S., and Crisanti, A., 1998, *Cryptosporidium parvum*: PCR-RFLP analysis of the TRAP-C1 (thrombospondin-related adhesive protein of *Cryptosporidium*-1) gene discriminates between two alleles differentially associated with parasite isolates of animal and human origin, *Exp. Parasitol.* **90**:195–198.

Steinberg, E. B., Mendoza, C. E., Glass, R., Arana, B., Lopez, M. B., Mejia, M., Gold, B. D., Priest, J. W., Bibb, W., Monroe, S. S., Bern, C., Bell, B. P., Hoekstra, R. M., Klein, R., Mintz, E. D., and Luby, S., 2004, Prevalence of infection with waterborne pathogens: A seroepidemiologic study in children 6–36 months old in San Juan Sacatepequez, Guatemala, *Am. J. Trop. Med. Hyg.* **70**:83–88.

Stinear, T., Matusan, A., Hines, K., and Sandery, M., 1996, Detection of a single viable *Cryptosporidium parvum* oocyst in environmental water concentrates by reverse transcription-PCR, *Appl. Environ. Microbiol.* **62**:3385–3390.

Straub, T. M., Daly, D. S., Wunshel, S., Rochelle, P. A., DeLeon, R., and Chandler, D. P., 2002, Genotyping *Cryptosporidium parvum* with an hsp70 Single-Nucleotide Polymorphism Microarray, *Appl. Environ. Microbiol.* **68**:1817–1826.

Strong, W. B., Gut, J., and Nelson, R. G., 2000, Cloning and sequence analysis of a highly polymorphic *Cryptosporidium parvum* gene encoding a 60-kilodalton glycoprotein and characterization of its 15- and 45-kilodalton zoite surface antigen products, *Infect. Immun.* **68**:4117–4134.

Sturbaum, G. D., Reed, C., Hoover, P. J., Jost, B. H., Marshall, M. M., and Sterling, C. R., 2001, Species-specific, nested PCR-restriction fragment length polymorphism detection of single *Cryptosporidium parvum* oocysts, *Appl. Environ. Microbiol.* **67**:2665–2668.

Sturbaum, G. D., Klonicki, P. T., Marshall, M. M., Jost, B. H., Clay, B. L., and Sterling, C. R., 2002, Immunomagnetic separation (IMS)-fluorescent antibody detection and IMS-PCR detection of seeded *Cryptosporidium parvum* oocysts in natural waters and their limitations, *Appl. Environ. Microbiol.* **68**:2991–2996.

Sturbaum, G. D., Jost, B. H., and Sterling, C. R., 2003, Nucleotide changes within three *Cryptosporidium parvum* surface protein encoding genes differentiate genotype I from genotype II isolates, *Mol. Biochem. Parasitol.* **128**:87–90.

Sulaiman, I. M., L., X., Yang, C., Escalante, L., Moore, A., Beard, C. B., Arrowood, M. J., and Lal, A. A., 1998, Differentiating human from animal isolates of *Cryptosporidium parvum.*, *Emerg. Infect. Dis.* **4**:681–685.

Sulaiman, I. M., Xiao, L. H., and Lal, A. A., 1999, Evaluation of *Cryptosporidium parvum* genotyping techniques, *Appl. Environ. Microbiol.* **65**:4431–4435.

Sulaiman, I. M., Morgan, U. M., Thompson, R. C., Lal, A. A., and Xiao, L., 2000, Phylogenetic relationships of *Cryptosporidium* parasites based on the 70-kilodalton heat shock protein (HSP70) gene, *Appl. Environ. Microbiol.* **66**:2385–2391.

Sulaiman, I. M., Lal, A. A., and Xiao, L., 2001, A population genetic study of the *Cryptosporidium parvum* human genotype parasites, *J. Eukaryot. Microbiol.* 24S–27S.

Sulaiman, I. M., Lal, A. A., and Xiao, L., 2002, Molecular phylogeny and evolutionary relationships of *Cryptosporidium* parasites at the actin locus, *J. Parasitol.* **88**:388–394.

Sulaiman, I. M., Hira, P. R., Zhou, L., Al Ali, F. M., Al-Shelahi, F. A., Shweiki, H. M., Iqbal, J., Khalid, N., and Xiao, L., 2005, Unique endemicity of cryptosporidiosis in Kuwaiti children, *J. Clin. Microbiol.* **43**:2805–2809.

Tamburrini, A., and Pozio, E., 1999, Long-term survival of *Cryptosporidium parvum* oocysts in seawater and in experimentally infected mussels (*Mytilus galloprovincialis*), *Int. J. Parasitol.* **29**:711–715.

Tangermann, R. H., Gordon, S., Wiesner, P., and Kreckman, L., 1991, An outbreak of cryptosporidiosis in a day-care center in Georgia, *Am. J. Epidemiol.* **133**:471–476.

Tanriverdi, S., Tanyeli, A., Baslamisli, F., Koksal, F., Kilinc, Y., Feng, X., Batzer, G., Tzipori, S., and Widmer, G., 2002, Detection and genotyping of oocysts of *Cryptosporidium parvum* by real-time PCR and melting curve analysis, *J. Clin. Microbiol.* **40**:3237–3244.

Taylor, J. P., Perdue, J. N., Dingley, D., Gustafson, T. L., Patterson, M., and Reed, L. A., 1985, Cryptosporidiosis outbreak in a day-care center, *Am. J. Dis. Child* **139**:1023–1025.

Teixidor, H. S., Godwin, T. A., and Ramirez, E. A., 1991, Cryptosporidiosis of the biliary tract in AIDS, *Radiology* **180**:51–56.

Thomson, M. A., Benson, J. W., and Wright, P. A., 1987, Two year study of *Cryptosporidium* infection, *Arch. Dis. Child* **62**:559–563.

Thurston-Enriquez, J. A., Watt, P., Dowd, S. E., Enriquez, R., Pepper, I. L., and Gerba, C. P., 2002, Detection of protozoan parasites and microsporidia in irrigation waters used for crop production, *J. Food Prot.* **65**:378–382.

Tiangtip, R., and Jongwutiwes, S., 2002, Molecular analysis of *Cryptosporidium* species isolated from HIV-infected patients in Thailand, *Trop. Med. Int. Health* **7**:357–364.

Tilley, M., Upton, S. J., and Freed, P. S., 1990, A comparative study of the biology of *Cryptosporidium serpentis* and *Cryptosporidium parvum* (Apicomplexa: Cryptosporidiidae, *J. Zoo Wildlife Med.* **21**:463–467.

Tilley, M., Upton, S. J., and Chrisp, C. E., 1991, A comparative study on the biology of *Cryptosporidium* sp. from guinea pigs and *Cryptosporidium parvum* (Apicomplexa), *Can. J. Microbiol.* **37**:949–952.

Traversa, D., Giangaspero, A., Molini, U., Iorio, R., Paoletti, B., Otranto, D., and Giansante, C., 2004, Genotyping of *Cryptosporidium* isolates from *Chamelea gallina* clams in Italy, *Appl. Environ. Microbiol.* **70**:4367–4370.

Tumwine, J. K., Kekitiinwa, A., Nabukeera, N., Akiyoshi, D. E., Rich, S. M., Widmer, G., Feng, X., and Tzipori, S., 2003, *Cryptosporidium parvum* in children with diarrhea in Mulago Hospital, Kampala, Uganda, *Am. J. Trop. Med. Hyg.* **68**:710–715.

Turkcapar, N., Kutlay, S., Nergizoglu, G., Atli, T., and Duman, N., 2002, Prevalence of *Cryptosporidium* infection in hemodialysis patients, *Nephron* **90**:344–346.

Tyzzer, E., 1907, A sporozoon found in the peptic glands of the common mouse, *Proc. Soc. Exp. Bio. Med.* **5**:12–13.

Tyzzer, E., 1910, An extracelluar coccidium, *Cryptosporidium muris* (gen. & sp. nov.), of the gastric glands of the common mouse, *J. Med. Res.* **18**:487–509.

Tyzzer, E., 1912, *Cryptosporidium parvum* (sp. Nov.), a coccidium found in the small intestine of the common mouse, *Arch. Protis.* **26**:394–412.

Tzipori, S., Angus, K. W., Campbell, I., and Gray, E. W., 1980, *Cryptosporidium*: Evidence for a single-species genus, *Infect. Immun.* **30**:884–886.

Tzipori, S., Angus, K. W., Campbell, I., and Sherwood, D., 1981a, Diarrhea in young red deer associated with infection with *Cryptosporidium, J. Infect. Dis.* **144**:170–175.

Tzipori, S., Angus, K. W., Gray, E. W., Campbell, I., and Allan, F., 1981b, Diarrhea in lambs experimentally infected with *Cryptosporidium* isolated from calves, *Am. J. Vet. Res.* **42**:1400–1404.

Tzipori, S., Angus, K. W., Campbell, I., and Gray, E. W., 1982, Experimental infection of lambs with *Cryptosporidium* isolated from a human patient with diarrhoea, *Gut* **23**: 71–74.

Ungar, B. L., 1990, Enzyme-linked immunoassay for detection of *Cryptosporidium* antigens in fecal specimens, *J. Clin. Microbiol.* **28**:2491–2495.

Upton, S. J., and Current, W. L., 1985, The species of *Cryptosporidium* (Apicomplexa: Cryptosporidiidae) infecting mammals, *J. Parasitol.* **71**:625–629.

Vakil, N. B., Schwartz, S. M., Buggy, B. P., Brummitt, C. F., Kherellah, M., Letzer, D. M., Gilson, I. H., and Jones, P. G., 1996, Biliary cryptosporidiosis in HIV-infected people after the waterborne outbreak of cryptosporidiosis in Milwaukee, *N. Engl. J. Med.* **334**:19–23.

Ventura, G., Cauda, R., Larocca, L. M., Riccioni, M. E., Tumbarello, M., and Lucia, M. B., 1997, Gastric cryptosporidiosis complicating HIV infection: Case report and review of the literature, *Eur. J. Gastroenterol. Hepatol.* **9**:307–310.

Vesey, G., Ashbolt, N., Fricker, E. J., Deere, D., Williams, K. L., Veal, D. A., and Dorsch, M., 1998, The Use Of a Ribosomal RNA Targeted Oligonucleotide Probe For Fluorescent Labelling Of Viable *Cryptosporidium parvum* Oocysts, *J. Appl. Microbiol.* **85**:429–440.

Vetterling, J. M., Jervis, H. R., Merrill, T. G., and Sprinz, H., 1971, *Cryptosporidium wrairi* sp. n. from the guinea pig Cavia porcellus, with an emendation of the genus, *J. Protozool.* **18**:243–247.

Wagner-Wiening, C., and Kimmig, P., 1995, Detection of viable *Cryptosporidium parvum* oocysts by PCR, *Appl. Environ. Microbiol.* **61**:4514–4516.

Wang, H. F., Swain, J. B., Besser, T. E., Jasmer, D., and Wyatt, C. R., 2003, Detection of antibodies to a recombinant *Cryptosporidium parvum* p23 in serum and feces from neonatal calves, *J. Parasitol.* **89**:918–923.

Ward, P. I., Deplazes, P., Regli, W., Rinder, H., and Mathis, A., 2002, Detection of eight *Cryptosporidium* genotypes in surface and waste waters in Europe, *Parasitology* **124**:359–368.

Ware, M. W., Wymer, L., Lindquist, H. D., and Schaefer, F. W., 2003, Evaluation of an alternative IMS dissociation procedure for use with Method 1622: Detection of *Cryptosporidium* in water, *J. Microbiol. Methods* **55**:575–583.

Weber, R., Bryan, R. T., Bishop, H. S., Wahlquist, S. P., Sullivan, J. J., and Juranek, D. D., 1991, Threshold of detection of *Cryptosporidium* oocysts in human stool specimens: Evidence for low sensitivity of current diagnostic methods, *J. Clin. Microbiol.* **29**:1323–1327.

Weber, R., Bryan, R. T., and Juranek, D. D., 1992, Improved stool concentration procedure for detection of *Cryptosporidium* oocysts in fecal specimens, *J. Clin. Microbiol.* **30**:2869–2873.

Webster, K. A., Pow, J. D., Giles, M., Catchpole, J., and Woodward, M. J., 1993, Detection of *Cryptosporidium parvum* using a specific polymerase chain reaction, *Vet. Parasitol.* **50**:35–44.

Webster, K. A., Smith, H. V., Giles, M., Dawson, L., and Robertson, L. J., 1996, Detection of *Cryptosporidium parvum* oocysts in faeces: Comparison of conventional coproscopical methods and the polymerase chain reaction, *Vet. Parasitol.* **61**:5–13.
Weinstein, P., Macaitis, M., Walker, C., and Cameron, S., 1993, Cryptosporidial diarrhoea in South Australia. An exploratory case-control study of risk factors for transmission, *Med. J. Aust.* **158**:117–119.
Widmer, G., 1998, Genetic heterogeneity and PCR detection of *Cryptosporidium parvum*, *Adv. Parasitol.* **40**:223–239.
Widmer, G., Orbacz, E. A., and Tzipori, S., 1999, beta-tubulin mRNA as a marker of *Cryptosporidium parvum* oocyst viability, *Appl. Environ. Microbiol.* **65**:1584–1588.
Widmer, G., Feng, X., and Tanriverdi, S., 2004, Genotyping of *Cryptosporidium parvum* With Microsatellite Markers, *Methods Mol. Biol.* **268**:177–188.
Wu, Z., Nagano, I., Matsuo, A., Uga, S., Kimata, I., Iseki, M., and Takahashi, Y., 2000, Specific PCR primers for *Cryptosporidium parvum* with extra high sensitivity, *Mol. Cell Probes* **14**:33–39.
Wu, Z., Nagano, I., Boonmars, T., Nakada, T., and Takahashi, Y., 2003, Intraspecies polymorphism of *Cryptosporidium parvum* revealed by PCR-restriction fragment length polymorphism (RFLP) and RFLP-single-strand conformational polymorphism analyses, *Appl. Environ. Microbiol.* **69**:4720–4726.
Xiao, L., and Ryan, U. M., 2004, Cryptosporidiosis: An update in molecular epidemiology, *Curr. Opin. Infect. Dis.* **17**:483–490.
Xiao, L. H., Escalante, L., Yang, C. F., Sulaiman, I., Escalante, A. A., Montali, R. J., Fayer, R., and Lal, A. A., 1999a, Phylogenetic analysis of *Cryptosporidium* parasites based on the small-subunit rRNA gene locus, *Appl. Environ. Microbiol.* **65**:1578–1583.
Xiao, L. H., Morgan, U. M., Limor, J., Escalante, A., Arrowood, M., Shulaw, W., Thompson, R. C. A., Fayer, R., and Lal, A. A., 1999b, Genetic diversity within *Cryptosporidium parvum* and related *Cryptosporidium* species, *Appl. Environ. Microbiol.* **65**:3386–3391.
Xiao, L., Alderisio, K., Limor, J., Royer, M., and Lal, A. A., 2000a, Identification of species and sources of *Cryptosporidium* oocysts in storm waters with a small-subunit rRNA-based diagnostic and genotyping tool, *Appl. Environ. Microbiol.* **66**:5492–5498.
Xiao, L., Morgan, U. M., Fayer, R., Thompson, R. C., and Lal, A. A., 2000b, *Cryptosporidium* systematics and implications for public health, *Parasitol. Today* **16**:287–292.
Xiao, L., Bern, C., Limor, J., Sulaiman, I., Roberts, J., Checkley, W., Cabrera, L., Gilman, R. H., and Lal, A. A., 2001a, Identification of 5 types of *Cryptosporidium* parasites in children in Lima, Peru, *J. Infect. Dis.* **183**:492–497.
Xiao, L., Limor, J., Bern, C., and Lal, A. A., 2001b, Tracking *Cryptosporidium parvum* by sequence analysis of small double-stranded RNA, *Emerg. Infect. Dis.* **7**:141–145.
Xiao, L., Singh, A., Limor, J., Graczyk, T. K., Gradus, S., and Lal, A., 2001c, Molecular characterization of *Cryptosporidium* oocysts in samples of raw surface water and wastewater, *Appl. Environ. Microbiol.* **67**:1097–1101.
Xiao, L., Bern, C., Sulaiman, I. M., and Lal, A. A., 2003, Molecular epidemiology of human cryptosporidiosis, In Thompson, R. C. A., Armson, A., and Ryan, U. M. (eds), *Cryptosporidium: From Molecules to Disease*, Elsevier, Amsterdam, New York, pp. 121–146.
Xiao, L., Fayer, R., Ryan, U., and Upton, S. J., 2004a, *Cryptosporidium* taxonomy: Recent advances and implications for public health, *Clin. Microbiol. Rev.* **17**:72–97.
Xiao, L., Lal, A. A., and Jiang, J., 2004b, Detection and differentiation of *Cryptosporidium* oocysts in water by PCR-RFLP, *Methods Mol. Biol.* **268**:163–176.
Xiao, L., Ryan, U. M., Graczyk, T. K., Limor, J., Li, L., Kombert, M., Junge, R., Sulaiman, I. M., Zhou, L., Arrowood, M. J., Koudela, B., Modry, D., and Lal, A. A., 2004c, Genetic

diversity of *Cryptosporidium* spp. in captive reptiles, *Appl. Environ. Microbiol.* **70**:891–899.

Yamamoto, N., Urabe, K., Takaoka, M., Nakazawa, K., Gotoh, A., Haga, M., Fuchigami, H., Kimata, I., and Iseki, M., 2000, Outbreak of cryptosporidiosis after contamination of the public water supply in Saitama Prefecture, Japan, in 1996, *Kansenshogaku Zasshi* **74**:518–526.

Yu, J. R., O'Hara, S. P., Lin, J. L., Dailey, M. E., and Cain, G., 2002, A common oocyst surface antigen of *Cryptosporidium* recognized by monoclonal antibodies, *Parasitol. Res.* **88**:412–420.

Zhou, L., Singh, A., Jiang, J., and Xiao, L., 2003, Molecular surveillance of *Cryptosporidium* spp. in raw wastewater in Milwaukee: Implications for understanding outbreak occurrence and transmission dynamics, *J. Clin. Microbiol.* **41**:5254–5257.

Zu, S. X., Li, J. F., Barrett, L. J., Fayer, R., Shu, S. Y., McAuliffe, J. F., Roche, J. K., and Guerrant, R. L., 1994, Seroepidemiologic study of *Cryptosporidium* infection in children from rural communities of Anhui, China, and Fortaleza, Brazil, *Am. J. Trop. Med. Hyg.* **51**:1–10.

CHAPTER 5

Toxoplasmosis

Ynes R. Ortega

5.1 PREFACE

Toxoplasma gondii is a coccidia that is the most widespread and prevalent parasite in the world. It can infect warm-blooded animals, including man. It is responsible for 20.7% of food-borne deaths due to known infectious agents. Waterborne outbreaks have also been associated with *Toxoplasma* in Canada and Brazil (Aramini *et al.*, 1999; Bahia-Oliveira *et al.*, 2003; Dubey, 2004). Toxoplasmosis can be asymptomatic or can cause abortion in humans if an acute infection develops during pregnancy. Healthy individuals may develop encephalitis. Serologically, *Toxoplasma* can be identified in as high as 85% of the population in some European countries, where meat is primarily eaten undercooked. In Paris, 84% of pregnant women have been exposed to *Toxoplasma*, as compared to 32% of pregnant women in New York.

Few cases of waterborne toxoplasmosis have been reported. It is estimated that most cases are transmitted via contaminated foods. In the United States, between 400 and 4000 cases of congenital toxoplasmosis occur annually. Of the 750 deaths attributed to toxoplasmosis each year, 50% are believed to be caused by eating contaminated meat, making toxoplasmosis the third leading cause of food-borne deaths in this country (Lopez *et al.*, 2000). These cases were assumed to have originated from mishandling in food service establishments and homes, not from food processing establishments. Acquisition of the parasite may be ingestion of raw or inadequately cooked infected meat or exposure to cat feces. Contamination may also occur from contact with soil when gardening or ingestion of unwashed fruits or vegetables contaminated with oocysts.

5.2 PARASITE DESCRIPTION

Toxoplasma gondii was described in the early 1900s. It has been identified as being able to infect over 300 species of mammals and 30 species of birds as intermediate hosts. Infection is acquired when a host ingests water or food which is contaminated with cat feces containing *Toxoplasma* oocysts. The oocysts excyst and the sporozoites migrate, and preferentially localize in muscle and the brain. The parasite can cross the placenta to infect the fetal tissues (Dubey, 1991).

A large variety of animals can acquire toxoplasmosis, but only cats (domestic and wild) are the definitive hosts. Outdoor cats are more likely to be infected with *Toxoplasma*.

The oocysts are highly resistant, even to desiccation, and can survive on dry surfaces for weeks or even months. The role of shellfish in parasite transmission is being studied (Arkush *et al.*, 2003) because shellfish can filter large volumes of water and concentrate viable *Toxoplasma* oocysts. Most marine mammals feed on mollusks and these mammals have a high mortality with meningoencephalitis caused by *Toxoplasma* (Dubey *et al.*, 2003e; Oksanen *et al.*, 1998).

In humans, most infections are asymptomatic; however, it can be fatal for immunocompromised individuals and the fetuses of women who acquire the infection during the first 4 to 5 months of pregnancy. Three different genotypes I, II, and III have been described in *T. gondii* (Howe and Sibley, 1995). Other strains fall into two classes: recombinant, which is closely related to the dominant types (I and III), and exotic. Type I is highly virulent in laboratory animals, whereas types II and III are non-virulent. In humans, Type II predominates in AIDS and congenital infections (encephalitis, pneumonitis, or disseminated infections). Type II has been isolated in about 75–80% of AIDS and non-AIDS immunocompromised patients. In Spain, the genotype I is more prevalent in congenital infections (Fuentes *et al.*, 2001). Ocular toxoplasmosis is a common sequelae of congenital toxoplasmosis, but can be dormant for years and emerge at adulthood, causing severe retinochoriditis. In these individuals type I, type IV, or novel types were frequently isolated.

Type I was implicated in outbreaks in Canada and Brazil and was characterized by severe ocular toxoplasmosis (Boothroyd and Grigg, 2002). Immunocompetent adults may also suffer from retinitis and enlarged lymph nodes.

5.3 LIFE CYCLE

Unsporulated oocysts are excreted in the feces of infected cats. Oocysts undergo sporulation for 24–48 h to become infectious. The oocysts are the environmentally resistant form and are excreted in the feces (Fig. 5.1). Other animals or humans can acquire the infection when oocysts are ingested via water, food, or soil.

Sporulated oocysts consist of two sporocysts, each containing four sporozoites. Once released from the sporocyst, the sporozoites penetrate the intestinal cells and lymph nodes, becoming tachyzoites. These multiply very fast and disperse throughout the body via blood or lymph where they multiply and can eventually encyst in the brain, liver, skeletal, and cardiac muscle. These cysts contain bradyzoites which

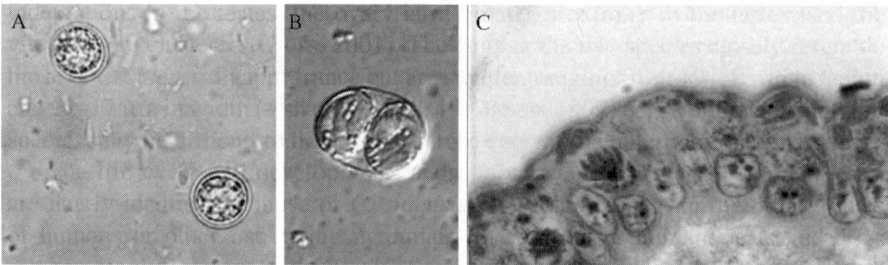

Figure 5.1. *Toxoplasma gondii* (a) unsporulated and (b) sporulated oocysts. (pictures obtained from http://www.dpd.cdc.gov/dpdx/HTML/ImageLibrary). (c) *Toxoplasma gondii* intracellular stages observed in cat intestinal tissue.

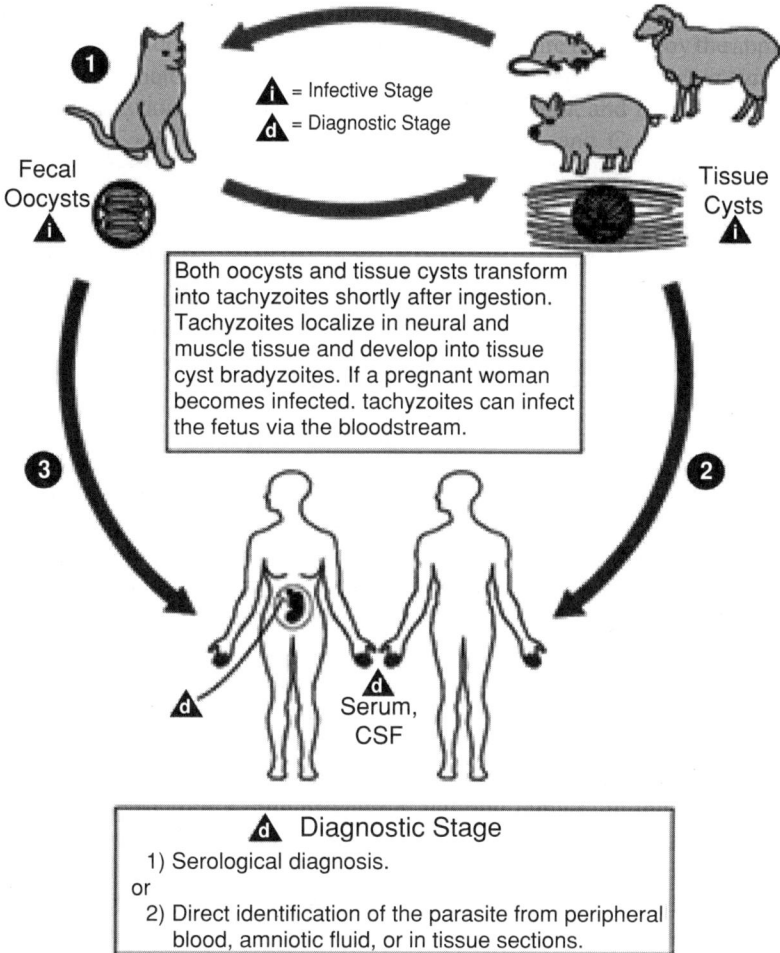

Figure 5.2. Life cycle of *Toxoplasma gondii*. Graph obtained from http:// www.dpd.cdc.gov/dpdx/HTML/ImageLibrary.

multiply slowly. Cysts may be viable for the duration of the hosts life. Tachyzoites are frequently present during the acute phase and bradyzoites in the chronic phase of the infection (Fig. 5.2).

When tissue cysts are ingested by a susceptible host, the cyst wall is digested by proteolytic enzymes. Bradyzoites are then released to infect the intestinal cells. Tachyzoites are dispersed and infect other tissues. If a cat ingests the tissues, tachyzoites infect the intestinal cells and begin asexual reproduction (schizogony) and sexual multiplication (gametogony). Macrogametocytes are fertilized by a male gamete (microgametocyte), forming the zygote which differentiates to become an oocyst, and is excreted in the feces.

A novel route of infection has been reported recently. *T. gondii* appeared in the bile and feces of interferon-gamma knockout (GKO) mice, but not wild mice after peroral infection with *T. gondii* cysts. The tachyzoite and bradyzoite specific mRNA were identified in bile and feces and was confirmed using the mouse infectivity assay (Piao *et al.*, 2005).

5.4 TRANSMISSION

Toxoplasmosis can be acquired by ingestion of contaminated water, food, or soil (Choi *et al.*, 1997; Coutinho *et al.*, 1982; Dubey, 2004; Ruiz *et al.*, 1973). Foodborne toxoplasmosis is most often acquired by consumption of raw or undercooked meats.

Congenital toxoplasmosis occurs when a pregnant female is exposed to *Toxoplasma* oocysts. *Toxoplasma* can cross the placenta and infect the fetus. This may result in diminished vision or blindness after birth. Symptoms include hydrocephaly, convulsion, and calcium deposits in the brain. Less frequently, toxoplasmosis is acquired by transfusion of blood or its components, or by organ transplantation. Latent toxoplasmosis can reactivate if the immune system of the host is compromised. Inhalation of aerosols containing oocysts from cat litters, farm animal feed, and bedding has also been suggested (Furuta *et al.*, 2001).

The first documented toxoplasmosis outbreak associated with a municipal water supply was described in 1995 in British Columbia, Canada. It was hypothesized that domestic cat or cougar feces contaminated a surface water reservoir with *T. gondii* oocysts. These animals were observed around the watershed and could shed oocysts in their feces near the waters' edge (Aramini *et al.*, 1999). During 1997–1999, a region of Brazil was surveyed for seropositivity to *Toxoplasma*. The survey population was selected randomly from schools, randomly chosen communities, and an army battalion. Out of 1436 persons tested, 84% of the population in the lower socioeconomic group was seropositive, compared with 62% and 23% of the middle and upper socioeconomic groups, respectively ($p < 0.001$). Multivariate analysis suggested that drinking unfiltered water increased the risk of seropositivity for the lower socioeconomic and middle socioeconomic populations (Bahia-Oliveira *et al.*, 2003).

5.5 IDENTIFICATION

Toxoplasma infections can be diagnosed by serological assays examining the antibody response toward the infection. Commercial agglutination and Elisa assays are available (Dubey *et al.*, 1995; Ryu *et al.*, 1996). Western blot assays have also been reported in the literature (Bessieres *et al.*, 1992; Saavedra and Ortega, 2004). *Toxoplasma* oocysts can be identified in the environment using conventional microscopy; however, the small number of parasites may be less sensitive. *Toxoplasma* oocysts have been isolated from mussels that could serve as paratenic hosts by concentrating oocysts (Arkush *et al.*, 2003). *Toxoplasma* oocysts can be identified from water samples using the current U.S. EPA method for concentration of *Cryptosporidium* (Isaac-Renton et al., 1998). Centrifugation and flocculation procedures using Aluminum sulfate and Ferric sulfate can also concentrate *Toxoplasma* oocysts. Sporulated oocysts were recovered more efficiently using aluminum sulfate while unsporulated oocysts could be better recovered using ferric sulfate (Kourenti and Karanis, 2004).

5.5.1 Molecular Assays

Toxoplasma oocysts are usually present in low numbers in contaminated water and foods. Rapid and sensitive detection methods are necessary. Tissue culture and animal models available for *Toxoplasma* are time-consuming, expensive, and labor-intensive. Therefore, Polymerase chain reaction (PCR) amplification has become the preferred method. Most PCR assays used for *Toxoplasma* identification use primers targeting the B1 gene. It is a 35-fold-repetitive gene that is highly specific and conserved among strains of *Toxoplasma* (Buchbinder et al., 2003). It also has a PCR amplification and detection method for *T. gondii* oocyst nucleic acid that incorporates uracil-N-glycosylase to prevent false-positive results, an internal standard control to identify false-negative results, and uses PCR product oligoprobe confirmation using a nonradioactive DNA hybridization immunoassay. This method can provide positive, confirmed results in less than 1 day and can detect less than 50 oocysts (Schwab and McDevitt, 2003).

Other assays have focused on the sensitivity of the assay. DNA was extracted with a modified Qiagen DNA Mini Kit method and was amplified by PCR using specific primers for the *T. gondii* B1 gene. *T. gondii* was detected correctly in 90% of the clinical specimens examined in less than 5 h, with a detection limit of two parasites/sample (Jalal et al., 2004).

Toxoplasma oocyst detection can be included as part of the waterborne parasite detection protocol. Water samples are filtered followed by a sucrose density gradient. Oocyst detection is done using PCR and bioassay. In an experimental seeding assay with 100 L of deionized water, a parasite density of one oocyst/L was successfully detected by PCR in 60% of cases and 10 oocysts/L was detected in 100% of cases. The sensitivity of the PCR assay varied from less than 10 to more than 1000 oocysts/L, depending on the sample source. PCR was always more sensitive than mouse inoculation. Out of 139 environmental water samples, 125 could be analyzed. *Toxoplasma* DNA was identified in 8% of the cases; however none were positive by mouse inoculation (Villena et al., 2004).

DNA amplification using the 18S-rRNA gene (MacPherson and Gajadhar, 1993) had a theoretical detection limit of 0.1 oocyst only when the samples were concentrated by aluminum sulfate flocculation (Kourenti and Karanis, 2004). TaqMan PCR assays using the B1 and ssrRNA genes have been successfully used in experimentally inoculated mussels.

The role of *Toxoplasma* on the morbidity and mortality of marine mammals has been studied extensively. Fatal meningoencephalitis has been reported in these animals. The sources of *T. gondii* oocysts in marine environments are unknown. However, bivalve shellfish have been demonstrated to serve as paratenic hosts by assimilation and concentration of infective cysts and oocysts. Therefore, *T. gondii* oocysts can be concentrated by shellfish, and sea otters could acquire the infection by eating the shellfish. A TaqMan PCR assay for detection of *T. gondii* SSrRNA was evaluated using experimentally spiked mussels with *T. gondii* oocysts. *T. gondii*-specific SSrRNA was detected in mussels as long as 21 days postinoculation. Detection was found more frequently in the digestive gland homogenate. Parasite infectivity was confirmed using a mouse bioassay (Arkush *et al.*, 2003).

A real-time PCR was developed in order to detect and quantify *T. gondii* B1 and bradizoite specific genes (SAG-4 and MAG-1) in serum and peripheral blood mononuclear cells specimens. The results were compared with those obtained with a nested PCR. Real-time PCR proved to be more sensitive than nested PCR for detection and quantification of either the B1 gene ($P < 0.001$) or the SAG-4/MAG-1 gene ($P < 0.05$). Real-time PCR has been shown to be particularly useful to accurately determine the parasite DNA load in follow-up specimens (Contini *et al.*, 2005).

Other targets used for *Toxoplasma* identification include a 529 bp sequence, which has 300 copies in the genome of *Toxoplasma*. This fragment was used for the development of a very sensitive and specific PCR for diagnostic purposes, and a quantitative competitive-PCR for the evaluation of cyst numbers in the brains of chronically infected mice. Polymerase chain reaction with the 529 bp fragment was more sensitive than with the 35-copy B1 gene. A highly significant correlation between visual counting of brain cysts and quantitative competitive-PCR was obtained in mice chronically infected with *Toxoplasma* (Homan *et al.*, 2000). Mobile genetic elements (MGE) that have 100–500 copies per cell were also used for design of assays for *Toxoplasma* identification. Two PCR-based strategies using specific primers amplified *T. gondii* MGEs; revealing information on element size and positional variation. The first PCR strategy involved the use of a standard two-primer PCR while the second strategy used a single specific primer in a step-up PCR protocol. The use of a standard two-primer PCR reaction revealed the presence of a virulence related marker in which all avirulent strains possessed an additional 688 bp band. The single primer PCR strategy demonstrated that all virulent strains had identical banding patterns suggesting invariance within this group of strains. However, all avirulent strains had different banding patterns indicating the presence of a number of individual lineages within this group (Terry *et al.*, 2001).

Single copy genes SAG1-4 and GRA4 genes have been used as targets for *Toxoplasma* characterization and identification. The genes SAG5A, SAG5B, and SAG5C were also examined to characterize strain virulence in the three major genotypes of *T. gondii*. Southern blot analysis using a SAG5-specific probe could differentiate between genotype I virulent strains from the avirulent strains of either genotype II or genotype III. A PCR-restriction fragment length polymorphism method based on the SAG5C gene can discriminate between strains of genotype I, II, and III using a single endonuclease digestion (Meisel *et al.*, 1996; Tinti *et al.*, 2003).

5.5.2 Riboprinting

Characterization of *Toxoplasma* isolates is achieved by using PCR amplified products digested with 13 enzymes. Discrimination between intracellular stages of coccidia in human tissues can be achieved using riboprints (through restriction enzyme analysis of the PCR-amplified small subunit rRNA gene). Together, the variation in riboprints and surface antigen gene structure reflects the phylogenetic diversity among these coccidia, and in addition, confirms the value of riboprinting in the identification of apicomplexan parasites such as *T. gondii* (Brindley et al., 1993). RFLP-PCR, RAPD, sequence length polymorphism, and sequencing has allowed for genotyping analysis (Aspinall et al., 2003; Bartova and Literak, 2004; Carme et al., 2002).

The coding region of GRA6 was amplified, sequenced, and compared for 30 *Toxoplasma* strains from eight different zymodemes (Z1–Z8). Sequence alignment demonstrated nucleotide polymorphisms. Types I, II, and III could be distinguished from each other. The large variety of amino acid changes supports the view that the GRA6 protein plays an important role in the antigenicity and pathogenicity of *T. gondii*. A PCR-RFLP method using MseI could differentiate the three *Toxoplasma* groups (Fazaeli et al., 2000).

5.6 PATHOGENICITY

Cell adhesion is a prerequisite for cell invasion. Various surface molecules are required in this process, including the SAG3 and SAG5 molecules. Once attachment occurs, tachyzoites release micronemal content, the conoid protrudes and forms an indentation in the host cell. Rhoptries containing proteins and lipids are released. A tight junction is formed between host cell and parasite. Tachyzoites multiply within the cell by binary fission. Multiplication continues until the host cell lyse and tachyzoites are released and can reinfect other cells. In chronic infections, bradyzoites are present and disease reactivation occurs when there is an impairment of the immune function. In murine models, tumor necrosis factor-α, interferon γ, and T cells are required to prevent disease reactivation (Roberts and McLeod, 2004).

5.7 EPIDEMIOLOGY

T. gondii can infect a variety of warm-blooded animals. The relevant species associated with transmission of *Toxoplasma* to humans will be described.

5.7.1 Humans
It has been estimated that 30–60% of adults in the United States have been exposed to *Toxoplasma* at some point in their lifetime (180, 185–191). A high seroprevalence of *Toxoplasma* in Europe and South American countries has also been reported. This may be due to the frequent consumption of raw meats.

According to the Third National Health and Nutrition Examination Survey in the United States (1988–1994), of 17,658 sera tested, the overall age-adjusted seroprevalence was 22.5%. Among women aged 15–44 years, seroprevalence was 15.0%. The seroprevalence in the Northeast was 29.2%, 22.8% in the South, 20.5% in the Midwest, and 17.5% in the West. Risks for acquiring *Toxoplasma* infection increased with age. It was higher among persons who were foreign-born, persons with a lower educational level, those who lived in crowded conditions, and those who worked in soil-related occupations. About 25% of adults and adolescents in the United States have been infected with *T. gondii* (Jones *et al.*, 2001). Sera collected in the National Health and Examination Survey (NHANES) from 1999–2000 was examined for seropositivity to *Toxoplasma*. Of 4234 persons 12–49 years of age, 15.8% were antibody positive; among women, 14.9% were seropositive. Prevalence was higher among non-Hispanic black persons (19.2%) than among non-Hispanic white persons (12.1%) (Jones *et al.*, 2003). A cross-sectional seroprevalence study in healthy adults in Maryland included Seventh Day Adventists who were vegetarians and control community volunteers who were not vegetarians. Overall, seroprevalence was 31% in the study group. People with *T. gondii* infection were less likely to be Seventh Day Adventists (24% versus 50%) than people without *T. gondii* infection (Roghmann *et al.*, 1999). In another study, seroprevalence of *Toxoplasma* in Yakima Indians (ages 1 to 66 years) was 20% (23/114) (Sturchler *et al.*, 1987).

The seroprevalence of *T. gondii* infection using an ELISA test was determined in primigravid women in India. Between August 1996 and September 1997, *Toxoplasma* seroprevalence was 41.75% in 503 women (Akoijam *et al.*, 2002). In Bombay, the seroprevalence of healthy adult voluntary blood donors (ages 13–50 years) was 30.9%, 67.8% in HIV infected hosts, and 28% in patients treated for cerebral tuberculoma or neurocysticercosis. *Toxoplasma* infection appears to be subclinical and prevalent throughout life but emerges as an important opportunistic infection in HIV/AIDS patients (Meisheri *et al.*, 1997). In Kashmir, 53.14% of 2371 women with recurrent abortions and 69.35% of 310 women with neonatal deaths tested positive for IgM antibody against *Toxoplasma*. Of the 177 women who received followed up visits, 94.26% of 122 women with recurrent abortions and 63.64% of 55 women with neonatal deaths delivered normal babies after they were treated with spiramycin during pregnancy (Zargar *et al.*, 1999). In Bangladesh, of 286 women examined by ELISA, 38.5% were seropositive for *Toxoplasma* IgG antibody, and of 88 randomly selected patients, 1.1% was positive for *Toxoplasma* IgM. The seroprevalence gradually increased with age and parity. The seroprevalence of antibody was higher among the poor women (53.0%) than the upper socio-economic class (22.0%) and among the women with jobs (55.0%) than the housewife group (35.0%) (Ashrafunnessa *et al.*, 1998). In Nepal, the seroprevalence of *T. gondii* infection in 404 apparently healthy subjects was 65.3% (Rai *et al.*, 1999).

In Korea, the seroprevalence of *Toxoplasma* in pregnant women was found to be low. Seropositivity for *Toxoplasma* was 0.79% in 5175 sera and 1.33% in 750 amniotic fluid samples (Song *et al.*, 2005). In Taiwan, a seroepidemiological survey of *T. gondii* infection among Atayal and Paiwan mountain aborigines and Southeast Asian laborers found that the overall seroprevalence of *T. gondii* infection was 19.4% for Atayal, 26.7% for Paiwan, 42.9% for Indonesian, 14.7% for Thai, and 11.3% for Filipinos. Atayal and Paiwan Indians with a history of eating raw meat seemed more susceptible to *T. gondii* infection than those who had never consumed raw meat (Fan *et al.*, 2002).

In Catania (Sicily), the seroprevalence of *T. gondii* in fertile women is 41.1% (Condorelli *et al.*, 1993). In the general northern Greek population, the prevalence of IgG-specific antibodies to *Toxoplasma* was 37, 29.9, and 24.1% in 1984, 1994, and 2004, respectively, and was 35.6, 25.6 and 20%, respectively, in women of reproductive age (15–39 years). The significant decline in prevalence, and the shift toward an older age group, observed during this period could be explained by the improved socio-economic situation (Diza et al., 2005).

Sera from 144 Ethiopian immigrants living in the Jezreel Valley were tested for antibodies against *T. gondii*. Of these, 34% of the immigrants were positive and the prevalence in the Ethiopians was higher than in Jewish kibbutz members (22.8%) and lower than in Arab villagers (55.8%). Prevalence increased from 0% in children less than 10-years old to 46% in individuals 40 years or older (Flatau et al., 1993). The incidence of toxoplasmosis in rural areas of Central African Republic on a healthy population was determined. About 40% of the adults had IgG antibodies against *T. gondii*, but in a pre-desert area 25% were positive (Dumas et al., 1990). In Rwanda, 50% of the adults of two communities had antibodies to *T. gondii*. Only 12% of the Ngenda population group of 14-years old was positive, whereas The Nyarutovu (NVU) population had a 31% positivity; suggesting that the Nyarutovu acquired the infection earlier in life (Gascon et al., 1989). In Dar es Salaam, Tanzania, the infection rate in normal pregnant women was 41.9%, in anemic women 52.5%, and 66.7% in individuals with hypertension (Gill and Mtimavalye, 1982). In four regions of Southern Africa (Natal, Eastern Cape, Western Cape, and South West Africa and Botswana) the overall seroprevalence was 20% (of 3379 sera tested), the highest prevalence occurring in Blacks (34%) and Indians (33%) of Natal, and the lowest in San (Bushmen) (9%) and Whites (12%) of South West Africa and Botswana (Jacobs and Mason, 1978). In Bamako, Mali, toxoplasmosis seroprevalence was 60% from AIDS patients, 22.6% from the HIV-seropositive blood donors, and 21% from the HIV-seronegative blood donors (Maiga et al., 2001).

The seroprevalence of *T. gondii* in 191 pregnant women was 25.7% in a Primary Care setting in Malaga, Spain. Significant associations using univariate and multivariate analysis was demonstrated in individuals with previous abortions and low economic status (Guerra and Fernandez, 1995). In another study, the seroprevalence in intravenous drug users was 47.6%, 12.2% in infants, and 30% in pregnant women (Gutierrez et al., 1996).

In Sweden, the seroprevalence of *Toxoplasma* for Swiss women was 46.0% and 45.8% for women of other nationalities at the time of delivery. The risk of seroconversion among seronegative women during their 9 months of pregnancy was 1.21% (Jacquier et al., 1995). Blood samples from more than 40,000 newborns, from two geographically different areas, were examined for the presence of IgG antibodies to *Toxoplasma* to determine the seroprevalence of *Toxoplasma* in their mothers. During a 16-month period between April 1997 and July 1998, the seroprevalence was 14.0% in Stockholm County and 25.7% in Skane County. The seroprevalence among women born in Stockholm was 11.1% and 24.9% in Skane. The corresponding figures for women born outside the Nordic countries were 24.3% and 29.4% (Petersson et al., 2000).

In the Netherlands, between 1995 and 1996, 7521 sera were tested and the national seroprevalence was found to be 40.5 %. Living in the Northwest, having professional contact with animals, living in a moderately urbanized area, being divorced or widowed, being born outside The Netherlands, frequent gardening and owning a cat were independently associated with *Toxoplasma* seropositivity. The seroprevalence among women aged 15–49 years was 35.2% in the study of 1995–1996 and was lower than in the pregnant women in the Southwest of The Netherlands in 1987–1988 (45.8 %). The steepest rise in seroprevalence occurred among women aged 25–44 years (Kortbeek *et al.*, 2004).

In Slovenia, during 1981 to 1994, a serological screening for toxoplasmosis was carried out on 20,953 pregnant women. Seropositivity decreased from 52% in the 1980s to 37% during 1991–1994, while during the same period, the incidence of suspected primary infections acquired in pregnancy rose from 0.33% to 0.75% (Logar *et al.*, 1995). Over a 12-month period, the incidence of congenital toxoplasmosis in 3959 pregnant women in Slovenia was 3/1000 (Logar *et al.*, 1992).

In South America, the prevalence of *Toxoplasma* has been studied in the past years. The prevalence of *T. gondii* in indigenous Brazilian tribes with different degrees of acculturation was studied. During 2000-2001, seroprevalence varied from 57.3 to 78.8%. Differential contact with soil-harboring oocysts from wild felines may be responsible for the variable seroprevalence in the different tribes (Sobral *et al.*, 2005). Also in Brazil, the seroprevalence of the Enawene-Nawe Indians of Mato Grosso was 80.4% (out of 148 samples). This community is isolated from non-Indians. They do not keep domestic animals, including cats. Their diet is based on insects, cassava, corn, honey, mushrooms, and fish. They do not consume other meats. Seropositivity increased significantly with age from 50 to 95%. Wild felines are considered a source of *Toxoplasma* which would contaminate soil, insect, and mushrooms (Amendoeira *et al.*, 2003). A serologic survey for *Toxoplasma* was done in Ticuna Indians from five villages in western Brazil and was compared with non-Indian inhabitants of the town of Codajas, Amazonas. Seroprevalence was 39% in the Ticuna population and 77% in the Codajas population (Lovelace *et al.*, 1978).

In the Yucpa community in Venezuela, the seroprevalence was 63% in 94 individuals (ages 3 months to 100 years) (az-Suarez *et al.*, 2003); whereas in Amerindians (aged 1–69 years) the overall prevalence of infection was 49.7% (of 447). A higher antibody rate was found in lowland settings compared to the mountain areas. No age-antibody association was detected in the mountain communities contrary to the lowland setting (83.3% in the oldest group). The results suggest that transmission by infective cat feces plays a predominant role in the spread of infection in this population (Chacin-Bonilla *et al.*, 2001). Another study conducted on 121 Amerindians of the Guajibo ethnic group, 4 to 45 years of age, found the overall prevalence to be 88% (de la *et al.*, 1999).

The wide variation in humans is thought to be a result of cultural habits, environmental conditions, socioeconomic status, and proximity to animals. A steady increase in prevalence with age was noted in all surveys.

5.7.2 Swine

An average of 29% of pigs worldwide is estimated to have *Toxoplasma*. The distribution of the parasite varies according to various regions and farm management.

The regional prevalence of *T. gondii* in pigs from 85 farms in five New England states was 47.4%. Herd prevalence rate was 90.6%. Within the herd, the seroprevalence ranged from 4 to 100%. All farms studied had one or more risk factors for exposure to *T. gondii*, suggesting that education on farm management practices should be targeted to include small producers (Gamble *et al.*, 1999). In Swedish pigs in 1999, 5.2% of 807 meat juice samples collected from 10 abattoirs in different parts of the country were positive. The seroprevalence was 3.3% in fattening pigs and 17.3% in adult swine (Lunden *et al.*, 2002). In Spain, seroprevalence of hunter-killed wild pigs between 1993 and 2004 from five geographic regions in the north and seven regions in the south was 38.4%. Seroprevalence was higher in pigs from high stocking per hectare. Sex, age, or hunting conditions (open or fenced) were not associated with high seroprevalence of *Toxoplasma* (Gauss *et al.*, 2005). In Portugal, antibodies to *T. gondii* were found in 15.6% of 333 pigs prior to slaughter. Viable *T. gondii* was isolated from 15 of 37 pigs. Using the SAG2 -RFLP and microsatellite analysis, 11 isolates were Type II and 4 were Type III (Sousa *et al.*, 2005). In Austria, blood samples were obtained from 4697 pigs. During a period of 10 years, the infection rate was reduced from 13.7 to 0.9%. Prevalence in breeding sows decreased from 43.4 to 4.3% and in fattening pigs 12.2 to 0.8% (Edelhofer, 1994). Under the Dutch field trial "Integrated Quality Control (IQC) for finishing pigs," 120 farms and 3 slaughterhouses were studied. The *Toxoplasma* seroprevalence was 2.1% in 23,348 serum samples. Seropositive animals were found from the earliest days of the finishing period. Housing and farm management play an important role in the prevention of *Toxoplasma* (Berends *et al.*, 1991).

In Serbia, during June 2002 to 2003, the seroprevalence was 76.3% in 611 cattle, 84.5% in 511 sheep, and 28.9% in 605 pigs. The risk factors for cattle were small herd size and farm location in Western Serbia, while housing in stables with access to outside pens was protective. In sheep, an increased risk of infection was found in ewes from state-owned flocks vs. private flocks. In pigs, the risk of infection was highly increased in adult animals and in those from finishing type farms (Klun *et al.*, 2005).

Some studies were also done in Asia. In northwestern Taiwan, in 1998, the overall seroprevalence of *T. gondii* infection was 28.8% among slaughtered pigs. No significant difference in seroprevalence was observed between male and female pigs (Fan *et al.*, 2004). In 1994, in Sumatra Indonesia, the seropositivity in two slaughter houses varied from 3.6 to 9.2% (Inoue *et al.*, 2001).

In Africa, *Toxoplasma* was also studied in domestic pigs. In Zimbabwe, *T. gondii* antibodies were found in 9.3% of 97 domestic pigs, 36.8% of 19 elands, 11.9% of 67 sables, 0% of 3 warthogs, 0% of 3 bushpigs, 50% of 2 white rhinos, 5.6% of 18 buffalos, 14.5% of 69 wildebeest, and 10.5% of 19 elephants examined (Hove and Dubey, 1999). In Ghana, the overall seroprevalence in pigs was 39%, and at different geographical locations, varied from 30.5, 42.5, and 43.9%. The age of the animal, the breed, the environmental conditions, and the management practices appeared to be the major determinants of prevalence of antibodies against *T. gondii*. Seroprevalence was significantly higher in crossbreed pigs (46.8%) than the Large White breed (38.8%) (rko-Mensah *et al.*, 2000).

Federally inspected abattoirs in Canada during 1991-1992 were sampled. Seroprevalence of the 2800 market-age pigs ranged from 3.5 to 13.2% in the different regions of the country. *T. gondii* ribosomal RNA was identified in 9 of 36 animals, but mouse bioassay testing was negative in all pig muscle samples. This suggests that serological evidence of *T. gondii* infection in pigs alone does not accurately assess the public health risks associated with consuming improperly cooked pork products (Gajadhar et al., 1998).

In the United States, the prevalence of *Toxoplasma* during 1983-1984 was 23.9% in 11,842 commercial pigs. Seroprevalence was 42% in breeder pigs, whereas in market pigs it was 23% (Dubey et al., 1991). In Oahu, Hawaii, sera from 509 pigs from 31 farms were examined. *T. gondii* antibodies were found in 48.5% of pigs. The prevalence of *T. gondii* antibodies in garbage-fed pigs was 67.3% (of 199 pigs) and 33.8% in grain-fed pigs (of 180 pigs) (Dubey et al., 1992). In Iowa, using the SAG2 loci, 83.7% of the isolates from pigs were Type II genotype. The type III genotype was identified in only 16.3% of the isolates. The distribution of these genotypes was similar to those observed in humans, but was different from those previously reported in animals. The type I genotype was not identified in the isolates from pigs, although these strains have previously been shown to account for approximately 10–25% of toxoplasmosis cases in humans (Mondragon et al., 1998). In Montana, seropositive animals at 1:16 or higher were 13.2% of sheep, 5.0% of pigs, and 22.7% of goats. Using the MAT, 3.2% of cattle, 3.1% of bison were positive, and none of the elk were positive (Dubey, 1985). Viable *T. gondii* were isolated from hearts and tongues of 51 out of 55 pigs from a farm in Massachusetts (Dubey et al., 2002). In Ossabaw Island, Georgia; a remote, barrier island, antibodies to *T. gondii* were found in 0.9% of 1264 pigs from the island. Of 170 feral pigs from mainland Georgia, 18.2% were seropositive. The markedly low prevalence of *T. gondii* on Ossabaw Island was attributed to the virtual absence of cats; only 1 domestic cat was known to be present (Dubey et al., 1997).

In certain regions of South America, the prevalence of toxoplasmosis in pigs was estimated. In Brazil, antibodies to *T. gondii* were found in 17% of 286 pigs prior to slaughter. Viable *T. gondii* was isolated from seven out of 28 pigs. RFLP analysis using products of the SAG2 locus identified two isolates of Type I and five of Type III (de et al., 2005). In Sao Paulo Brazil, in 5-month-old pigs obtained at abattoirs, 9.6% were seropositive, which was lower than the same age animals in Lima, Peru (32.3%) (Suarez-Aranda et al., 2000). Another study showed a seroprevalence of 27.7% in 137 pigs at a slaughter house in Peru (Saavedra and Ortega, 2004). In Argentina, antibodies to *T. gondii* were detected in 37.8% of 230 slaughter sows belonging to 83 farms distributed in 5 provinces. Distribution among provinces varied from 3.3 to 62.8%. Monthly evaluation of pigs from an intensive management indoor farm demonstrated 4.5% seropositivity. A cross-sectional study in an outdoor farm demonstrated 40.2% seropositivity. This prevalence was related to the facilities and management of the farm (Venturini et al., 2004). Another study in Argentina from September 1991 to May 1992 demonstrated 11% seropositivity in 109 pigs at 1:1024 or higher serum dilutions, and 36.7% at 1:16 or less. *Toxoplasma* was isolated in 14 pigs using the mouse bioassay. The authors suggest that antibody production in infected pigs is apparently dependent on the pathogenicity of the parasite strain (Omata et al., 1994).

5.7.3 Poultry
Toxoplasma was isolated in 0.4% of 716 Croatian chicken brain tissues using the mouse bioassay (Kuticic and Wikerhauser, 2000). In Egypt, the seroprevalence of

Toxoplasma was 18.7% in 150 chickens. Of these, 10% in house-bred chickens and 11.1% of farm-bred chickens were positive. Tissue cysts of *T. gondii* were demonstrated in 78.6% of the positive chickens (Deyab and Hassanein, 2005). Further studies included not only in determining the seroprevalence, but also in determining the genotypes using the SAG2 locus, and their differences on virulence in various areas in the world. In the United States, the prevalence of *T. gondii* was determined in 118 free-range chickens from 14 counties in Ohio and in 11 chickens from a pig farm in Massachusetts. *T. gondii* antibodies were found in 20 of 118 chickens from Ohio. Viable parasites were isolated in 19 chickens and isolates were avirulent for mice. Five isolates were type II and 14 were type III (Dubey *et al.*, 2003b). In Granada, West Indies, 52% of 102 free-range chickens were seropositive for *Toxoplasma* and parasites were isolated from 36 chickens. All were avirulent for mice. Of these chicken, 29 were Type III, 5 were Type I, 1 was Type II, and 1 had both Type I and III (Dubey *et al.*, 2005). In Brazil, 16 of 40 free range chickens were seropositive from a rural area. Parasites were isolated in 81% of 16 seropositive chickens. Of these seven isolates were type I and six were type III (Dubey *et al.*, 2003c). In Argentina, 65% of 29 free-range chickens were seropositive for *Toxoplasma* and parasites were isolated from 9 of 19 seropositive chickens. One was type I, 1 was type II, and 7 were type III (Dubey *et al.*, 2003d). In Mexico, seroprevalence was 6.2% in 208 free-range chickens. *T. gondii* was isolated from 6 of 13 seropositive chickens. All were avirulent for mice, 5 were type III, and 1 was type I (Dubey *et al.*, 2004b). In a commercial farm in Israel, antibodies to *Toxoplasma* were found in 45 of 96 free-range chickens. *T. gondii* was isolated in 42.2% of seropositive chickens and of these, 17 were type II, and 2 were type III (Dubey *et al.*, 2004d). In a rural area surrounding Giza in Egypt, seroprevalence of *T. gondii* was 40.4% in 121 free range chickens and 15.8% of 19 ducks. Of 20 chicken isolates, 17 were type III and three were type II. The duck isolate of *T. gondii* was type III. None of the isolates were lethal for mice (Dubey *et al.*, 2003a). *Toxoplasma* has also been found in other bird and animal species. In the United States, *T. gondii* type III has been isolated from skunks, lories, and goose, and Type II has been isolated in cats. All Type III isolates were mouse virulent (Dubey *et al.*, 2004c). In Poland, the prevalence of *T. gondii* in chicks of wild birds and captive individuals was detected in 5.8% of 205 white stork chicks and 13.6% of 44 adult storks (Andrzejewska *et al.*, 2004). The high prevalence in chicken may be associated with chicken feed from the ground.

5.7.4 Sheep and Goats

Congenital transmission in pedigree Charollais and outbred sheep flocks has been reported. Overall rates of transmission per pregnancy, as determined by PCR based diagnosis, were consistent over time in a commercial sheep flock (69%) and in sympatric (60%) and allopatric (41%) populations of Charollais sheep. The result of this was that 53.7 % of lambs were acquiring an infection prior to birth: 46.4% of live lambs and 90.0% of dead lambs (Williams *et al.*, 2005). In Worcestershire, UK, sheep flocks were examined. Significant differences in the frequency of abortion between sheep families ranged between 0% and 48%, and infection frequencies with *T. gondii* for different families varied between 0% and 100% (Morley *et al.*, 2005). In Morocco, 27.6% of 261 sheep intended for consumption in Marrakech were seropositive for IgG specific anti-*Toxoplasma* (Sawadogo *et al.*, 2005). In the southeastern region of Brazil 34.7% of the samples were seropositive (Figliuolo *et al.*, 2004). In Italy, during the period 1999-2002, specific IgG antibodies were detected in 2048 (28.4%) sheep and 302 (12.3%) goats, and specific IgM antibodies were found in 652 (9%) sheep and 139 (5.6%) goats. From a total of 2471 ovine and 362 caprine fetal samples, 271 (11.1%) ovine and 23 (6.4%) caprine samples were positive by PCR (Masala *et al.*, 2003). The seroprevalence of antibodies to *T. gondii* in goats of Satun Province in Thailand was 27.9% in 631 goats. Female goats were 1.73 times more likely than male to be seropositive and dairy goats were more seropositive than meat goats (Jittapalapong *et al.*, 2005).

Serum samples from 4339 wild cervids collected in Norway were tested for antibodies against *T. gondii* using the direct agglutination test. Positive titers were found in 33.9% of 760 roe deer, 12.6% of 2142 moose, 7.7% of 571 red deer, and 1.0% of 866 reindeer. Significant factors such as age were relevant in roe deer, moose, and red deer. Sex was significant in moose, but not for roe deer or red deer, and geographic regions were significant in only roe deer and male moose (Vikoren *et al.*, 2004). In the United States, *T. gondii* was isolated from white-tailed deer from Mississippi, raccoons, bobcats, gray fox, red fox, coyote from Georgia, and black bears from Pennsylvania. All three genotypes of *T. gondii* based on the SAG2 locus were circulating among wildlife (Dubey *et al.*, 2004a). Llamas in the Peruvian Andean region also have been seropositive to *Toxoplasma*. Using the IFAT assay, 55.8% of 43 llamas and 5.5% of 200 vicunas tested positive (Chavez-Velasquez *et al.*, 2005).

5.7.5 Other Animal Species

Animals other than pigs have also been studied to determine their role in parasite transmission to humans. Serologic surveys indicate that *T. gondii* infections are common worldwide from Alaska to Australia in wild carnivores, including pigs, bears, felids, fox, raccoons, and skunks. Clinical and subclinical toxoplasmosis has been reported from wild cervids, ungulates, marsupials, monkeys, and marine mammals (Hill *et al.*, 2005).

Overall, *Toxoplasma* in cattle is suspected to be 25% worldwide. *Toxoplasma* parasites present in milk does not seem to be a high risk factor in parasite transmission. Chickens can get infected with *Toxoplasma*, but chickens are not usually eaten raw; therefore the risk of acquiring the infection is reduced. Although seroprevalence in animals is high, infectious parasites have been demonstrated in a few animal species; including swine and chickens. Food-borne outbreaks following ingestion of raw meats were described in France (Choi *et al.*, 1997; Vaillant *et al.*, 2005). Few cases associated with drinking unpasteurized goat milk (Sacks *et al.*, 1982), and eating raw meats and organs from wild boars, seal, caribou, and lamb have been described.

Sera was obtained from 12,628 clinically ill, client-owned cats in the United States. Overall, 31.6% of the cats were seropositive for *T. gondii*-specific IgM, IgG, or both. Seroprevalence increases as cats age and is higher in male and domestic shorthair cats, compared with females and other breeds (Vollaire *et al.*, 2005). In Brazil, antibodies to *T. gondii* were found in 40% of stray cats, and 50.5% in stray dogs (Meireles *et al.*, 2004). In Austria, using the IFAT, 35% of foxes and 26% of the dogs examined were positive (Wanha et al., 2005). In the UK, lung fluid from over 549 foxes was examined using IFAT. Of these, 20% were seropositive to *T. gondii* (Hamilton et al., 2005). In Southern Argentina, 20% of 84 free-ranging foxes had antibodies to *Toxoplasma* (Martino et al., 2004).

Marine mammals are also susceptible to *Toxoplasma* infection. Using an IFA, 36% of 80 California sea otters and 38% of 21 Washington sea otters examined were seropositive for *T. gondii*. None of 65 Alaskan sea otters examined had antibodies to *Toxoplasma* (Miller et al., 2002). Another study reported *T. gondii* in 77% of 115 dead, and in 60% of 30 apparently healthy sea otters, in 16% of 311 Pacific harbor seals, 42% of 45 sea lions, 16% of 32 ringed seals, and 50% of 8 bearded seals, 11.1% of 9 spotted seals, 98% of 141 Atlantic bottlenose dolphins, and 6% of 53 walruses (Dubey et al., 2003e).

Toxoplasma gondii antibodies were present in 13.9% of 961 Polish farmed mink. On large farms, the seropositivity was lower (2.9%), than on small farms (26.33%). On farms feeding fish, percentage of seropositivity was lower (2.2%), than on farms based on non-frozen slaughter offal (43.4%) (Smielewska-Los and Turniak, 2004).

Rodents have also been reported as having *Toxoplasma* antibodies. In 456 wild rabbits, the prevalence was 14.2%. Prevalence of infection was significantly higher in wild rabbits from northeast Spain (53.8%), where rabbits lived in a forest. In other areas with drier conditions, prevalence ranged from 6.1 to 14.6% (Almeria et al., 2004). Capybaras, the largest rodent used for meat in South and Central America were seropositive by IFAT (69.8%) and with the MAT 63 (42.3%) (Canon-Franco et al., 2003). Raccoons from Fairfax County, Virginia were also surveyed. Out of 256 racoons, 84.4% had been exposed to *T. gondii* (Hancock et al., 2005). In Sao Paulo Brazil, antibodies to *T. gondii* were found in 82 (20.4%) of the 396 opossums using the MAT assay, and using the IFAT 148 of 396 were positive (Yai et al., 2003).

In Zimbabwe, wild animals have also been examined for the presence of *Toxoplasma* antibodies. Significantly high seroprevalence were found in the felidae (92% of 26), bovidae (55.9% of 34), and farm-reared struthionidae (48% of 50). The nyala had the highest seroprevalence at 90% (9/10). Low anti-*Toxoplasma* antibody prevalence was found in greater kudu (20% of 10), giraffe (10% of 10), and elephant (10% of 20). No antibodies were detected in the wild African suidae and bushpig (Hove and Mukaratirwa, 2005). In Thailand, 45.5% of 156 captive elephants were positive by MAT. In the same region, 14.09% of 447 dairy cattle on 14 dairy farms were also positive for *Toxoplasma*. Coinfections of *Neospora* and *Toxoplasma* were identified in 4.76% of the cattle (Gondim et al., 1999).

5.8 TREATMENT

Individuals with acute toxoplasmosis and congenital infections can be treated with pyrimethamine and sulfadiazine. These drugs are effective with tachyzoites, but not with the bradizoites in mature cysts. Spiramycin and clindamicin are effective, but have serious side effects (Alves and Vitor, 2005; Chakraborty et al., 1997; Djurkovic-Djakovic et al., 2005; guirre-Cruz et al., 1998; Lescano et al., 2004; Schmidt et al., 2005; Sordet et al., 1998; Tabbara et al., 2005).

Five hundred forty women of child bearing age with still births and spontaneous abortions in their obstetrical history were tested serologically for anti-*Toxoplasma* antibody using microlatex agglutination test. Maximum prevalence (10.2%) and highest titer of anti-*Toxoplasma* antibodies were observed in women of 35–42 years age group. The overall prevalence of toxoplasmosis in these women was 7.7%. Seropositive pregnant women were treated using a combined regimen of sulfadiazine and pyrimethamine. Incidence of toxoplasmosis in women is low because of infrequent and uncommon practices, such as a substantial number of the population surveyed ingested undercooked or uncooked food stuff, especially meat (Chakraborty *et al.*, 1997).

Mice intraperitonealy inoculated with *Toxoplasma* tachizoites were treated with nifurtimox alone or in combination with pyrimethamine. Nifurtimox alone was not significantly effective against murine toxoplasmosis. However, when combined with pyrimethamine, a strong anti-*Toxoplasma* effect was obtained in comparison with survival rates associated with pyrimethamine or nifurtimox alone (guirre-Cruz *et al.*, 1998).

The *in vitro* activity of atovaquone-loaded nanocapsules against tachyzoites of *T. gondii* was comparable to atovaquone suspension form. The sensibility of *T. gondii* to atovaquone varies according to the strains, and the activity of atovaquone in the treatment of toxoplasmosis is enhanced when administered in nanoparticular form (Sordet *et al.*, 1998).

The efficacy of prolonged administration of azithromycin and pyrimethamine was evaluated in mice experimentally infected with a cystogenic strain of *T. gondii*. Mice started an oral treatment of 120 days, 20 days post infection. The association of both drugs provided the best results by diminishing the cyst count in the brain of the animals (Lescano *et al.*, 2004).

The tolerability and efficacy of pyrimethamine and sulfadiazine in children with congenital toxoplasmosis was evaluated. Anemia or thrombocytopenia was not observed in treated children; however, progression of eye lesions was observed during the follow-up period. Although treatment was well tolerated in 86% (25/29) of the children and did not affect their weight gain, drug effectiveness at recommended concentrations was limited (Schmidt *et al.*, 2005).

5.9 INACTIVATION

Cats, mice, and chickens have been used to determine infectivity and viability of *Toxoplasma* (Hellesnes and Mohn, 1977; Hiramoto *et al.*, 2001; Lindsay *et al.*, 2005; Piao *et al.*, 2005). Whether this infectivity is selective to certain genotypes is still under study. *Toxoplasma* tachyzoites can be propagated using the MRC-5 cell line and most other fibroblast cell lines.

Cysts produced in mouse brains were used to experimentally spike milk and prepare homemade cheese. Cysts were infectious for 20 days at refrigeration temperature and survived the production process of homemade fresh cheese and storage for a period of 10 days. These findings support the importance of milk pasteurization before any processing or ingestion (Hiramoto *et al.*, 2001).

High pressure processing (HPP) is an effective non-thermal method of eliminating non-spore forming bacteria. *Toxoplasma* oocysts were exposed from 100 to 550 MPa for 1 min in the HPP unit. Oocysts treated with 550 to 340 MPa were rendered noninfectious for mice. These results suggest that HPP technology may be useful in the removal of *T. gondii* oocysts from food products (Lindsay *et al.*, 2005).

Oocysts remain viable when stored at 10–25°C for 200 days. At 35°C, oocysts were infective for 32 days, but not for 62 days; at 40°C they were still infective for 9 days, but not for 28 days. At 45°C, oocysts were non-infectious for a 2 day incubation. At 60°C, oocysts were rendered non infectious after 1 min. At 4°C, oocysts can be infectious for up to 54 months. At freezing temperatures, oocysts were still infectious at −5°C and −10°C after 106 days of storage. Sporulated oocysts are highly resistant and can survive freezing at −20°C. Freezing to −12°C and cooking to an internal temperature of 67°C can kill *Toxoplasma* cysts in meats (Dubey, 1996).

Unsporulated oocysts irradiated at 0.4 to 0.8 kGy sporulated, but were not infective to mice. Sporulated oocysts irradiated at ≥ 0.4 kGy were able to excyst, and sporozoites were infective, but not capable of inducing a viable infection in mice. *T. gondii* was detected in histologic sections of mice up to 5 days, but not at 7 days after feeding oocysts irradiated at 0.5 kGy. Raspberries inoculated with sporulated *T. gondii* oocysts were rendered innocuous after irradiation at 0.4 kGy (Dubey *et al.*, 1998).

Toxoplasma gondii cysts were stored at −21°C for various periods of time and then inoculated into mice. Parasite cysts were rendered inactive only after freezing for 5 h or longer (Hellesnes and Mohn, 1977).

Pigs acquire *Toxoplasma* infection more commonly after birth than via transplacental infection. In most pigs, toxoplasmosis presents subclinically, whereas in young pigs clinical toxoplasmosis is observed. Tissue cysts persist in brain, heart, and tongue for several months (Dubey, 1986).

REFERENCES

Akoijam, B. S., Shashikant, Singh, S., and Kapoor, S. K., 2002, Seroprevalence of *Toxoplasma* infection among primigravid women attending antenatal clinic at a secondary level hospital in North India, *J. Indian Med. Assoc.* **100**:591–596, 602.

Almeria, S., Calvete, C., Pages, A., Gauss, C., and Dubey, J. P., 2004, Factors affecting the seroprevalence of *Toxoplasma gondii* infection in wild rabbits (Oryctolagus cuniculus) from Spain, *Vet. Parasitol.* **123**:265–270.

Alves, C. F., and Vitor, R. W., 2005, Efficacy of atovaquone and sulfadiazine in the treatment of mice infected with *Toxoplasma gondii* strains isolated in Brazil, *Parasite* **12**:171–177.

Amendoeira, M. R., Sobral, C. A., Teva, A., de Lima, J. N., and Klein, C. H., 2003, Serological survey of *Toxoplasma gondii* infection in isolated Amerindians, Mato Grosso, *Rev. Soc. Bras. Med. Trop.* **36**:671–676.

Andrzejewska, I., Tryjanowski, P., Zduniak, P., Dolata, P. T., Ptaszyk, J., and Cwiertnia, P., 2004, *Toxoplasma gondii* antibodies in the white stork Ciconia ciconia, *Berl. Munch. Tierarztl. Wschr.* **117**:274–275.

Aramini, J. J., Stephen, C., Dubey, J. P., Engelstoft, C., Schwantje, H., and Ribble, C. S., 1999, Potential contamination of drinking water with *Toxoplasma gondii* oocysts, *Epidemiol. Infect.* **122**:305-315.

Arkush, K. D., Miller, M. A., Leutenegger, C. M., Gardner, I. A., Packham, A. E., Heckeroth, A. R., Tenter, A. M., Barr, B. C., and Conrad, P. A., 2003, Molecular and bioassay-based detection of *Toxoplasma gondii* oocyst uptake by mussels (Mytilus galloprovincialis), *Int. J. Parasitol.* **33**:1087-1097.

Ashrafunnessa, Khatun, S., Islam, M. N., and Huq, T., 1998, Seroprevalence of *Toxoplasma* antibodies among the antenatal population in Bangladesh, *J. Obstet. Gynaecol. Res.* **24**:115-119.

Aspinall, T. V., Guy, E. C., Roberts, K. E., Joynson, D. H., Hyde, J. E., and Sims, P. F., 2003, Molecular evidence for multiple *Toxoplasma gondii* infections in individual patients in England and Wales: Public health implications, *Int. J. Parasitol.* **33**:97-103.

az-Suarez, O., Estevez, J., Garcia, M., Cheng-Ng, R., Araujo, J., and Garcia, M., 2003, Seroepidemiology of toxoplasmosis in a Yucpa Amerindian community of Sierra de Perija, Zulia State, Venezuela, *Rev. Med. Chil.* **131**:1003-1010.

Bahia-Oliveira, L. M., Jones, J. L., zevedo-Silva, J., Alves, C. C., Orefice, F., and Addiss, D. G., 2003, Highly endemic, waterborne toxoplasmosis in north Rio de Janeiro state, Brazil, *Emerg. Infect. Dis.* **9**:55-62.

Bartova, E., and Literak, I., 2004, K24 *T. gondii* isolate is a hybrid and has the virulence of lineage I isolates, *Parasite* **11**:183-188.

Berends, B. R., Smeets, J. F., Harbers, A. H., van, K. F., and Snijders, J. M., 1991, Investigations with enzyme-linked immunosorbent assays for Trichinella spiralis and *Toxoplasma gondii* in the Dutch "Integrated Quality Control for finishing pigs" research project, *Vet. Q.* **13**:190-198.

Bessieres, M. H., Le, B. S., and Seguela, J. P., 1992, Analysis by immunoblotting of *Toxoplasma gondii* exo-antigens and comparison with somatic antigens, *Parasitol. Res.* **78**:222-228.

Boothroyd, J. C., and Grigg, M. E., 2002, Population biology of *Toxoplasma gondii* and its relevance to human infection: Do different strains cause different disease? *Curr. Opin. Microbiol.* **5**:438-442.

Brindley, P. J., Gazzinelli, R. T., Denkers, E. Y., Davis, S. W., Dubey, J. P., Belfort, R., Jr., Martins, M. C., Silveira, C., Jamra, L., and Waters, A. P., 1993, Differentiation of *Toxoplasma gondii* from closely related coccidia by riboprint analysis and a surface antigen gene polymerase chain reaction, *Am. J. Trop. Med. Hyg.* **48**:447-456.

Buchbinder, S., Blatz, R., and Rodloff, A. C., 2003, Comparison of real-time PCR detection methods for B1 and P30 genes of *Toxoplasma gondii*, *Diagn. Microbiol. Infect. Dis.* **45**:269-271.

Canon-Franco, W. A., Yai, L. E., Joppert, A. M., Souza, C. E., D'Auria, S. R., Dubey, J. P., and Gennari, S. M., 2003, Seroprevalence of *Toxoplasma gondii* antibodies in the rodent capybara (Hidrochoeris hidrochoeris) from Brazil, *J. Parasitol.* **89**:850.

Carme, B., Bissuel, F., Ajzenberg, D., Bouyne, R., Aznar, C., Demar, M., Bichat, S., Louvel, D., Bourbigot, A. M., Peneau, C., Neron, P., and Darde, M. L., 2002, Severe acquired toxoplasmosis in immunocompetent adult patients in French Guiana, *J. Clin. Microbiol.* **40**:4037-4044.

Chacin-Bonilla, L., Sanchez-Chavez, Y., Monsalve, F., and Estevez, J., 2001, Seroepidemiology of toxoplasmosis in amerindians from western Venezuela, *Am. J. Trop. Med. Hyg.* **65**:131-135.

Chakraborty, P., Sinha, S., Adhya, S., Chakraborty, G., and Bhattacharya, P., 1997, Toxoplasmosis in women of child bearing age and infant follow up after in-utero treatment, *Indian J. Pediatr.* **64**:879-882.

Chavez-Velasquez, A., varez-Garcia, G., Gomez-Bautista, M., Casas-Astos, E., Serrano-Martinez, E., and Ortega-Mora, L. M., 2005, *Toxoplasma gondii* infection in adult llamas (Lama glama) and vicunas (Vicugnavicugna) in the Peruvian Andean region, *Vet. Parasitol.* **130**:93–97.

Choi, W. Y., Nam, H. W., Kwak, N. H., Huh, W., Kim, Y. R., Kang, M. W., Cho, S. Y., and Dubey, J. P., 1997, Food-borne outbreaks of human toxoplasmosis, *J. Infect. Dis.* **175**:1280–1282.

Condorelli, F., Scalia, G., Stivala, A., Costanzo, M. C., Adragna, A. D., Franceschino, C., Santagati, M. G., Furneri, P. M., Marino, A., and Castro, A., 1993, Seroprevalence to some TORCH agents in a Sicilian female population of fertile age, *Eur. J. Epidemiol.* **9**:341–343.

Contini, C., Seraceni, S., Cultrera, R., Incorvaia, C., Sebastiani, A., and Picot, S., 2005, Evaluation of a Real-time PCR-based assay using the lightcycler system for detection of *Toxoplasma gondii* bradyzoite genes in blood specimens from patients with toxoplasmic retinochoroiditis, *Int. J. Parasitol.* **35**:275–283.

Coutinho, S. G., Lobo, R., and Dutra, G., 1982, Isolation of *Toxoplasma* from the soil during an outbreak of toxoplasmosis in a rural area in Brazil, *J. Parasitol.* **68**:866–868.

de la, R. M., Bolivar, J., and Perez, H. A., 1999, *Toxoplasma gondii* infection in Amerindians of Venezuelan Amazon, *Medicina (B Aires)* **59**:759–762.

de, A. D. S. C., de Carvalho, A. C., Ragozo, A. M., Soares, R. M., Amaku, M., Yai, L. E., Dubey, J. P., and Gennari, S. M., 2005, First isolation and molecular characterization of *Toxoplasma gondii* from finishing pigs from Sao Paulo State, Brazil, *Vet. Parasitol.* **131**:207–211.

Deyab, A. K., and Hassanein, R., 2005, Zoonotic toxoplasmosis in chicken, *J. Egypt. Soc. Parasitol.* **35**:341–350.

Diza, E., Frantzidou, F., Souliou, E., Arvanitidou, M., Gioula, G., and Antoniadis, A., 2005, Seroprevalence of *Toxoplasma gondii* in northern Greece during the last 20 years, *Clin. Microbiol. Infect.* **11**:719–723.

Djurkovic-Djakovic, O., Nikolic, A., Bobic, B., Klun, I., and Aleksic, A., 2005, Stage conversion of *Toxoplasma gondii* RH parasites in mice by treatment with atovaquone and pyrrolidine dithiocarbamate, *Microbes. Infect.* **7**:49–54.

Dubey, J. P., 1985, Serologic prevalence of toxoplasmosis in cattle, sheep, goats, pigs, bison, and elk in Montana, *J. Am. Vet. Med. Assoc.* **186**:969–970.

Dubey, J. P., 1986, A review of toxoplasmosis in pigs, *Vet. Parasitol.* **19**:181–223.

Dubey, J. P., 1991, Toxoplasmosis–an overview, *Southeast Asian J. Trop. Med. Public Health*, **22**(Suppl):88–92.

Dubey, J. P., 1996, Strategies to reduce transmission of *Toxoplasma gondii* to animals and humans, *Vet. Parasitol.* **64**:65–70.

Dubey, J. P., 2004, Toxoplasmosis–a waterborne zoonosis, *Vet. Parasitol.* **126**:57–72.

Dubey, J. P., Leighty, J. C., Beal, V. C., Anderson, W. R., Andrews, C. D., and Thulliez, P., 1991, National seroprevalence of *Toxoplasma gondii* in pigs, *J. Parasitol.* **77**:517–521.

Dubey, J. P., Gamble, H. R., Rodrigues, A. O., and Thulliez, P., 1992, Prevalence of antibodies to *Toxoplasma gondii* and Trichinella spiralis in 509 pigs from 31 farms in Oahu, Hawaii, *Vet. Parasitol.* **43**:57–63.

Dubey, J. P., Weigel, R. M., Siegel, A. M., Thulliez, P., Kitron, U. D., Mitchell, M. A., Mannelli, A., Mateus-Pinilla, N. E., Shen, S. K., and Kwok, O. C., 1995, Sources and reservoirs of *Toxoplasma gondii* infection on 47 swine farms in Illinois, *J. Parasitol.* **81**:723–729.

Dubey, J. P., Rollor, E. A., Smith, K., Kwok, O. C., and Thulliez, P., 1997, Low seroprevalence of *Toxoplasma gondii* in feral pigs from a remote island lacking cats, *J. Parasitol.* **83**:839–841.

Dubey, J. P., Thayer, D. W., Speer, C. A., and Shen, S. K., 1998, Effect of gamma irradiation on unsporulated and sporulated *Toxoplasma gondii* oocysts, *Int. J. Parasitol.* **28**:369–375.

Dubey, J. P., Gamble, H. R., Hill, D., Sreekumar, C., Romand, S., and Thuilliez, P., 2002, High prevalence of viable *Toxoplasma gondii* infection in market weight pigs from a farm in Massachusetts, *J. Parasitol.* **88**:1234–1238.

Dubey, J. P., Graham, D. H., Dahl, E., Hilali, M., El-Ghaysh, A., Sreekumar, C., Kwok, O. C., Shen, S. K., and Lehmann, T., 2003a, Isolation and molecular characterization of *Toxoplasma gondii* from chickens and ducks from Egypt, *Vet. Parasitol.* **114**: 89–95.

Dubey, J. P., Graham, D. H., Dahl, E., Sreekumar, C., Lehmann, T., Davis, M. F., and Morishita, T. Y., 2003b, *Toxoplasma gondii* isolates from free-ranging chickens from the United States, *J. Parasitol.* **89**:1060–1062.

Dubey, J. P., Navarro, I. T., Graham, D. H., Dahl, E., Freire, R. L., Prudencio, L. B., Sreekumar, C., Vianna, M. C., and Lehmann, T., 2003c, Characterization of *Toxoplasma gondii* isolates from free range chickens from Parana, Brazil, *Vet. Parasitol.* **117**:229–234.

Dubey, J. P., Venturini, M. C., Venturini, L., Piscopo, M., Graham, D. H., Dahl, E., Sreekumar, C., Vianna, M. C., and Lehmann, T., 2003d, Isolation and genotyping of *Toxoplasma gondii* from free-ranging chickens from Argentina, *J. Parasitol.* **89**:1063–1064.

Dubey, J. P., Zarnke, R., Thomas, N. J., Wong, S. K., Van, B. W., Briggs, M., Davis, J. W., Ewing, R., Mense, M., Kwok, O. C., Romand, S., and Thulliez, P., 2003e, *Toxoplasma gondii*, Neospora caninum, Sarcocystis neurona, and Sarcocystis canis-like infections in marine mammals, *Vet. Parasitol.* **116**:275–296.

Dubey, J. P., Graham, D. H., De Young, R. W., Dahl, E., Eberhard, M. L., Nace, E. K., Won, K., Bishop, H., Punkosdy, G., Sreekumar, C., Vianna, M. C., Shen, S. K., Kwok, O. C., Sumners, J. A., Demarais, S., Humphreys, J. G., and Lehmann, T., 2004a, Molecular and biologic characteristics of *Toxoplasma gondii* isolates from wildlife in the United States, *J. Parasitol.* **90**:67–71.

Dubey, J. P., Morales, E. S., and Lehmann, T., 2004b, Isolation and genotyping of *Toxoplasma gondii* from free-ranging chickens from Mexico, *J. Parasitol.* **90**:411–413.

Dubey, J. P., Parnell, P. G., Sreekumar, C., Vianna, M. C., De Young, R. W., Dahl, E., and Lehmann, T., 2004c, Biologic and molecular characteristics of *Toxoplasma gondii* isolates from striped skunk (Mephitis mephitis), Canada goose (Branta canadensis), black-winged lory (Eos cyanogenia), and cats (Felis catus), *J. Parasitol.* **90**:1171–1174.

Dubey, J. P., Salant, H., Sreekumar, C., Dahl, E., Vianna, M. C., Shen, S. K., Kwok, O. C., Spira, D., Hamburger, J., and Lehmann, T. V., 2004d, High prevalence of *Toxoplasma gondii* in a commercial flock of chickens in Israel, and public health implications of free-range farming, *Vet. Parasitol.* **121**:317–322.

Dubey, J. P., Bhaiyat, M. I., de, A. C., Macpherson, C. N., Sharma, R. N., Sreekumar, C., Vianna, M. C., Shen, S. K., Kwok, O. C., Miska, K. B., Hill, D. E., and Lehmann, T., 2005, Isolation, tissue distribution, and molecular characterization of *Toxoplasma gondii* from chickens in Grenada, West Indies, *J. Parasitol.* **91**:557–560.

Dumas, N., Cazaux, M., Meunier, D. M., Seguela, J. P., and Charlet, J. P., 1990, Toxoplasmosis in the Central African Republic. Complementary study in a rural area, *Bull. Soc. Pathol. Exot.* **83**:342–348.

Edelhofer, R., 1994, Prevalence of antibodies against *Toxoplasma gondii* in pigs in Austria—an evaluation of data from 1982 and 1992, *Parasitol. Res.* **80**:642–644.

Fan, C. K., Su, K. E., Wu, G. H., and Chiou, H. Y., 2002, Seroepidemiology of *Toxoplasma gondii* infection among two mountain aboriginal populations and Southeast Asian laborers in Taiwan, *J. Parasitol.* **88**:411–414.

Fan, C. K., Su, K. E., and Tsai, Y. J., 2004, Serological survey of *Toxoplasma gondii* infection among slaughtered pigs in northwestern Taiwan, *J. Parasitol.* **90**:653–654.

Fazaeli, A., Carter, P. E., Darde, M. L., and Pennington, T. H., 2000, Molecular typing of *Toxoplasma gondii* strains by GRA6 gene sequence analysis, *Int. J. Parasitol.* **30**:637–642.

Figliuolo, L. P., Kasai, N., Ragozo, A. M., de, P., V, Dias, R. A., Souza, S. L., and Gennari, S. M., 2004, Prevalence of anti-*Toxoplasma gondii* and anti-Neospora caninum antibodies in ovine from Sao Paulo State, Brazil, *Vet. Parasitol.* **123**:161–166.

Flatau, E., Nishri, Z., Mates, A., Qupty, G., Reichman, N., and Raz, R., 1993, Seroprevalence of antibodies against *Toxoplasma gondii* among recently immigrating Ethiopian Jews, *Isr. J. Med. Sci.* **29**:395–397.

Fuentes, I., Rubio, J. M., Ramirez, C., and Alvar, J., 2001, Genotypic characterization of *Toxoplasma gondii* strains associated with human toxoplasmosis in Spain: Direct analysis from clinical samples, *J. Clin. Microbiol.* **39**:1566–1570.

Furuta, T., Une, Y., Omura, M., Matsutani, N., Nomura, Y., Kikuchi, T., Hattori, S., and Yoshikawa, Y., 2001, Horizontal transmission of *Toxoplasma gondii* in squirrel monkeys (Saimiri sciureus), *Exp. Anim.* **50**:299–306.

Gajadhar, A. A., Aramini, J. J., Tiffin, G., and Bisaillon, J. R., 1998, Prevalence of *Toxoplasma gondii* in Canadian market-age pigs, *J. Parasitol.* **84**:759–763.

Gamble, H. R., Brady, R. C., and Dubey, J. P., 1999, Prevalence of *Toxoplasma gondii* infection in domestic pigs in the New England states, *Vet. Parasitol.* **82**:129–136.

Gascon, J., Torres-Rodriguez, J. M., Soldevila, M., and Merlos, A. M., 1989, Seroepidemiology of toxoplasmosis in 2 communities of Rwanda (Central Africa), *Rev. Inst. Med. Trop. Sao Paulo*, **31**:399–402.

Gauss, C. B., Dubey, J. P., Vidal, D., Ruiz, F., Vicente, J., Marco, I., Lavin, S., Gortazar, C., and Almeria, S., 2005, Seroprevalence of *Toxoplasma gondii* in wild pigs (Sus scrofa) from Spain, *Vet. Parasitol.* **131**:151–156.

Gill, H. S., and Mtimavalye, L. A., 1982, Prevalence of *Toxoplasma* antibodies in pregnant African women in Tanzania, *Afr. J. Med. Med. Sci.* **11**:167–170.

Gondim, L. F., Sartor, I. F., Hasegawa, M., and Yamane, I., 1999, Seroprevalence of Neospora caninum in dairy cattle in Bahia, Brazil, *Vet. Parasitol.* **86**:71–75.

Guerra, G. C., and Fernandez, S. J., 1995, Seroprevalence of *Toxoplasma gondii* in pregnant women, *Aten. Primaria* **16**:151–153.

guirre-Cruz, L., Velasco, O., and Sotelo, J., 1998, Nifurtimox plus pyrimethamine for treatment of murine toxoplasmosis, *J. Parasitol.* **84**:1032–1033.

Gutierrez, J., Roldan, C., and Maroto, M. C., 1996, Seroprevalence of human toxoplasmosis, *Microbios* **85**:73–75.

Hamilton, C. M., Gray, R., Wright, S. E., Gangadharan, B., Laurenson, K., and Innes, E. A., 2005, Prevalence of antibodies to *Toxoplasmagondii* and Neosporacaninum in red foxes (Vulpesvulpes) from around the UK, *Vet. Parasitol.* **130**:169–173.

Hancock, K., Thiele, L. A., Zajac, A. M., Elvingert, F., and Lindsay, D. S., 2005, Prevalence of antibodies to *Toxoplasma gondii* in raccoons (Procyon lotor) from an urban area of Northern Virginia, *J. Parasitol.* **91**:694–695.

Hellesnes, I., and Mohn, S. F., 1977, Effects of freezing on the infectivity of *Toxoplasma gondii* cysts for white mice, *Zentralbl. Bakteriol. [Orig. A]* **238**:143–148.

Hill, D. E., Chirukandoth, S., and Dubey, J. P., 2005, Biology and epidemiology of *Toxoplasma gondii* in man and animals, *Anim. Health Res. Rev.* **6**:41–61.

Hiramoto, R. M., Mayrbaurl-Borges, M., Galisteo, A. J., Jr., Meireles, L. R., Macre, M. S., and Andrade, H. F., Jr., 2001, Infectivity of cysts of the ME-49 *Toxoplasma gondii* strain in bovine milk and homemade cheese, *Rev. Saude Publica* **35**:113–118.

Homan, W. L., Vercammen, M., De, B. J., and Verschueren, H., 2000, Identification of a 200- to 300-fold repetitive 529 bp DNA fragment in *Toxoplasma gondii*, and its use for diagnostic and quantitative PCR, *Int. J. Parasitol.* **30**:69–75.

Hove, T., and Dubey, J. P., 1999, Prevalence of *Toxoplasma gondii* antibodies in sera of domestic pigs and some wild game species from Zimbabwe, *J. Parasitol.* **85**:372–373.

Hove, T., and Mukaratirwa, S., 2005, Seroprevalence of *Toxoplasma gondii* in farm-reared ostriches and wild game species from Zimbabwe, *Acta Trop.* **94**:49–53.

Howe, D. K., and Sibley, L. D., 1995, *Toxoplasma gondii* comprises three clonal lineages: Correlation of parasite genotype with human disease, *J. Infect. Dis.* **172**:1561–1566.

Inoue, I., Leow, C. S., Husin, D., Matsuo, K., and Darmani, P., 2001, A survey of *Toxoplasma gondii* antibodies in pigs in Indonesia, *Southeast Asian J. Trop. Med. Public Health* **32**:38–40.

Isaac-Renton, J., Bowie, W. R., King, A., Irwin, G. S., Ong, C. S., Fung, C. P., Shokeir, M. O., and Dubey, J. P., 1998, Detection of *Toxoplasma gondii* oocysts in drinking water, *Appl. Environ. Microbiol.* **64**:2278–2280.

Jacobs, M. R., and Mason, P. R., 1978, Prevalence of *Toxoplasma* antibodies in Southern Africa, *S. Afr. Med. J.* **53**:619–621.

Jacquier, P., Hohlfeld, P., Vorkauf, H., and Zuber, P., 1995, Epidemiology of toxoplasmosis in Switzerland: National study of seroprevalence monitored in pregnant women 1990–1991, *Schweiz. Med. Wochenschr. Suppl.* **65**:29S–38S.

Jalal, S., Nord, C. E., Lappalainen, M., and Evengard, B., 2004, Rapid and sensitive diagnosis of *Toxoplasma gondii* infections by PCR, *Clin. Microbiol. Infect.* **10**:937–939.

Jittapalapong, S., Sangvaranond, A., Pinyopanuwat, N., Chimnoi, W., Khachaeram, W., Koizumi, S., and Maruyama, S., 2005, Seroprevalence of *Toxoplasma gondii* infection in domestic goats in Satun Province, Thailand, *Vet. Parasitol.* **127**:17–22.

Jones, J. L., Kruszon-Moran, D., Wilson, M., McQuillan, G., Navin, T., and McAuley, J. B., 2001, *Toxoplasma gondii* infection in the United States: Seroprevalence and risk factors, *Am. J. Epidemiol.* **154**:357–365.

Jones, J. L., Kruszon-Moran, D., and Wilson, M., 2003, *Toxoplasma gondii* infection in the United States, 1999-2000, *Emerg. Infect. Dis.* **9**:1371–1374.

Klun, I., Djurkovic-Djakovic, O., Katic-Radivojevic, S., and Nikolic, A., 2005, Cross-sectional survey on *Toxoplasma gondii* infection in cattle, sheep, and pigs in Serbia: Seroprevalence and risk factors, *Vet. Parasitol.* **135**:121–131.

Kortbeek, L. M., De Melker, H. E., Veldhuijzen, I. K., and Conyn-Van Spaendonck, M. A., 2004, Population-based *Toxoplasma* seroprevalence study in The Netherlands, *Epidemiol. Infect.* **132**:839–845.

Kourenti, C., and Karanis, P., 2004, Development of a sensitive polymerase chain reaction method for the detection of *Toxoplasma gondii* in water, *Water Sci. Technol.* **50**:287–291.

Kuticic, V., and Wikerhauser, T., 2000, A survey of chickens for viable toxoplasms in Croatia, *Acta Vet. Hung.* **48**:183–185.

Lescano, S. A., Amato, N., V, Chieffi, P. P., Bezerra, R. C., Gakiya, E., Ferreira, C. S., and Braz, L. M., 2004, Evaluation of the efficacy of azithromycin and pyrimethamine, for treatment of experimental infection of mice with *Toxoplasma gondii* cystogenic strain, *Rev. Soc. Bras. Med. Trop.* **37**:460–462.

Lindsay, D. S., Collins, M. V., Jordan, C. N., Flick, G. J., and Dubey, J. P., 2005, Effects of high pressure processing on infectivity of *Toxoplasma gondii* oocysts for mice, *J. Parasitol.* **91**:699–701.

Logar, J., Novak-Antolic, Z., Zore, A., Cerar, V., and Likar, M., 1992, Incidence of congenital toxoplasmosis in the Republic of Slovenia, *Scand. J. Infect. Dis.* **24**:105–108.

Logar, J., Novak-Antolic, Z., and Zore, A., 1995, Serological screening for toxoplasmosis in pregnancy in Slovenia, *Scand. J. Infect. Dis.* **27**:163–164.
Lopez, A., Dietz, V. J., Wilson, M., Navin, T. R., and Jones, J. L., 2000, Preventing congenital toxoplasmosis, *MMWR Recomm. Rep.* **49**:59–68.
Lovelace, J. K., Moraes, M. A., and Hagerby, E., 1978, Toxoplasmosis among the Ticuna Indians in the state of Amazonas, Brazil, *Trop. Geogr. Med.* **30**:295–300.
Lunden, A., Lind, P., Engvall, E. O., Gustavsson, K., Uggla, A., and Vagsholm, I., 2002, Serological survey of *Toxoplasma gondii* infection in pigs slaughtered in Sweden, *Scand. J. Infect. Dis.* **34**:362–365.
MacPherson, J. M., and Gajadhar, A. A., 1993, Sensitive and specific polymerase chain reaction detection of *Toxoplasma gondii* for veterinary and medical diagnosis, *Can. J. Vet. Res.* **57**:45–48.
Maiga, I., Kiemtore, P., and Tounkara, A., 2001, Prevalence of anti *Toxoplasma* antibodies in patients with acquired immunodeficiency syndrome and blood donors in Bamako, *Bull. Soc. Pathol. Exot.* **94**:268–270.
Martino, P. E., Montenegro, J. L., Preziosi, J. A., Venturini, C., Bacigalupe, D., Stanchi, N. O., and Bautista, E. L., 2004, Serological survey of selected pathogens of free-ranging foxes in southern Argentina, 1998–2001, *Rev. Sci. Tech.* **23**:801–806.
Masala, G., Porcu, R., Madau, L., Tanda, A., Ibba, B., Satta, G., and Tola, S., 2003, Survey of ovine and caprine toxoplasmosis by IFAT and PCR assays in Sardinia, Italy, *Vet. Parasitol.* **117**:15–21.
Meireles, L. R., Galisteo, A. J., Jr., Pompeu, E., and Andrade, H. F., Jr., 2004, *Toxoplasma gondii* spreading in an urban area evaluated by seroprevalence in free-living cats and dogs, *Trop. Med. Int. Health* **9**:876–881.
Meisel, R., Stachelhaus, S., Mevelec, M. N., Reichmann, G., Dubremetz, J. F., and Fischer, H. G., 1996, Identification of two alleles in the GRA4 locus of *Toxoplasma gondii* determining a differential epitope which allows discrimination of type I versus type II and III strains, *Mol. Biochem. Parasitol.* **81**:259–263.
Meisheri, Y. V., Mehta, S., and Patel, U., 1997, A prospective study of seroprevalence of Toxoplasmosis in general population, and in HIV/AIDS patients in Bombay, India, *J. Postgrad. Med.* **43**:93–97.
Miller, M. A., Gardner, I. A., Packham, A., Mazet, J. K., Hanni, K. D., Jessup, D., Estes, J., Jameson, R., Dodd, E., Barr, B. C., Lowenstine, L. J., Gulland, F. M., and Conrad, P. A., 2002, Evaluation of an indirect fluorescent antibody test (IFAT) for demonstration of antibodies to *Toxoplasma gondii* in the sea otter (*Enhydra lutris*), *J. Parasitol.* **88**:594–599.
Mondragon, R., Howe, D. K., Dubey, J. P., and Sibley, L. D., 1998, Genotypic analysis of *Toxoplasma gondii* isolates from pigs, *J. Parasitol.* **84**:639–641.
Morley, E. K., Williams, R. H., Hughes, J. M., Terry, R. S., Duncanson, P., Smith, J. E., and Hide, G., 2005, Significant familial differences in the frequency of abortion and *Toxoplasma gondii* infection within a flock of Charollais sheep, *Parasitology* **131**:181–185.
Oksanen, A., Tryland, M., Johnsen, K., and Dubey, J. P., 1998, Serosurvey of *Toxoplasma gondii* in North Atlantic marine mammals by the use of agglutination test employing whole tachyzoites and dithiothreitol, *Comp. Immunol. Microbiol. Infect. Dis.* **21**:107–114.
Omata, Y., Dilorenzo, C., Venturini, C., Venturini, L., Igarashi, I., Saito, A., and Suzuki, N., 1994, Correlation between antibody levels in *Toxoplasma gondii* infected pigs and pathogenicity of the isolated parasite, *Vet. Parasitol.* **51**:205–210.
Petersson, K., Stray-Pedersen, B., Malm, G., Forsgren, M., and Evengard, B., 2000, Seroprevalence of *Toxoplasma gondii* among pregnant women in Sweden, *Acta Obstet. Gynecol. Scand.* **79**:824–829.

Piao, L. X., Aosai, F., Mun, H. S., and Yano, A., 2005, Peroral infectivity of *Toxoplasma gondii* in bile and feces of interferon-gamma knockout mice, *Microbiol. Immunol.* **49**:239–243.

Rai, S. K., Matsumura, T., Ono, K., Abe, A., Hirai, K., Rai, G., Sumi, K., Kubota, K., Uga, S., and Shrestha, H. G., 1999, High *Toxoplasma* seroprevalence associated with meat eating habits of locals in Nepal, *Asia Pac. J. Public Health* **11**:89–93.

rko-Mensah, J., Bosompem, K. M., Canacoo, E. A., Wastling, J. M., and Akanmori, B. D., 2000, The seroprevalence of toxoplasmosis in pigs in Ghana, *Acta Trop.* **76**:27–31.

Roberts, C., and McLeod, R., 2004, *Toxoplasma gondii*. In Gorbach, S. L., Bartlett, J. G., and Blacklow, N. R. (eds), *Infections Diseases*, Vol. 282, Lippincott Williams & Wilkins, Philadelphia, PA, pp. 2334–2339.

Roghmann, M. C., Faulkner, C. T., Lefkowitz, A., Patton, S., Zimmerman, J., and Morris, J. G., Jr., 1999, Decreased seroprevalence for *Toxoplasma gondii* in Seventh Day Adventists in Maryland, *Am. J. Trop. Med. Hyg.* **60**:790–792.

Ruiz, A., Frenkel, J. K., and Cerdas, L., 1973, Isolation of *Toxoplasma* from soil, *J. Parasitol.* **59**:204–206.

Ryu, J. S., Min, D. Y., Ahn, M. H., Choi, H. G., Rho, S. C., Shin, Y. J., Choi, B., and Joo, H. D., 1996, *Toxoplasma* antibody titers by ELISA and indirect latex agglutination test in pregnant women, *Korean J. Parasitol.* **34**:233–238.

Saavedra, G. M., and Ortega, Y. R., 2004, Seroprevalence of *Toxoplasma gondii* in swine from slaughterhouses in Lima, Peru, and Georgia, U.S.A, *J. Parasitol.* **90**:902–904.

Sacks, J. J., Roberto, R. R., and Brooks, N. F., 1982, Toxoplasmosis infection associated with raw goat's milk, *JAMA* **248**:1728–1732.

Sawadogo, P., Hafid, J., Bellete, B., Sung, R. T., Chakdi, M., Flori, P., Raberin, H., Hamouni, I. B., Chait, A., and Dalal, A., 2005, Seroprevalence of T. *gondii* in sheep from Marrakech, Morocco, *Vet. Parasitol.* **130**:89–92.

Schmidt, D. R., Hogh, B., Andersen, O., Hansen, S. H., Dalhoff, K., and Petersen, E., 2005, Treatment of infants with congenital toxoplasmosis: Tolerability and plasma concentrations of sulfadiazine and pyrimethamine, *Eur. J. Pediatr.* **165**:19–25.

Schwab, K. J., and McDevitt, J. J., 2003, Development of a PCR-enzyme immunoassay oligoprobe detection method for *Toxoplasma gondii* oocysts, incorporating PCR controls, *Appl. Environ. Microbiol.* **69**:5819–5825.

Smielewska-Los, E., and Turniak, W., 2004, *Toxoplasma gondii* infection in Polish farmed mink, *Vet. Parasitol.* **122**:201–206.

Sobral, C. A., Amendoeira, M. R., Teva, A., Patel, B. N., and Klein, C. H., 2005, Seroprevalence of infection with *Toxoplasma gondii* in indigenous Brazilian populations, *Am. J. Trop. Med. Hyg.* **72**:37–41.

Song, K. J., Shin, J. C., Shin, H. J., and Nam, H. W., 2005, Seroprevalence of toxoplasmosis in Korean pregnant women, *Korean J. Parasitol.* **43**:69–71.

Sordet, F., Aumjaud, Y., Fessi, H., and Derouin, F., 1998, Assessment of the activity of atovaquone-loaded nanocapsules in the treatment of acute and chronic murine toxoplasmosis, *Parasite* **5**:223–229.

Sousa, S. D., Ajzenberg, D., Canada, N., Freire, L., Costa, J. M., Darde, M. L., Thulliez, P., and Dubey, J. P., 2005, Biologic and molecular characterization of *Toxoplasma gondii* isolates from pigs from Portugal, *Vet. Parasitol.* **135**:133–136.

Sturchler, D., DiGiacomo, R. F., and Rausch, L., 1987, Parasitic infections in Yakima Indians, *Ann. Trop. Med. Parasitol.* **81**:291–299.

Suarez-Aranda, F., Galisteo, A. J., Hiramoto, R. M., Cardoso, R. P., Meireles, L. R., Miguel, O., and Andrade, H. F., Jr., 2000, The prevalence and avidity of *Toxoplasma gondii* IgG antibodies in pigs from Brazil and Peru, *Vet. Parasitol.* **91**:23–32.

Tabbara, K. F., Hammouda, E., Tawfik, A., Al-Omar, O. M., and bu El-Asrar, A. M., 2005, Azithromycin prophylaxis and treatment of murine toxoplasmosis, *Saudi. Med. J.* **26**:393–397.

Terry, R. S., Smith, J. E., Duncanson, P., and Hide, G., 2001, MGE-PCR: A novel approach to the analysis of *Toxoplasma gondii* strain differentiation using mobile genetic elements, *Int. J. Parasitol.* **31**:155–161.

Tinti, M., Possenti, A., Cherchi, S., Barca, S., and Spano, F., 2003, Analysis of the SAG5 locus reveals a distinct genomic organisation in virulent and avirulent strains of *Toxoplasma gondii*, *Int. J. Parasitol.* **33**:1605–1616.

Vaillant, V., de, V. H., Baron, E., Ancelle, T., Colin, P., Delmas, M. C., Dufour, B., Pouillot, R., Le, S. Y., Weinbreck, P., Jougla, E., and Desenclos, J. C., 2005, Food-borne infections in France, *Foodborne Pathog. Dis.* **2**:221–232.

Venturini, M. C., Bacigalupe, D., Venturini, L., Rambeaud, M., Basso, W., Unzaga, J. M., and Perfumo, C. J., 2004, Seroprevalence of *Toxoplasma gondii* in sows from slaughterhouses and in pigs from an indoor and an outdoor farm in Argentina, *Vet. Parasitol.* **124**:161–165.

Vikoren, T., Tharaldsen, J., Fredriksen, B., and Handeland, K., 2004, Prevalence of *Toxoplasma gondii* antibodies in wild red deer, roe deer, moose, and reindeer from Norway, *Vet. Parasitol.* **120**:159–169.

Villena, I., Aubert, D., Gomis, P., Ferte, H., Inglard, J. C., is-Bisiaux, H., Dondon, J. M., Pisano, E., Ortis, N., and Pinon, J. M., 2004, Evaluation of a strategy for *Toxoplasma gondii* oocyst detection in water, *Appl. Environ. Microbiol.* **70**:4035–4039.

Vollaire, M. R., Radecki, S. V., and Lappin, M. R., 2005, Seroprevalence of *Toxoplasma gondii* antibodies in clinically ill cats in the United States, *Am. J. Vet. Res.* **66**:874–877.

Wanha, K., Edelhofer, R., Gabler-Eduardo, C., and Prosl, H., 2005, Prevalence of antibodies against Neospora caninum and *Toxoplasma gondii* in dogs and foxes in Austria, *Vet. Parasitol.* **128**:189–193.

Williams, R. H., Morley, E. K., Hughes, J. M., Duncanson, P., Terry, R. S., Smith, J. E., and Hide, G., 2005, High levels of congenital transmission of *Toxoplasma gondii* in longitudinal and cross-sectional studies on sheep farms provides evidence of vertical transmission in ovine hosts, *Parasitology* **130**:301–307.

Yai, L. E., Canon-Franco, W. A., Geraldi, V. C., Summa, M. E., Camargo, M. C., Dubey, J. P., and Gennari, S. M., 2003, Seroprevalence of Neospora caninum and *Toxoplasma gondii* antibodies in the South American opossum (Didelphis marsupialis) from the city of Sao Paulo, Brazil, *J. Parasitol.* **89**:870–871.

Zargar, A. H., Wani, A. I., Masoodi, S. R., Laway, B. A., Kakroo, D. K., Thokar, M. A., Sofi, B. A., and Bashir, M. I., 1999, Seroprevalence of toxoplasmosis in women with recurrent abortions/neonatal deaths and its treatment outcome, *Indian J. Pathol. Microbiol.* **42**:483–486.

CHAPTER 6

Food-Borne Nematode Infections

Charles R. Sterling

6.1 PREFACE

For most of human evolution, man lived in widely dispersed, small nomadic groups that subsisted on hunting and gathering. While this lifestyle undoubtedly exposed individuals to the risk of occasional zoonotic disease transmission, population size likely minimized the effect of exposure to such diseases. The Neolithic period, which saw the advent of the agricultural revolution and the eventual domestication of certain animal species, not only brought about a rapid increase and concentration in human population but also fostered an environment in which substantial increases in infectious and nutritional diseases could rapidly spread (Armelagos, 1991). In addition, man's descent from the trees, subsequent eccrine evolution, and the development of agriculture forced him to become "water-bound," thereby putting him in contact with a group of infectious agents that can be described as "water associated" (Desowitz, 1981). It is within this context that many zoonotic diseases, including those of nematode origin that are discussed in this chapter, may have gained a foothold in human populations.

When one thinks about food-borne nematode infections, the first parasites that usually come to mind are *Trichinella* and *Anisakis*. While these two are certainly important in the overall context of food-borne infections caused by nematodes, since they are directly transmitted via infected food to susceptible individuals, one also has to consider other less frequently encountered nematode infections and infections caused by rather common nematodes, but which are transferred to humans via indirect food-borne routes. The overall goal of this chapter is to provide a rather broad, general perspective of the important nematode parasites that may find their way into our food chain, either directly or indirectly, and thus may serve as a potential source of infection and disease.

6.2 *TRICHINELLA* SPP.

6.2.1 Background

The credit for the discovery of *Trichinella* infection in humans is credited to Paget who in 1835 encountered cysts in the diaphragm muscle of an Italian man who had died of tuberculosis. It was Owen (1835), however, who publicized this finding and provided the name *Trichina spiralis* for the worm encountered in the tissues of this patient. Interestingly, earlier anatomical observations conducted by others in Europe indicate that calcifications and concretions encountered within muscle tissues of deceased patients may also have been *Trichinella* cysts, but were not adequately identified (Gould, 1970). The discovery of cysts in the extensor muscles of the thigh of a hog by Leidy (1846) helped pave the way for the discovery of the main features of the life cycle of this parasite by Leuckart (1859) and especially Virchow (1860). The first case of fatal trichinellosis in man was described by Zenker (1860), while the first instance in which this disease was clinically diagnosed during the acute phase of infection was reported by Friedriech (1862). An excellent overview of the history of *Trichinella* and trichinellosis is provided by Campbell (1983).

Man, pigs, and *Trichinella* have likely had a long-standing relationship. Just how long is anybody's guess, but encysted larvae have been recovered from a 3200-year-old Egyptian mummy (Carvalho-Gonçalves *et al.*, 2003). The origin of domestic pigs can be traced to the Eurasian wild boar, *Sus scofa*. Subspecies of this ancestor likely diverged some 500,000 years ago, providing ancestral stock for pigs of Asian and European origin. Pig domestication occurred about 9000 years ago in China, with more recent introgression of the Asian subspecies into European domestic breeds in the eighteenth and nineteenth centuries, giving us the present day domestic pig (Giuffra *et al.*, 2000). Given that pigs are highly omnivorous and that numerous animal species can harbor *Trichinella*, it is likely that sylvatic cycles preceded the inclusion of man into the category of animal species that could harbor this infection. Predation by early man on wild pigs and other wild game, and certainly domestication of pigs, closed the zoonotic loop on this parasite. Today, trichinellosis is maintained in sylvatic and urban cycles and man fits into the epidemiological picture of this infection within both of these cycles.

6.2.2 Speciation

Trichinella was considered to be a single species until the discovery in the early 1970s of distinctive biological variants and nonencapsulated forms (Gajadhar and Gamble, 2000). Today, as a result of molecular, biochemical, and experimental studies, eight distinct species are recognized and include *T. spiralis*, *T. nativa*, *T. brivoti*, *T. murrelli*, *T. nelsoni*, *T. pseudospiralis*, *T. papuae*, and *T. zimbabwensis* (Bruschi and Murrell, 2002; Murrell *et al.*, 2000; Pozio *et al.*, 2002). The first five constitute the so-called encapsulated species, while the latter three are nonencapsulated species. In addition, three distinctive genotypes, *Trichinella* T6, T8, and T9, have been identified from distinct geographical regions in carnivores, but have not yet been classified as species (Kapel, 2000a; Murrell and Pozio, 2000; Pozio, 2000a). From a human perspective, only *T. spiralis* is maintained in an urban domestic cycle (Pozio, 2000a). All other recognized species and genotypes are transmitted and maintained in somewhat distinctive sylvatic cycles, however, this certainly has not precluded their transmission to humans; as will be discussed later. In addition, sylvatic reservoirs frequently become synanthropic, thus bringing infections in contact with domestic animals and humans. Collectively, *Trichinella* spp. have a wide cosmopolitan distribution and infect a very broad range of mammals, primarily occurring in those with scavenging and carnivorous habits. In humans, *Trichinella* prevalence has been estimated at perhaps as many as 11,000,000 infections (Dupouy-Camet, 2000).

6.2.3 Life Cycle

The complete life cycle of *Trichinella* spp. occurs within the same mammalian host and involves enteral and parenteral phases. There are, however, some exceptions to this in which some species appear to have limited development in some hosts

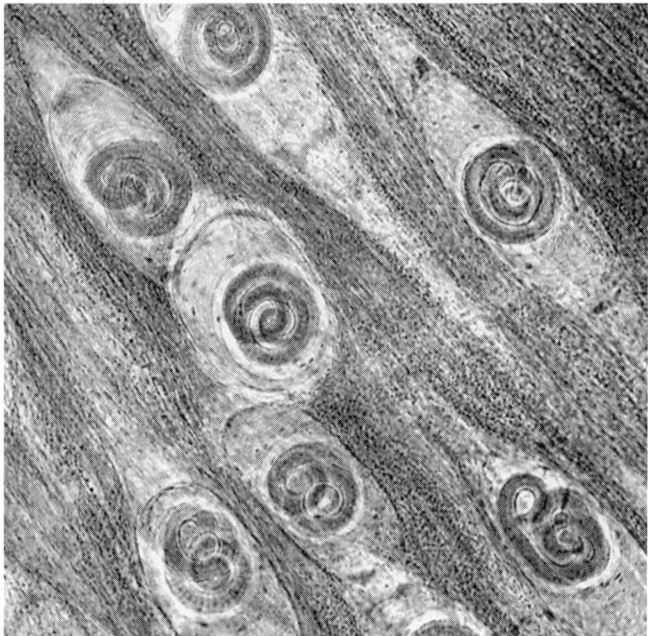

Figure 6.1. Encysted muscle larvae of *Trichinella spiralis*.

(Kapel, 2000a). Infection and the enteral phase commence when first stage larvae contained within muscle tissues (Fig. 6.1) are ingested by another host. The larvae are released from the "nurse cell" by the action of gastric fluids and enzymes and pass into the small intestine, where they penetrate the epithelium to become intra-multicellular organisms (Despommier *et al.*, 1978). Larvae molt four times at this site within 30 h to become sexually active adults. Mating ensues and female worms give birth to an estimated 500 to 1500 live newborn larvae (ovoviparous) until host immunity brings about their expulsion (Wakelin and Denham, 1983). The parenteral phase begins with larval entry into the circulatory system and distribution to multiple body organs. Only larvae that invade muscle cells are destined to survive and do so by establishing a unique, albeit highly modified, intracellular niche. This may take from 3 to 4 weeks to complete, at which time they are infective to another host. The complex of parasite and host cell becomes a remarkably stable unit termed the nurse cell (Despommier, 1990, 1998), which in some instances may survive for years. This may especially be true in natural hosts. In three species, *T. pseudospiralis*, *T. papuae*, and *T. zimbabwensis*, however, the larvae do not induce cyst or capsular formation and it is not known how long such larvae can survive in host muscle tissue.

6.2.4 Epidemiology

There is no easy way to discuss the overall epidemiology of *Trichinella* without providing information on each of the species involved in this complex picture. Likewise, it has to be kept in mind that most of these species have the potential of infecting man and that in many instances, the species responsible for initiating an infection or outbreak has never been determined. In addition, those species occurring in sylvatic cycles can exist in areas where the domestic cycle involving *T. spiralis* predominates, further complicating the epidemiological picture. Because of this, it is appropriate to discuss the biology and epidemiology unique to each *Trichinella* sp. and then put it in the context of human infection and the overall epidemiological picture. Overviews of these species are provided in the reviews of Kapel (2000a) and Pozio (2000a; 2001a).

6.2.4.1 *Trichinella spiralis*

Trichinella spiralis is responsible for the domestic cycle of trichinellosis because of its primary association with the domestic pig. It also occurs in sylvatic pigs and other synanthropic animals such as rats. Infections occur in many countries of temperate and tropical regions but are unlikely to occur in colder regions because muscle larvae are incapable of surviving in frozen animal carcasses (Pozio, 2001a). Exceptions to this, however, might occur in situations where wildlife may harbor this infection near human settlements. Geographic distribution of this species within temperate and tropical zones, therefore, is frequently determined by the extent to which domestic, synanthropic, and potentially sylvatic cycles may overlap. *T. spiralis* is highly pathogenic to humans and is more infective to pigs and other domestic animal species such as the horse, cow, sheep, and goats than are other *Trichinella* spp. or genotypes (Kapel, 2000a; Kapel *et al.*, 1998). This species is infrequently encountered in wild animals. For these reasons, and because of human influences, *T. spiralis* has become the predominant species encountered within humans. Epidemiologically, transmission routes of *T. spiralis* in farm animals such as the pig are usually the result of poor breeding and rearing habits instituted by humans. Identified transmission routes include the feeding of pigs with infected scraps of meat from other pigs, tail-biting among pigs, ingestion of feces from pigs fed infected meat, ingestion of infected rats on poorly managed pig farms where effective barriers of rodent control have not been installed, and ingestion of other synanthropic or sylvatic animals under similar circumstances (Pozio, 2000a).

6.2.4.2 *Trichinella nativa*

This species is interesting because of its geographic distribution, being restricted to arctic and subarctic regions where it is largely a parasite of sylvatic and scavenging carnivores such as bears, foxes, wolves, and mustelids. Biological features that help distinguish this species include the variable freeze tolerance of muscle larvae and the inability to maintain long-term infections in pigs, boars, and other livestock (Kapel, 2000a). The most freeze-resistant larvae, which can remain viable for years in some instances, are found in animals such as the polar bear and arctic fox that live at the more northern latitudes of this parasite's range (Dick and Pozio, 2001; Kapel *et al.*, 1999). *T. nativa* is highly pathogenic in humans and occurs where sylvatic animals are hunted and consumed as a food source.

6.2.4.3 *Trichinella brivoti*

Trichinella brivoti is a sylvatic species infecting wolves, foxes, raccoon dogs, mustelids, and wild boars in more temperate parts of Europe and Asia. It has also been encountered in synanthropic rats, domestic pigs, and horses in mainland Europe and other parts of Asia. Encysted larvae have been reported to survive up to 6 months in the frozen ($-20°C$) muscle tissues of carnivores (Pozio et al., 1989). When encountered in wild boars and domestic pigs, its prevalence is always lower than for *T. spiralis* (Pozio, 2001a). Infectivity is comparable in the wild boar and domestic pig and the former has been cited as a frequent source of infection to humans where infection is moderately pathogenic (Kapel, 2000a, 2000b).

6.2.4.4 *Trichinella murrelli*

This parasite appears to have a predominantly sylvatic cycle in carnivores such as foxes, coyotes, raccoons, black bears, and bobcats of the more temperate regions of the United States. Very low infectivity to noninfectivity has been reported from domestic pigs (Kapel and Gamble 2000; Yao et al., 1997). Muscle larvae are moderately tolerant to freezing, but less so than for *T. brivoti* (Pozio et al., 1994a). Infections in humans have been reported in France following the consumption of raw horsemeat imported from the United States (Ancelle et al., 1988; Dupouy-Camet et al., 1994).

6.2.4.5 *Trichinella nelsoni*

Trichinella nelsoni infections predominate in scavenging carnivores, south of the Sahara region in Africa. Jackals, hyenas, cheetahs, lions, leopards, and wild pigs have been found to harbor infections. The latter are thought to be a common source of human infection (Pozio et al., 1994b). Infectivity is moderate for both domestic pigs and wild boar (Kapel and Gamble, 2000). Interestingly, this genotype is very sensitive to freezing, yet somewhat resistant to elevated temperatures such as might be encountered in decaying meat in the tropics of Africa (Sokolova, 1979). This genotype is not highly pathogenic in humans (Bura and Willett, 1977).

6.2.4.6 *Trichinella pseudospiralis*

A distinguishing feature of this species is that muscle larvae are not encapsulated. In addition, this species can infect birds as well as mammals. A broad diversity of animal species, including wild carnivores, meat eating birds, marsupials, rodents, pigs, and man have been found to harbor infection with *T. pseudospiralis* (Kapel, 2000a; Pozio, 2001a). It appears to be a very biologically diverse species with molecular and biochemical differences having been detected in isolates from varying regions (La Rosa et al., 2001). Larvae are more difficult to detect on necropsy and in digests than for the encapsulated species, so this species may be underreported. This species does not appear to tolerate heat or freezing as well as encapsulating species (Sokolova, 1979). Severe disease has been reported in humans from Australia, Asia, and Europe. In each instance, infection from wild pigs was implicated (Kapel, 2000a).

6.2.4.7 *Trichinella papuae*

Trichinella papuae, another nonencapsulated species, has recently been described from wild and domestic pigs of Papua New Guinea (Pozio et al., 1999). The distribution of this parasite is unknown; as are reservoirs. It is not infective to birds, as is *T. pseudospiralis*, has an intermediate sized larvae, and has a low infectivity to mice (Pozio et al., 1999). This parasite has recently also proven infective to several reptilian species (Pozio et al., 2004). A focus of human infection with this parasite has been described from New Guinea (Pozio, 2001b).

6.2.4.8 Trichinella zimbabwensis

The third nonencapsulated *Trichinella* species was recently described from farmed crocodiles in Zimbabwe and is similar in morphology to *T. papuae* (Pozio et al., 2002). This species is also infective to mammals. It represents the first nonencapsulated species described from Africa and because it can infect both reptiles and mammals suggests a more ancient lineage for this parasite than previously thought (Pozio et al., 2002). Infections have not been reported in humans to date.

In addition to the above-recognized species, the following genotypes have also been described but not yet classified.

6.2.4.9 Trichinella T6

This genotype is related to *T. nativa* and has been detected in numerous carnivores (bears, wolves, foxes, mountain lions) of subarctic regions of North America. Like *T. nativa*, this genotype can survive in frozen tissue for several years (Worley et al., 1990). Human infections have resulted from eating game meat (Dworkin et al., 1996). *Trichinella* T6 and *T. nativa* can interbreed and live in sympatry in Arctic wolf populations (La Rosa et al., 2003). This has led to the suggestion that these two genotypes may have diverged and are still diverging as a result of glaciation episodes in North America.

6.2.4.10 Trichinella T8 and T9

These two genotypes are related to *T. brivoti*, but with disparate geographical distributions. The former has been isolated from carnivores of South Africa and Namibia whereas the latter has only been seen in a dog and bear from Japan (Kapel, 2000a). Very little is known of the biological characteristics of these two genotypes.

6.2.5 Human Trichinellosis–Epidemiology

In the not too distant past, all trichinellosis was thought to be caused by a single species, *T. spiralis*. This view had been propagated in many circles because of the well-known and ingrained association of this infection with the domestic pig. The advent of molecular tools and biochemical methods capable of distinguishing species has greatly impacted our knowledge of *Trichinella* as a species complex and has contributed to a more thorough understanding of the epidemiology of this parasite.

The modern day epidemiology of human trichinellosis is governed by complex interactions involving changes in human activity as well as by interactions that bring humans into contact with sylvatic species of the disease. Most human trichinellosis is still caused by *T. spiralis* and is maintained in the domestic cycle because of human activity linked to pig breeding and rearing. Transmission routes of epidemiological importance that likely impact the domestic cycle include: (1) pigs eating scraps composed of other infected pigs, (2) predation on other infected synanthropic animals, such as rats, (3) pig habits such as tail-biting and coprophagy where infections are endemic, and (4) occasional ingestion of infected sylvatic animals; although this latter route does not lead to persistent cycling because the domestic pig, in many instances, becomes a dead end host (Pozio, 2000a). *T. spiralis*, however, can easily cross back into the sylvatic habitat and become a persistent problem where poor management practices prevail (Murrell and Pozio, 2000).

The problem of human infection with *T. spiralis* can usually be linked to factors related to either nonexistent public health or veterinary standards that might help prevent disease transmission, or to a breakdown in such standards. In China and many areas of Latin America, pigs have become a part of the cash economy for the family unit. As such, they are often butchered in backyard operations and their meat is sold without benefit of any formal inspection. The lack of health education and an awareness of *Trichinella* itself in such environments contribute to a sustained spread of disease (Pozio, 2000a). In eastern Europe and parts of the former USSR, political change and economic destabilization have resulted in a diminished quality of veterinary care and controls that have led to a sharp rise in the prevalence of domestic trichinellosis. Pig production on large well-controlled collective farms has diminished significantly and fallen back to small individual farms that lack adequate feed, technology, and veterinary services. In such situations, the prevalence of *Trichinella* infection in pigs is frequently 10 times higher than on the collective farms (Bessonov, 1994), ultimately resulting in more human infections. In countries such as Lithuania and Romania, the incidence of trichinellosis has increased ninefold and 17-fold, respectively, over the past two decades, making it a truly reemergent disease problem (Olteanu, 1997; Rockiene and Rocka, 1997).

Globalization of trade is another risk factor that has resulted in the spread of trichinellosis from endemic to largely nonendemic regions. This type of transference results in meat products ending up in markets where inspection services may not be familiar with this disease problem. Nowhere has this been more evident than in France and northern Italy, where infected horsemeat from North America, Mexico, and especially eastern European countries has contributed to more than 3300 human infections (Pozio, 2000b). *T. spiralis* infection in horses carries with it the implication that such noncarnivorous animals are being fed infected food products and again underscores the need for adequate veterinary control practices and regulations (Dupouy-Camet *et al.*, 1994).

In countries such as the United States, where pig production has increased and control programs and surveillance have demonstrated a dramatic reduction in the annual number of cases of trichinellosis in humans due to *T. spiralis*, there has been an increase of cases associated with consumption of infected game meat (Moorhead *et al.*, 1999; Murrell and Pozio, 2000). Many of these new infections have been attributed to *T. nativa*, *T. murrelli*, or *Trichinella* T6 (Gajadhar and Gamble, 2000). Regions of Canada, and particularly the Northwest territories and Quebec, have experienced human rates of infection 200-fold over that of the general population in spite of a decline in domestic trichinellosis, once again underscoring the likely connection with consumption of infected game meat (Murrell and Pozio, 2000). In many countries the impact of sylvatic *Trichinella* transmission to humans cannot be adequately addressed because of the lack of appropriate diagnostic tests.

In many instances, human behavior influences the transmission of certain *Trichinella* genotypes within defined habitats. International travel and the fashionable consumption of undercooked meat can explain outbreaks in certain population groups (Dupouy-Camet, 2000). Likewise, outbreaks have occurred in defined ethnic communities within countries where the overall prevalence rates are low, but where in their native countries the prevalence rates are much higher. This situation is likely linked to customs of food preparation (Pozio, 2000a; Shantz and McAuley, 1991). In Arctic regions, hunting habits of native peoples not only contribute to high rates of infection among human populations, but help sustain infection in wildlife populations as well. Among the Inuit culture, sled dogs are fed the remains of killed animals including polar bears, which have some of the highest rates of trichinellosis recorded because of their carnivorous and scavenging habits. Dogs that die of trichinellosis are left as carcasses that become available to infect other scavenging carnivores (Kapel et al., 1997).

Ecological modifications can also impact the distribution of *Trichinella* genotypes. A decrease in the number of farms and reforestation in parts of Europe has led to an increase in the wild boar population and maintenance of sylvatic trichinellosis (Pozio et al., 1996). Likewise, food access of pigs to potential wildlife reservoirs is likely to increase as human populations expand and with it the need for increased food production in certain areas of the world. Such situations have already been implicated as a major risk factor for transmission to sylvatic animals (Dame et al., 1987; Murrell et al., 1987).

Another factor that likely plays a role in the overall epidemiology of trichinellosis and certainly in the reemergence of this disease problem is the issue of misdiagnosis. Clinical manifestations of disease such as fever, myalgia, and fatigue, particularly in the early stages of infection, can easily be mistaken as influenza. Other features of infection, such as the appearance of facial edema, could also be mistaken as manifestations of an allergic response brought on by the flu. Because of this issue, the disease is likely to be underreported in many areas where health care systems are incapable of providing optimal service, especially with respect to infectious disease diagnosis.

6.2.6 Clinical Manifestations

Clinical aspects of trichinellosis have been aptly described and are summarized in excellent reviews by Capo and Despommier (1996) and Bruschi and Murrell (2002). The course of disease depends, in part, on the infecting species and may result in mild, to moderate, to severe symptoms. The incubation period may be variable, ranging from 7 to 30 days. During the enteral phase, which usually lasts about 6 weeks, patients may exhibit upper abdominal pain, diarrhea or constipation, vomiting, malaise, and low-grade fever. Variations depend on the severity of the infection, age, sex, ethnicity, and immune status of the infected individual. Such symptoms, which are not uncommon with other enteral infections, often make diagnosis difficult at this time during the infection. Gastrointestinal symptoms may overlap with the parenteral phase of the infection during which time the larvae are migrating and invading muscle tissue. This phase, which may begin as early as 2 weeks following infection, is characterized by inflammatory and allergic responses that may be manifest as facial edema, diffuse myalgia, conjunctivitis, fever, headache, and urticaria. The most severe cases usually serve as index cases of an outbreak and prompt epidemiological investigations. Today, death occurs in only rare instances because of improved therapies, and the long-term consequences of harboring larvae are somewhat controversial since many will ultimately undergo complete calcification.

6.2.7 Diagnosis and Treatment

An important aspect of dealing effectively with trichinellosis is to affect a diagnosis as early as possible during the course of an infection. As already mentioned, clinical signs of disease early in the infection can often be misleading. Obtaining a history of food consumption can play an important role in suspecting trichinellosis and this should be followed up with blood work to demonstrate a hypereosinophilia and an elevation of certain muscle enzymes, such as creatine phosphokinase and lactate dehydrogenase. Taken collectively, the food consumption history and laboratory findings are highly pathognomonic for the disease (Bruschi and Murrell, 2002; Despommier *et al.*, 2000). A definitive diagnosis can be made by finding encysted muscle larvae or by using immunodiagnostic procedures, but the latter are not likely to be positive until several weeks into the course of an infection, by which time the larvae are fully ensconced in the muscle tissue (Bruschi and Murrell, 2002).

An important goal of early treatment is to limit muscle invasion by larvae and barring that, to limit muscle damage after larvae have invaded. Drugs such as mebendazole, albendazole, and thiabendazole, members of the benzimidazole family, work quite effectively against intestinal worms, although the latter has been noted to have frequent side effects (Kociecka, 2000). These drugs are not likely to control infections once they have reached the muscle encapsulated stage and in severe cases, hospitalization is required (Pozio *et al.*, 2001). In severe cases, treatment may involve administration of glucocorticosteroids, analgetic drugs, or even immunomodulating drugs, depending on the individual patient (Kociecka, 2000).

6.2.8 Prevention and Control

Effective prevention and control of trichinellosis depends not only on reducing transmission of the disease to humans via livestock and wild game, but also on reducing transmission of infection between sylvatic and synanthropic animals and domestic livestock. In highly industrialized countries where pig farming is more tightly regulated and where inspection or processing methods are in place to insure trichina free meat, trichinellosis has been more or less confined to the few who acquire infection via consumption of wild game, or who because of culinary habits have acquired a taste for food prepared in ways inadequate to insure safety. In many parts of the world, however, economic problems, the erosion or lack of appropriate veterinary and medical infrastructures, and poor education translate into virtually no control of this disease at the level of the farm and abattoir and into no general information about the disease being disseminated to the general public or to the medical profession (van Knapen, 2000).

The discovery of the association of trichinellosis in humans with the consumption of infected pork in the nineteenth century led to the advent of microscopy and later trichinoscopy as a means of detecting infection in pigs. Virchow, who did much to advance our knowledge of *Trichinella*, has been credited with initiating inspection of pork for the control of trichinellosis in provincial provinces of Germany in the mid-1860s (Gould, 1970; Nöckler et al., 2000). Trichinoscopy, which relies on the postmortem examination of select muscle samples by compression between glass slides, and is viewed as having rather limited sensitivity, served as the mainstay of diagnosing infected meat from its inception until 1978 when the pooled artificial digestion method was introduced (Nöckler et al., 2000). This technique, which utilizes pepsin digestion of select muscles to release encysted larvae, results in a threefold or greater increase in sensitivity when compared to trichinoscopy (Forbes et al., 2003). The pooled digestion assay permits testing of up to 100 carcasses at the same time and for sensitivity in the range of 1 larva per gram of muscle tissue, digestion of at least 5 grams of muscle tissue is recommended (Gamble, 1996; Gamble et al., 2000; Nöckler et al., 2000). Identification of muscle to be tested may vary because of differences in predilection sites by the various *Trichinella* species and because of infected host differences. In addition, not all *Trichinella* species encyst and thus trichinoscopy may be unable to detect such infections (Nöckler et al., 2000). In spite of this, trichinoscopy is still in use, particularly in the examination of muscle tissue from wildlife in epidemiological studies. The International Commission on Trichinellosis has strongly urged that all laboratories that test for the presence of Trichinella adapt quality assurance guidelines and the Canadian Food Inspection Agency's Centre for Animal Parasitology has taken the lead in this regard (Gajadhar and Forbes, 2002).

The use of immunological techniques applicable to the serodiagnostic detection of Trichinella infections in pigs and other animals has gained in popularity, but have yet to become a permanent diagnostic fixture because they have not been standardized (Gamble et al., 2004). Such techniques, and particularly the enzyme-linked immunsorbent assay (ELISA), have the advantage of permitting testing of pre and postmortem animals and with greater sensitivity. However, the ELISA may not be able to detect very early or late stage infections and for this reason has not replaced digestion testing (Nöckler et al., 2000). Still, the ELISA has proven invaluable as a test for pig herd surveillance and for detection of infection in humans (Gamble, 1996; Gamble et al., 2004).

Processing methods to control trichinellosis have also been recommended. Such methods are particularly applicable where meat from pigs or other animals are not tested by recommended inspection procedures, these include cooking, freezing, and irradiation (Gamble et al., 2000). Cooking of meat to an internal temperature of 71°C (160°F) should be sufficient to kill any *Trichinella* larvae present. In the absence of temperature monitoring, meat texture and color should be monitored. Freezing guidelines call for cuts of meat up to 15 cm thick to be frozen solid for no less than 3 weeks, rising to 4 weeks for meat up to 69 cm thick. Irradiation, where permitted, at 0.3 kGy has proven to inactivate *Trichinella* larvae and is recommended for sealed packaged food only. Interestingly, irradiation was reported to be effective in killing *Trichinella* cysts in pork by scientists of the USDA as early as 1921 (Steele, 2000). The key element associated with these recommendations is good consumer education.

In addition to the above, prevention and control of trichinellosis requires utilization of strict rules of good production practices on farms along with good veterinary practices to minimize the transmission risks of *Trichinella* spp. The following summarizes guidelines required for *Trichinella* free pig production (Gamble *et al.*, 2000; Murrell and Pozio, 2000; van Knapen, 2000):

- Construction of animal containment facilities that will exclude rodents and wildlife.
- Provisions for feed storage that do not permit rodent or animal entry and purchase of food from approved facilities.
- Adherence to garbage feeding regulations and cooking of meat products to inactivate any larvae that may be present.
- Maintenance of effective rodent and bird control programs.
- Purchase of new animals only from certified *Trichinella*-free farms.
- Absence of garbage dumps within a 2 km radius of the farm.
- Appropriate disposal of pig and other animal carcasses to minimize infection risks on the farm and with other animals.

Eradication of *Trichinella* from wild animal populations may never be possible due to predatory and scavenging habits, but certainly could be impacted upon if hunters would properly dispose of offals and carcasses.

Finally, pig vaccination against *T. spiralis* using antigens of the newborn larvae of the same or different species of *Trichinella* has been attempted and could have applications in situations where the above-mentioned management practices are not used or have not been successful (Marincculic *et al.*, 1991; Marti *et al.*, 1987; Smith, 1987). Such an approach, however, is likely to be very expensive and have limited use.

6.3 *ANISAKIS SIMPLEX* AND RELATED SPECIES

6.3.1 Background

Worms parasitizing fish were recorded as early as the thirteenth century. It was not until 1845, however, that Dujardin described the taxonomic position of *Anisakis* from dolphins, a common worm parasite encountered in fish with marine mammals as definitive hosts (Bouree *et al.*, 1995). This nematode and the relatives *Pseudoterranova decipiens* and *Contracaecum* spp. are all anisakids belonging to the same superfamily of worms as the common human intestinal roundworm, *Ascaris lumbricoides*. All are likely to cause human infection through the consumption of raw or undercooked fish (Sakanari and McKerrow, 1989). Additional anisakid worms have been implicated in causing human disease but case reports are rare. *A. simplex* and *P. decipiens* are the only anisakids reported from human cases in the United States http://vm.cfsan.fda.gov/~mow/chap25.html.

The first case involving human infection with *Anisakis* was recorded in Holland from an eosinophilic intestinal lesion in a patient with severe abdominal pain. This case was important because it linked the eating of infected raw herring with observed symptoms (van Thiel *et al.*, 1960). Endoscopy was used to directly view larvae of *Anisakis* in an infected individual in 1968 (Namiki, 1989) and in 1975 the first case of human infection was recorded from the United States (Pinkus *et al.*, 1975).

6.3.2 Life Cycle

The life cycles of the various anisakids usually involve two intermediate hosts with the final definitive host being a marine mammal such as dolphins, sea lions, and whales (Bouree et al., 1995; Sakanari and Mckerrow, 1989; Smith, 1983). Adult worms reside within the stomach of their definitive host and eggs are passed in the feces to embryonate in seawater. Eggs containing first stage larvae can hatch or be ingested. In either case, planktonic crustaceans ingest the larvae and serve as first intermediate hosts for this infection. Larvae develop to the second stage in these intermediate hosts, which are then consumed by fish where they become infective third stage larvae. Distribution of these third stage larvae within the fish hosts can vary from the body cavity to encystment in various tissues. In the normal life cycle, marine mammals consume infected fish at which point the larvae molt within the stomach to become adult worms. In some instances, it is possible for a larger predatory fish or animal, such as man, to eat a smaller infected fish in which case the life cycle is completed. In these hosts, the worms do not mature. It is also theoretically possible for marine mammals to be infected via ingestion of infected crustaceans, which might result in larval development in their body cavities or tissues, but this has not been reported.

6.3.3 Epidemiology

Anisakiasis, or disease caused by anisakid worms, is frequently reported in the literature as simply due to *Anisakis* spp. This is probably because of the fact that attempts are often not made to distinguish between the infecting anisakid species. There is little doubt that more attention needs to be paid to which distinct species is causing infection and to try to discern which fish species is responsible for the cause of infection so that the epidemiology of these infections can be better defined. Having made this statement, there is no doubt that the transmission of these anisakid infections to humans is associated with the food-borne route via the consumption of raw and undercooked fish. Fish dishes from various parts of the world that are considered to pose high risk of infection include sushi, sashimi, pickled or smoked herring, dry cured salmon, lomi-lomi, cebiche, and pickled anchovies (Audicana et al., 2002).

Epidemiological studies from Japan, where 95% of the cases of anisakiasis have been reported from, indicate that the main sources of *A. simplex* are the spotted chub mackerel and Japanese flying squid. In Western Europe, *A. simplex* is principally found in herring and in Spain, anchovies and sardines are the main culprits (Audicana et al., 2002). Fish found in North American waters that may harbor *A. simplex* infections include wild caught and farm reared salmon (Deardorff and Kent, 1989), Pacific herring (Moser and Hsieh, 1992), and Atlantic cod (Chandra and Khan, 1988). Halibut, mackerel, rockfish, and squid may also be infected (Sakanari and Mckerrow, 1989). *A. simplex* has also been reported from ahi off Hawaiian shores (Deardorff et al., 1991). Recently, *A. simplex* was reported for the first time from the American shad in Oregon rivers (Shields et al., 2002). This report is significant because they are an introduced species that was derived from stock from the Hudson River in New York where they are normally not infected. In addition, infected fish were found in rivers distant from ocean sources, suggesting an ecological expansion of this infection into a new environment. Interestingly, *A. simplex* and other anisakids have recently been reported from river otters in the Pacific Northwest at considerable distances from the ocean (Hoberg *et al.*, 1997). Otters have been observed feeding on living and dead shad during their spawning runs and outmigrations (Shields *et al.*, 2002). Taken collectively, these observations suggest that this new host—parasite relationship within the Northwestern United States has led to infections spreading within a new habitat that also pose risks to certain wildlife species and potentially to humans as well. It is also a reminder that the introduction of stock fish species into new areas, whether infected or not, can alter host—parasite relationships.

Pseudoterranova decipiens, the other anasakid associated with human infection in North America, is most commonly encountered in cod and can be found in cod worldwide (Oshima, 1987). Red snapper, another popular fish consumed in the United States, has been thought to be the main source of codworm anisakiasis (Oshima, 1987). In Mexico, where cebiche is a popular dish, five fish species commonly used in its preparation were found to be frequently infected with this parasite. They included the lane snapper, yellow fin mojarra, barracuda, red grouper, and white grunt (Laffon-Leal *et al.*, 2000).

As of 2002, a total of ~14,000 cases of anisakiasis have been reported, with Japan accounting for about 2000 new cases annually (Audicana *et al.*, 2002). Epidemiological studies conducted in Japan, have shown that cases of anisakiasis were more likely to be encountered in coastal areas where individuals were involved in the fish industry (Asaishi *et al.*, 1980). Cases from Europe, the United States and elsewhere also appear to be on the rise, but are more likely the result of culinary habits associated with ethnic groups or restaurants. The increased recognition of anisakiasis as a problem is probably due in large part to an increased awareness of this infection and the means by which to diagnose it.

6.3.4 Clinical Manifestations

The clinical symptoms of acute anisakiasis include sudden epigastric pain, nausea, vomiting, diarrhea, and frequently urticaria (Audicana *et al.*, 2002; Bouree *et al.*, 1995; Sakanari and Mckerrow, 1989;). In fact, most of the symptoms are probably the result of digestive tract allergic reactions (Asaishi *et al.*, 1980; Audicana *et al.*, 2002; Moreno-Ancillo *et al.*, 1997). Onset of symptoms is usually rapid and may persist for 1–5 days. Infections may frequently be misdiagnosed at this time. Chronic anisakiasis results from larval invasion that causes abscess or eosinophilic granulomas. This form of disease can mimic other conditions such as appendicitis, gastroduodenal ulcer, colitis, inflammatory bowel disease, and intestinal obstruction (Moreno-Ancillo *et al.*, 1997). Some *Anisakis* allergens are heat and freeze resistant. Cooking, therefore, which kills the parasite, might not prevent potent allergic responses (Audicana *et al.*, 1997; Moreno-Ancillo *et al.*, 1997). In some individuals, sensitization may even lead to severe anaphylactic shock (Audicana *et al.*, 2002; Moreno-Ancillo *et al.*, 1997).

6.3.5 Diagnosis and Treatment

Vague symptoms during the early stages of the disease can often lead to a misdiagnosis (Sakanari and Mckerrow, 1989). Because clinical manifestations of allergy often accompany anisakiasis, it is clinically relevant to link consumption of fish with the attendant disease symptoms and then to rule out allergic responses to fish as the likely cause of disease. The diagnosis of allergy brought on by an anisakid infection can be assessed using the following criteria: (1) a history of urticaria, angioedema or anaphylaxis following fish consumption, (2) a positive skin-prick test with a somatic extract of larvae, (3) a radioimmunoassay which detects *Anisakis*-specific IgE responses, and (4) a lack of reaction to proteins from the ingested fish host (Audicana *et al.*, 2002). Immunoblotting techniques that detect the presence of antigen specific antibody responses may also be of use (Del Pozo *et al.*, 1996; Montoro *et al.*, 1997). The value of immunoassays is likely increased in populations previously sensitized to *Anisakis* antigens. Many serologic assays, such as Ouchterlony tests, indirect haemagglutination, complement fixation, and ELISA can yield false positive reactions due to the cross-reactivity with other parasite antigens (Kennedy *et al.*, 1988). It must also be kept in mind that many of these assays are of limited use in the case of acute disease.

Endoscopy, applied early during the course of infection, is still one of the most important tools in both the diagnosis and treatment of anisakiasis (Bouree *et al.*, 1995; Sakanari and Mckerrow, 1989). Worms cannot only be visualized, but removed with the biopsy forceps for identification. Worm removal also minimizes the chance of allergic responsiveness characteristic of more chronic infections. Even in chronic cases, this technique may identify larval cuticular debris within granulomatous lesions (Gomez *et al.*, 1998). Chemotherapeutic treatments have not been described for anisakiasis.

6.3.6 Prevention and Control

Visual examination of fish is required in Europe with extraction of visible parasites and removal of heavily parasitized fish from the markets (Audicana *et al.*, 2002). Evisceration of fish soon after they are caught is recommended since larvae are known to leave the digestive tract and go into the muscles (Declerck, 1988; Smith and Wooten, 1975). Salting, smoke curing, and marinating are not likely to affect larval viability (Bouree *et al.*, 1995). The United States Food and Drug Administration has listed guidelines for seafood not cooked to temperatures above 140° F throughout, which includes blast freezing to $-31°F$ or below for 15 h, or regular freezing to $-10°F$ for 7 days (Deardorff *et al.*, 1991). These measures should reduce the risk of anasakid infection but may not totally eliminate the risk of allergic responses. For that, one would have to consider a fish-free diet!

6.4 *ANGIOSTRONGYLUS CANTONENSIS* AND *ANGIOSTRONGYLUS COSTARICENSIS*

Angiostrongylus cantonensis and *Angiostrongylus costaricensis* are metastrongyle nematodes. As a group, the metastrongyles usually occur in the lungs of mammals. *A. cantonensis*, a parasite of the lungs of rats, was first observed in China in 1935 (Chen, 1935). The first reported human infection described a 15-year-old boy in Taiwan in 1945 with symptoms of meningitis, and from whom worms were recovered from cerebrospinal fluid (Beaver and Rosen, 1964). Humans are accidental hosts for this infection, as the worms never reach maturity. It is now recognized as a major cause of eosinophilic meningitis worldwide, although most of the cases still occur in areas around Southeast Asia (Alicata, 1991; Cross, 1987; Lo Re and Gluckman, 2001; Rosen et al., 1967). Snails and slugs serve as intermediate hosts, while shrimp, crabs, and land planarians serve as transport hosts (Alicata, 1991). Oysters and clams have been experimentally infected and from which juvenile worms produced infections in rats (Cheng, 1965). Adult worms reside within the pulmonary arteries of rodents where females lay their eggs. Eggs hatch at this site and first stage larvae work their way into the lungs, up the trachea, are swallowed, and subsequently passed in the feces. Snails and slugs consume these larvae, which then undergo two molts to become infective third stage larvae. Rodents and man get infected by ingesting raw snails, vegetables contaminated with snail or slug slime, or transport hosts that have consumed infected snails or slugs (Lo Re and Gluckman, 2001). In rodents, larvae migrate to the lungs to become adults, while in man the larvae make it as far as the central nervous system where they die and produce inflammation. Increased dispersal of infected rodents worldwide via shipborne routes is likely contributing to a spread of this disease (Kliks and Palumbo, 1992). Infections have recently been reported in humans in the United States and Caribbean (New et al., 1995; Slom et al., 2002) and the parasite has been reported as endemic in Louisiana wildlife (Kim et al., 2002). Clinical manifestations develop from 2 to 35 days after infection and most frequently present as headache (Lo Re and Glickman, 2001). Fever and vomiting are less frequently observed. Abnormal skin sensations of the extremities, trunk, or face are also frequently encountered and can persist for weeks. Diagnosis is difficult at best and usually involves a history of exposure in an endemic area along with a clinical presentation suggestive of infection with this parasite. An elevated CSF eosinophilia along with serologic testing will confirm infection (Lo Re and Glickman, 2001). Unfortunately, commercially available serologic tests are not available. Therapy is controversial and in some instances treatments have resulted in inflammatory reactions due to responses directed against antigens of dying worms (Pien and Pien, 1999). Use of glucocorticosteroids may reduce illness duration (Chau, 2003). Control is difficult because attempts to eliminate snail hosts may harm native or beneficial snail species (Alicata, 1991). Avoidance of eating uncooked or undercooked snails and transport hosts is the best way to prevent infection.

Angiostrongylus costaricensis is also a parasite of rodents, but adult worms reside within the mesenteric arteries of the ileocecal region (Morera, 1973; Waisberg et al., 1999). Abdominal angiostrongyliasis, due to infection with this worm, was first reported from humans in 1967 in Costa Rico (Morera, 1967; Céspedes et al., 1967) and is endemic in Central and South America where it primarily infects children, perhaps due to their play habits (Loría-Cortés and Lobo-Sanahuja, 1980; Morera, 1973; Morera et al., 1982). Unlike *A. cantonensis*, this parasite can mature to adults in humans. Infections lead to inflammation and thrombosis in the terminal branches of the superior mesenteric artery, and possibly to perforation and peritonitis (Waisberg et al., 1999). The life cycle of this parasite parallels that of *A. cantonensis* with the exception of adult worm site location in the definitive host. Adult worms residing within the mesenteric arteries deposit eggs that hatch in the feces; releasing first stage larvae. These are ingested by land slugs in which they undergo two molts to become infective third stage larvae. These larvae pass from the slugs in mucous secretions that may contaminate food or other surfaces. Accidental human infection results from eating inadequately cleaned and/or cooked vegetables and contact with surfaces contaminated with slug mucous containing infective larvae (Morera, 1973). Experimental infections conducted in dogs have led to the suggestion that they may serve as reservoir hosts of this infection (Rodriquez et al., 2002). In human infections, adult worms damage the vascular endothelium causing thrombosis and necrosis. Eggs, larvae, and excretory/secretory products produced by adult worms also contribute to inflammatory processes in infected individuals (Morera, 1988). The clinical picture is one of ileitis, often mimicking acute appendicitis (Waisberg et al., 1999). Included in this picture are abdominal pain, fever, anorexia, nausea, and vomiting. Continued low-grade fever and pain may persist for several weeks and infections may become chronic. Leucocytosis and eosinophilia are prominent features of infection (Morera, 1988) and as for *A. cantonensis*, the clinical diagnosis often depends on the clinical picture being tied to a history of travel to an endemic area. Ectopic localization of adult worms and eggs has also been seen in children, leading to visceral larval migrans-like symptoms (Morera et al., 1982). Sensitive ELISA serologic tests do exist, but they lack high specificity (Geiger et al., 2001; Hulbert et al., 1992). An outbreak involving 22 cases of abdominal angiostrongyliasis was linked epidemiologically to the consumption of mint used in preparing ceviche in Guatemala (Kramer et al., 1998). This was the first instance linking a specific food item to this disease and points to the importance of thorough washing or cooking of all food items that may have been contaminated by slug secretions. Cases of angiostrongyliasis have been reported from the United States but they are rare and usually linked to travel to an endemic region (Wu et al., 1997). The parasite, however, is enzootic in Texas cotton rat populations (Ubelaker and Hall, 1979). Therapy for this infection is also controversial since it may result in worm migration that could aggravate lesions (Waisberg et al., 1999). A laparotomy may have to be performed to rule out suspicion of neoplasms. In severe cases, surgery may have to be performed. Preventive measures include thorough cooking of mollusk hosts and foods that may have been contaminated by slug secretions. Rodent control might be helpful, but is frequently difficult to maintain in endemic areas.

6.5 *GNATHOSTOMA* SPP.

Gnathostoma spp. normally occur as parasites in the stomach of carnivorous mammals. Human disease may result from the ingestion of an infected fish or other intermediate host or from ingestion of infected paratenic hosts that have not been adequately cooked (Rojekittikhun et al., 2002). The highest prevalence of disease is seen in several areas of Asia, but recently gnathostomiasis has been reported with increasing frequency in Mexico and other parts of South America (Miyazaki, 1966; Nawa, 1991; Pelaez and Perez-Reyes, 1970; Rojas-Molina et al., 1999). In this life cycle, adults live in the stomach of infected hosts and pass eggs which embryonate in water. Free-swimming first stage larvae are then ingested by copepods where they mature to second stage larvae in the body cavity. Fish or other hosts then ingest these first intermediate hosts. In these hosts, the parasites develop into the third stage infective larval form in the musculature or connective tissue. The final host acquires the infection by eating infected fish or other second intermediate or paratenic hosts (Miyazaki, 1954). Humans behave like paratenic hosts in that consuming raw or undercooked infected foods leads to no further development of the larvae. This is true whether the consumed host is intermediate or paratenic. More than 125 species of hosts serve as second intermediate hosts and more than 35 species of paratenic hosts are known. These include crustaceans, fish, amphibians, reptiles, birds, many mammals, and man (Miyazaki, 1966; Rusnak and Lucey, 1993). One of the interesting aspects of this life cycle is that many hosts seem to have the ability to serve as both intermediate and paratenic hosts. In man, the larvae tend to migrate through subcutaneous tissue or other organs, producing symptoms dependent on site location. Symptoms may develop within 24 h of larval ingestion and include many of the common indications of intestinal upset. Eosinophilia accompanied by cutaneal or visceral migrans symptoms are also seen (Crowley, 1995; Rusnak and Lucey, 1993). Cutaneal gnathostomiasis, which is frequently observed, is likely to manifest as a creeping eruption commonly associated with hookworm infections (Magana *et al.*, 2004). Visceral gnathostomiasis is usually more severe with symptoms dictated by the site location of migrating larvae (Rusnak and Lucey, 1993). Infections in humans involving persistence of juvenile worms for up to 10 years in humans have been reported (Taniguchi *et al.*, 1991). Outbreaks of disease involving consumption of ceviche or raw fish have recently been reported in Myanmar and Mexico (Chai *et al.*, 2003; Diaz Camacho *et al.*, 2003). Rarely are cases seen in the United States and Europe, but when reported, they are likely the result of travel to endemic countries where raw or undercooked fish was consumed (Menard *et al.*, 2003). Clinical presentation that involves creeping eruption along with a travel history that involves consumption of raw fish are early indications of gnathosomiasis. ELISA and western blot follow-ups are also useful, but as for many parasitic infections, absolute confirmation usually depends on worm isolation and identification (Rojas-Molina *et al.*, 1999). Most anthelminthic treatments have proven ineffective (Ruiz-Maldonado and Mosqueda-Cabrera, 1999), although albendazole may cause worms to migrate to the skin where they can be extracted (Ogata *et al.*, 1998). In the absence of success using chemotherapy, surgical removal would be the treatment of choice (Feinstein and Rodriguez-Valdes, 1984). In the context of many other parasite problems, this is clearly a localized emerging infection. Cultural food practices no doubt impact disease prevalence in certain areas. Indigenous peoples and travelers to endemic areas need to be educated to the threat and better understand good cooking and eating practices.

6.6 *GONGYLONEMA* SPP.

Gongylonema pulchrum has been the main species associated with the 50+ human infections reported in the literature (Gutierrez, 1999). This parasite normally occurs in ruminants and swine, with man being an accidental host. Interestingly, another related parasite, *G. neoplasticum*, which normally occurs in rodents, was associated with tumor research at the beginning of the twentieth century that led to a Nobel Prize (Campbell, 1997). Adult worms live in the esophageal epithelium of their normal hosts. Eggs passed in the feces are consumed by cockroaches and dung beetles where larvae mature to the infective third stage. Infection of the definitive host occurs after ingestion of the insect host (Waite and Gorrie, 1935). As unthinkable as it may sound, it has been suggested that most human infections follow ingestion of cockroaches (Beaver *et al.*, 1984). It could also follow that infection might occur while drinking water in which larvae had been released from disintegrating insect intermediate hosts. Unbeknownst to most of us, we do consume insect parts on a regular basis, and the USDA even has guidelines for allowable numbers of insect parts, eggs, etc., that can be present in food (http://www.cfsan.fda.gov/~dms/dalbook.html). In case report studies on human infections, the patient usually complains of sensations associated with something moving in the skin around the mouth area. Worms can be extracted following local anesthesia and albendazole therapy is usually prescribed (Eberhard and Busillo, 1999; Wilson *et al.*, 2001). In most cases, patients do not recall knowingly ingesting insects and in many there may have been travel to areas where sanitation might have played a role in transmission of this disease. In the cases reported from the United States the majority occurred in individuals living in southeastern states, although a few occurred in individuals from large northeastern cities (Wilson *et al.*, 2001). It is obvious that risk factors for acquiring this disease are not fully understood.

6.7 OTHER NEMATODE INFECTIONS WITH FOOD-BORNE ASSOCIATIONS

Nematodes, as a group, cause more human infections that any other group of parasites (Cromptom, 1999; Stoll, 1999) and the Disability Adjusted Life Years (DALYs) lost due to the three major ones (Ascaris, Trichuris, and the hookworms) surpass that due to malaria infections (Chan, 1997). While this chapter has primarily focused on nematode infections acquired directly from food sources, it would be a mistake to neglect infections that may be acquired indirectly from food or via other nonconventional routes considered to be food sources. In the latter category could be included potential transmammary routes of infection as have been suggested for *Ancylostoma duodenale* (Navitsky *et al.*, 1998; Nwosu, 1981; Prociv and Luke, 1995) and *Strongyloides fuelleborni* (Brown and Girardeau; 1977). While larvae of these parasites were not found in mothers' milk, the frequency of infection in mothers and their nursing children highly suggests lactogenic infections. It also has to be kept in mind that nematodes are very prodigious egg producers and that in many parts of the world, human feces and raw sewage is used as fertilizer to grow crops for human consumption (Cai *et al.*, 1988; Choi and Chang, 1967; Needham *et al.*, 1998). In such instances, it is certainly feasible for eggs to end up on vegetables that if consumed raw or poorly cooked could contribute to a substantial number of human infections. One also has to consider the role of flies as transport hosts in disseminating helminth eggs to food that people might consume. Flies have recently gained a good deal of notoriety in transmitting protozoan infections (Graczyk *et al.*, 1999a, 1999b), but their role in the transport of helminth eggs requires study. Lastly, there was a recent report on the zoonotic risk of *Toxocara canis* infection through consumption of pig or poultry viscera (Taira *et al.*, 2004). While this study was an assessment of the potential risk using pigs as a model, it does point to the possibility that human infection could occur via ingestion of undercooked foods from these sources that may harbor *T. canis* larvae. As we gain knowledge of more of these infections and as more epidemiological studies are undertaken to determine where they come from, it is likely we will become more respectful of food-borne nematode infections.

REFERENCES

Alicata, J. E., 1991, The discovery of *Angiostrongylus cantonensis* as a cause of human eosinophilic meningitis, *Parasitol. Today* **7**:151–153.

Ancelle, T., Dupouy-Camet, J., Bougnoux, M. E., Fourestie, V., Petit, H., Mougeot, G., Nozais, J. P., and Lapierre, J., 1988, Two outbreaks of trichinosis caused by horse meat in France in 1985, *Am. J. Epidemiol.* **127**:1302–1311.

Armelagos, G. J., 1991, Human evolution and the evolution of human disease, *Ethn.Dis.* **1**:21–26.

Asaishi, K., Nishino, C., Ebata, T., Totsuka, M., Hayasaka, H., and Suzuki, T., 1980, Studies on the etiologic mechanism of anisakiasis. 1. Immunological reactions of digestive tract induced by *Anisakis* larva, *Gastroenterol. Jpn.* **15**: 120–127.

Asaishi K., Nishino, C., Totsuka, M., Hayasaka, H., and Suzuki, T., 1980, Studies on the etiologic mechanism of anisakiasis. 2. Epidemiologic study of inhabitants and questionaire survey in Japan, *Gastroenterol. Jpn.* **15**:128–134.

Audicana, L., Audicana, M. T., Fernandez de Corres, L., and Kennedy, M. W., 1997, Cooking and freezing may not protect against allergenic reactions to ingested *Anisakis simplex* antigens in humans. *Vet. Rec.* **140**:235.

Audicana, M. T., Ansotegui, I. J., Fernandez de Corres, L., and Kennedy, M. W., 2002, *Anisakis simplex*: Dangerous- dead and alive? *Trends Parasitol.* **18**:20–25.

Beaver, P. C., and Rosen, L., 1964, Memorandum on the first report of *Angiostrongylus* in man, by Nomura and Lin, 1945. *Am. J. Trop. Med. Hyg.* **13**:589–590.

Beaver, P. C., Yung, R. C., and Cupp, E. W., 1984, *Clinical Parasitology*, 9th edn. Lea & Febiger, Philadelphia, pp. 345–346.

Bessonov, A. S., 1994, Trichinellosis in the former USSR. Epidemic situation (1988–1992). In Campbell, W. C., Pozio, E., and Bruschi, F. (eds), *Trichinellosis*, Instituto Superiore di Sanità Press, Rome, Italy, pp. 505–510.

Bouree, P., Paugam, A., and Petithory, J.-C., 1995, Anisakidosis: Report of 25 cases and a review of the literature, *Comp. Immun. Microbiol. Infect. Dis.* **18**:75–84.

Brown, R. C., and Girardeau, H. R., 1977, Transmammary passage of *Strongyloides* sp. larvae in the human host, *Am. J. Trop. Med. Hyg.* **26**:215–219.

Bruschi, F., and Murrell, K. D., 2002, New aspects of human trichinellosis: The impact of new *Trichinella* species. *Postgrad. Med. J.* **78**:15–22.

Bura, M. W. T., and Willett, W. C., 1977, An outbreak of trichinosis in Tanzania, *East Afr. Med. J.* **54**:185–193.

Cai, S. W., Zhou, S. Y., Wang, J. Q., Li, S. Y., Zhu, X. L., Wang, J. J., and Xue, J. R., 1988, A bacteriological and helminthological investigation of a sewage-irrigate area in a Beijing suburb. *Biomed. Environ. Sci.* **1**:332–338.

Campbell, W. C., 1983, Historical introduction, In Campbell, W. C. (ed), *Trichinella and Trichinellosis*, Plenum Press, New York pp. 1–30.

Campbell, W. C., 1997, The worm and the tumor: Reflections on Fibiger's nobel prize, *Prospectives Biol. Med.* **40**:498–504.

Capo, V., and Despommier, D. D., 1996, Clinical aspects of infection with *Trichinella* spp., *Clin. Microbiol. Rev.* **9**:47–54.

Céspedes, R., Salas, J., Mekbel, S., Troper, L., Müllner, F., and Morera, P., 1967, Granulomas entéricos y linfácticos con intensa eosinophilia tisular producidos por un estrongilídeo (*Strongylata*), *Acta Méd. Costarric.* **10**:235–255.

Chai, J. Y., Han, E. T., Shin, E. H., Park, J. H., Chu, J. P., Hirota, M., Nakamura-Uchiyama, F., and Nawa, Y., 2003, An outbreak of gnathostomiasis among Korean emigrants in Myanmar, *Am. J. Trop. Med. Hyg.* **69**:67–73.

Chan, M. -S., 1997, The global burden of intestinal nematode infections—fifty years on, *Parasitol. Today* **13**:438–443.

Chandra, C. V., and Khan, R. A., 1988, Nematode infestation of fillets from Atlantic cod, *Gadus morhua*, off eastern Canada, *J. Parasitol.* **74**:1038–1040.

Chau, T. T. H., Thwaites, G. E., Chuong, L. V., Sinh, D. X., and Farrar, J. J., 2003, Headache and confusion: The dangers of a raw snail dinner, *The Lancet* **361**:1866.

Chen, H. T., 1935, Un nouveau nematode pulmonaire: *Pulmonema cantonensis* n.g.n.sp., des rats de Canton, *Ann. Parasitol. Hum. Comp.* **13**:312–317.

Cheng, T. C., 1965, The American oyster and clam as experimental intermediate hosts of *Angiostrongylus cantonensis*, *J. Parasitol.* **51**:296.

Choi, W. Y., and Chang, K., The incidence of parasites found on vegetables, *Kisaengchunghak Chapchi.* **5**:153–158.

Cromptom, D. W., 1999. How much human helminthiasis is there in the world? *J. Parasitol.* **85**:397–403.

Cross, J. H., 1987, Public health importance of *Angiostrongylus cantonensis* and its relatives, *Parasitol. Today* **3**:367–369.

Crowley, J. J., 1995, Cutaneous Gnathostomiasis, *J. Am. Acad. Dermatol.* **33**:825–828.

Dame J. B., Murrell, K. D., Worley, D. E., and Schad, G. A., 1987, *Trichinella spiralis*: Genetic evidence for synanthropic subspecies in sylvatic hosts, *Exp. Parasitol.* **64**:195–203.

De Boni U, Lenczner, M. M., and Scott, J. W., 1977, Autopsy of an Egyptian mummy (Nakht-ROM I). 6. *Trichinella spiralis* cysts, *Can Med. Assoc. J*. **117**:472.

del Pozo, M. D., Moneo, I., de Corres, L. R., Audicana, M. T., Munoz, D., Fernandez, E., Navarro, J. A., and Garcia, M., 1996, Laboratory determinations in *Anisakis simplex* allergy, *J. Allergy Clin. Immunol.* **97**:977–984.

Deardorff, T. L., and Kent, M. L., 1989, Prevalence of larval *Anisakis simplex* in pen-reared and wild-caught salmon (Salmonidae) from Puget Sound, Washington, *J. Wildl. Dis.* **25**:416–419.

Deardorff, T. L., Kayes, S. G., and Fukumura, T., 1991, Human anisakiasis transmitted by marine food products, *Hawaii Med. J.* **50**:9–16.

Declerck, D., 1988, Presence de larves de *Anisakis simplex* dans le hareng (*Clupea harangus* L.), *Rev. Agric.* **41**:971–980.

Desowitz, R. S., 1981, *New Guinea Tapeworms and Jewish Grandmothers*. WW Norton & Company, New York.

Despommier, D. D., 1990, *Trichinella spiralis*: The worm that would be virus, *Parasitol. Today* **6**:193–196.

Despommier, D. D., 1998, How does *Trichinella spiralis* make itself at home? *Parasitol. Today* **14**:318–323.

Despommier, D. D., Sukhdeo, M., and Meerovitch, E., 1978, *Trichinella spiralis*: Site selection of the larva during the enteral phase of the infection in mice, *Exp. Parasitol.* **44**:209–215.
Despommier, D. D., Gwadz, R. W., Hotez, P. J., and Knirsch, C. A., 2000, *Trichinella spiralis*, In Despommier, D. D., Gwadz, R. W., Hotez, P. J., and Knirsch, C. A. (eds), *Parasitic Diseases*, 4th edn. Apple Tree Publications, New York, pp. 124–132.
Diaz Camacho, S. P., Willms, K., de la Cruz Otero Mdel, C., Zazueta Ramos, M. L., Bayliss, G. S., Castro Velazquez, R., Osuna Ramirez, I., Bojorquez Contreras, A., Torres Montoya, E. H., and Sanchez Gonzales, S., 2003, Acute outbreak of gnathostomiasis in a fishing community in Sinaloa, Mexico, *Parastiol. Int.* **52**:133–40.
Dick, T. A., and Pozio, E., 2001, *Trichinella* spp. and trichinellosis, In Samuel, W. M., Pybus, M. J., and Kocau, A. A. (eds), *Parasitic Diseases of Wild Mammals*, 2nd edn. Iowa State University Press, Ames, IA, pp. 380–396.
Dupouy, J., 2000, Trichinellosis: A worldwide Zoonosis, *Vet. Parasitol.* **93**:191–200.
Dupouy-Camet, J., Soule, C., and Ancelle, T., 1994, Recent news on trichinellosis: Another outbreak due to horse meat consumption in France in 1993, *Parasite* **1**:99–103.
Dworkin, M. S., Gamble, H. R., Zarlenga, D. S., and Tennican, P. O., 1996, Outbreak of trichinellosis associated with eating cougar jerky, *J. Inf. Dis.* **174**:663–666.
Eberhard, M. L., and Busillo, C., 1999, Human *Gongylonema* infection in a resident of New York City, *Am. J. Trop. Med. Hyg.* **61**:51–52.
Feinstein, R. J., and Rodriguez-Valdes, J., 1984, Gnathostomiasis, or larva migrans Profundus, *J. Am. Acad. Dermatol.* **11**:738–740.
Forbes, L. B., Parker, S., and Scandrett, W. B., 2003, Comparison of a modified digestion assay with trichinoscopy for the detection of *Trichinella* larvae in pork, *J. Food Prot.* **66**:1043–1046.
Friedriech, N., 1862, Ein Beitrag zur Pathologie der Trichinenkrankheit beim Menschen, *Virchow Arch. Path. Anat.* **25**:399–413.
Gajadhar, A. A., and Gamble, H. R., 2000, Historical perspectives and current global challenges of *Trichinella* and trichinellosis, *Vet. Parsitol.* **93**:183–189.
Gajadhar, A. A., and Forbes, L. B., 2002, An internationally recognized quality assurance system for diagnostic parasitology in animal health and food safety, with example data on trichinellosis, *Vet. Parasitol.* **103**:133–140.
Gamble, H. R., 1996, Detection of trichinellosis in pigs by artificial digestion and enzyme immunoassay, *J. Food Prot.* **59**:295–298.
Gamble, H. R., Bessonov, A. S., Cuperlovic, K., Gajadhar, A. A., van Knapen, F., Noeckler, K., Schenone, H., and Zhu, X., 2000, International Commission on Trichinellosis: Recommendations on methods for the control of *Trichinella* in domestic and wild animals intended for human consumption, *Vet. Parasitol.* **93**:393–408.
Gamble, H. R., Pozio, E., Bruschi, R., Nöckler, K., Kapel, C. M., and Gajadhar, A. A., 2004, International Commission on Trichinellosis: Recommendations on the use of serological tests for the detection of *Trichinella* infections in animals and man, *Parasite* **11**:3–13.
Geiger, S. M., Laitano, A. C., Sievers-Tostes, C., Agostini, A. A., Schulz-Key, H., and Graeff-Teixeira, C., 2001, Detection of the acute phase of abdominal angiostrongyliasis with parasite-specific IgG enzyme linked immunosorbent assay, *Mem. Inst. Oswaldo Cruz* **96**:515–518.
Giuffra, E., Kijas, J. M. H., Amarger, V., Carlborg, O., Jeon, J. –T., and Andersson, L., 2000, The origin of the domestic pig: Independent domestication and subsequent introgression, *Genetics* **154**:1785–1791.
Gomez, B., Tabar, A. I., Tunon, T., Larrinaga, B., Alvarez, M. J., Garcia, B. E., and Olaguibel, J. M., 1998, Eosinophilic gastroenteritis and *Anisakis*, *Allergy* **53**:1148–1154.

Gould, S. E., 1970, History, In Gould, S. E. (ed), *Trichinosis in Man and Animals*, Charles C. Thomas, Springfield, IL, pp. 3–18.

Graczyk T. K., Cranfield M. R., Fayer, R., and Bixler, H., 1999a, House flies (*Musca domestica*) as transport hosts of *Cryptosporidium parvum*, *Am. J. Trop. Med. Hyg.* **61**:500–504.

Graczyk T. K., Fayer, R., Cranfield, M. R., Mhangami-Ruwende, B., Knight, R., Trout, J. M., and Bixler, H., 1999b, Filth flies are transport hosts of *Cryptosporidium parvum*, *Emerg. Infect. Dis.* **5**:726–727.

Gutierrez, Y., 1999, Other tissue nematode infections, In Guerrant, R. L., Walker, D. H., and Weller, P. F. (eds), *Tropical Infectious Diseases: Principles, Pathogens and Practice*, Churchill Livingstone, New York, pp. 933–948.

Hoberg, E. P., Henny, C. J., Hedstrom, O. R., and Grove, R. A., 1997, Intestinal helmints of river otters (*Lutra Canadensis*) from the Pacific Northwest, *J. Parasitol.* **83**:105–110.

Hulbert, T. V., Larsen, R. A., and Chandrasoma, P. T., 1992, Abdominal angiostrongyliasis mimicking appendicitis and Meckels' diverticulum: Report of a case in the United states and review, *Clin. Infect. Dis.* **14**:836–840.

Kapel, C. M. O., 2000a, Host diversity and biological characteristics of the *Trichinella* genotypes and their effect on transmission, *Vet. Parasitol.* **93**:263–278.

Kapel, C. M. O., 2000b, Experimental infection with sylvatic and domestic *Trichinella* spp. in wild boars infectivity, muscle distribution, and antibody response, *J. Parasitol.* **93**:263–278.

Kapel, C. M. O., and Gamble, H. R., 2000, Infectivity, persistence, and antibody response to domestic and sylvatic *Trichinella* spp. in experimentally infected pigs, *Int. J. Parasitol.* **30**:215–221.

Kapel, C. M. O., Henriksen, S. S., and Nansen, P., 1997, Prevalence of *Trichinella nativa* in Greenland according to zoogeography, In Ortega-Pierres, H. R., Gamble, F. Van Knapen and Wakelin, D. (eds), *Trichinellosis ICT9*, Centro de Investigacion y Estudios Avanzados del Instituto Politecnico Nacional, Mexico, D.F., Mexico, pp. 591–597.

Kapel, C. M. O., Webster, P., Lind, P., Pozio, E., Henriksen, S. A., Murrell, K. D., and Nansen, P., 1998, *Trichinella spiralis*, *Trichinella brivoti*, and *Trichinella nativa*: Infectivity, larval distribution in muscle, and antibody response afger experimental infection of pigs, *Parasitol. Res.* **84**:264–271.

Kapel, C. M. O., Pozio, E., Sacchi, L., and Prestrud, P., 1999, Freeze tolerance, morphology, and RAPD-PCR identification of *Trichinella nativa* in naturally infected arctic foxes, *J. Parasitol.* **85**:144–147.

Kennedy, M. W., Tierney, J., Ye, P., McMonagle, F. A., McIntosh, A., McLaughlin, D., and Smith, J. W., 1988, The secreted and somatic antigens of the third stage larva of *Anisakis simplex*, and antigenic relationship with *Ascaris suum*, *Ascaris lumbricoides*, and *Toxocara canis*, *Mol. Biochem. Parasitol.* **31**:35–46.

Kim, D. Y., Stewart, T. B., Bauer, R. W., and Mitchell, M., 2002, *Parastrongylus* (=*Angiostrongylus*) *cantonensis* now endemic in Louisiana wildlife, *J. Parasitol.* **88**:1024–1026.

Kliks, M. M., and Palumbo, N. E., 1992, Eosinophilic meningitis beyond the Pacific basin: The global dispersal of a peridomestic zoonosis caused by *Angiostrongylus cantonensis*, the nematode lungworm of rats, *Soc. Sci. Med.* **34**:199–212.

Kociecka, W., 2000, Trichinellosis: Human disease, diagnosis, and treatment, *Vet. Parasitol.* **93**:365–383.

Kramer, M. H., Greer, G. J., Quinonez, J. F., Padilla, N. R., Hernandez, B., Arana, B. A., Lorenzana, R., Morera, P., Hightower, A. W., Eberhard, M. L., and Herwaldt, B. L., 1998, First reported outbreak of abdominal angiostrongyliasis, *Clin. Infect. Dis.* **26**:365–372.

La Rosa, G., Marucci, G., Zarlenga, D. S., and Pozio, E., 2001, *Trichinella pseudospiralis* populations of the Palearctic region and their relationship with populations of the Nearctic and Australian regions, *Int. J. Parasitol.* **31**:297–305.

La Rosa, G., Marucci, G., Zarlenga, D. S., Casulli, A., Zarnke, R. L., and Pozio, E., 2003, Molecular identification of natural hybrids between *Trichinella nativa* and *Trichinella* T6 provides evidence of gene flow and ongoing genetic divergence, *Int. J. Parasitol.* **33**:209–216.

Laffon-Leal, S. M., Vidal-Martinez, V. M., and Arjona-Torres, G., 2000, 'Cebiche'—a potential source of human anisakiasis in Mexico? *J. Helminthol.* **74**:151–154.

Leidy, J., 1846, Remarks on trichina, *Proc. Acad. Anat. Sci.* **3**:107–108.

Leuckart, R., 1859, Untersuchungen uber *Trichina spiralis*. Zugleich um Beitrag zur Kenntnis der Wurmkrankenheiten, Winter, Leipzig.

Lo Re, V., and Gluckman, J., 2001, Eosinophilic meningitis due to *Angiostrongylus cantonensis* in a returned traveler: Case report and a review of the literature, *Clin. Inf. Dis.* **33**:e112–e115.

Loría-Cortés, R. M., and Lobo-Sanahuja, J. F., 1980, Clinical abdominal angiostrongylosis. A study of 116 children with intestinal eosinophilic granuloma caused by *Angiostrongylus costaricensis, Am. J. Trop. Med. Hyg.* **29**:538–544.

Magana, M., Messina, M., Bustamante, F., and Cazarin, J., 2004, Gnathostomiasis: Clinicopathologic study, *Am. J. Dermatopathol.* **26**:91–95.

Marinculic A., Gamble, H. R., Urban, J. F., Rapic, D., Zivicnjak, T., Smith, H. J., and Murrell, K. D., 1991, Immunity in swine inoculated with larvae or extracts of a pig isolate and a sylvatic isolate of *Trichinella spiralis, Am. J. Vet. Res.* **52**:754–758.

Marti, H. P., Murrell, K. D., and Gamble, H. R., 1987, *Trichinella spiralis*: Immunization of pigs with newborn larval antigens, *Exp. Parasitol,* **63**:68–73.

Menard, A., Dos Santos, G., Dekumyoy, P., Ranque, S., Delmont, J., Danis, M., Bricaire, F., and Caumes, E., 2003, Imported cutaneous gnathostomiasis: A report of five cases, *Trans. R. Soc. Trop. Med. Hyg.* **97**:200–202.

Miyazaki, I., 1954, Studies on *Gnathostoma* occurring in Japan (Nematoda: Gnathostomidae). 2. Life history of *Gnathostoma* and morphological comparison of its larval forms, *Kyushu Mem. Med. Sci.* **5**:123–140.

Miyazaki, I., 1966, *Gnathostoma* and gnathostomiasis in Japan. In Morishita, K., Komiya, Y., and Matsubayashi, H. (eds), *Progress of Medical Parasitology in Japan 3*, Meguro Parasitological Museum, Tokyo, Japan, pp. 529–586.

Montoro, A., Perteguer, M. J., Chivato, T., Laguna, R., and Cuellar, C., 1997, Recidivous acute urticaria caused by *Anisakis simplex, Allergy* **52**:985–991.

Moorhead, A., Grunenwald, P. E., Dietz, V. J., and Schantz, P. M., 1999, Trichinellosis in the United States. 1991–1996: Declining but not gone, *Am. J. Trop. Med. Hyg.* **60**:66–69.

Moreno-Ancillo, A., Caballero, M. T., Cabañas, R., Contreras, J., Martin-Barroso, J. A., Barranco, P., and López-Serrano, M. C., 1997, Allergic reactions to *Anisakis simplex* parasitizing seafood, *Ann. Allergy, Asthma Immunol.* **79**:246–250.

Morera, P., 1967, Granulomas entéricos y linfáticos con intensa eosinophilia tisular producidos por un estrongilídeo (*Strongylata*; Raillet y Henry, 1913). 2. Aspecto parasitológico (nota previa), *Acta Méd. Costarric.* **10**:257–265.

Morera, P., 1973, Life history and redescription of *Angiostrongylus costaricensis* Morera and Céspedes, 1971, *Am. J. Trop. Med. Hyg.* **22**:613–621.

Morera, P., 1988, Angiostrongilíase abdominal. Um problema de saúde pública? *Rev. Soc. Bras. Med. Trop.* **21**:81–83.

Morera, P., Perez, F., Mora, F., and Castro, L., 1982, Visceral larva migrans-like syndrome caused by *Angiostrongylus costaricensis, Am. J. Trop. Med. Hyg.* **31**:67–70.

Moser, M., and Hsieh, J., 1992, Biological tags for stock separation in Pacific herring *Clupea harengus pallasi* in California, *J. Parasitol.* **78**:54–60.

Murrell, K. D., and Pozio, E., 2000, Trichinellosis: The Zoonosis that won't go quietly, *Int. J. Parasitol.* **30**:1339–1349.

Murrell, K. D., Stringfellow, F., Dame, J. B., Leiby, D. A., Duffy, C., and Schad, G. A., 1987, *Trichinella spiralis* in an agricultural ecosystem. 2. Evidence for natural transmission of *Trichinella spiralis spiralis* from domestic swine to wildlife, *J. Parasitol.* **73**:103–109.

Murrell, K. D., Lichtenfels, R. J., Zarlenga, D. S., and Pozio, E., 2000, The systematics of the genus *Trichinella* with a key to species, *Vet. Parasitol.* **93**:293–307.

Namiki, M., 1989,. A history of research into gastric anisakiasis in Japan, In Ishikura, H. and Namiki, N. (eds), *Gastric Anisakiasis in Japan*, Springer-Verlag, New York, pp. 13–18.

Navitsky, R. C., Dreyfuss, M. L., Shrestha, J., Khatry, S. K., Stoltzfus, R. J., and Albonico, M., 1998, *Ancylostoma duodenale* is responsible for hookworm infections among pregnant women in the rural plains of Nepal, *J. Parasitol.* **84**:647–651.

Nawa, Y., 1991, Historical review and current status of gnathostomiasis in Asia, *Southeast Asian J. Trop. Med. Public Health.* **22**(Suppl):217–219.

Needham, C., Kim, H. T., Hoa, N. V., Cong, L. D., Michael, E., Drake, L., Hall, A., and Bundy, D. A., 1998, Epidemiology of soil-transmitted nematode infections in Ha Nam province, Vietnam, *Trop. Med. Int. Health* **3**:904–912.

New, D., Little, M. D., and Cross, J., 1995, *Angiostrongylus cantonensis* infection from eating raw snails, *N. Engl. J. Med.* **332**:1105–1106.

Nöckler, K., Pozio, E., Voigt, W. P., and Heidrich, J., 2000, Detection of *Trichinella* infection in food animals, *Vet. Parasitol.* **93**:335–350.

Nwosu, A. B., 1981, Human neonatal infections with hookworms in an endemic area of Southern Nigeria. A possible transmammary route, *Trop. Geogr. Med.* **33**:105–111.

Olteanu, G. H., 1997, New studies on epidemiology and control of trichinellosis in Romania, In Ortega-Pierres, Gamble, H. R., Van Knapen, F., and Wakelin, D. (eds), *Trichinellosis ICT9*, Centro de Investigacion y Estudios Avanzados del Instituto Politecnico Nacional, Mexico, D.F., Mexico, pp. 517–531.

Ogata, K., Nawa, Y., Akahane, H., Diaz Camacho, S. P., Lamothe-Argumedo, R. and Cruz-Reyes, A., 1998, Short Report: Gnathostomiasis in Mexico, *Am. J. Trop. Med. Hyg.* **58**:316–318.

Oshima, T., 1987, Anisakiasis–Is the sushi bar guilty? *Parasitol. Today* **3**:44–48.

Owen, R., 1835, Description of a microscopic entozoan infesting the muscles of the human body, *Trans. Zool. Soc.* **1**:315–324.

Pelaez, D., and Perez-Reyes, R., 1970, Gnatostomiasis humana en America, *Rev. Latinoma. Microbiol.* **12**:83–91.

Pien, F. D., and Pien, B. C., 1999, *Angiostrongylus cantonensis* eosinophilic meningitis, *Int. J. Infect. Dis.* **3**:161–163.

Pinkus, G. S., Coolidge, C., and Little, M. D., 1975, Intestinal anisakiasis. First report from North America, *Am. J. Med.* **59**:114–120.

Pozio, E., 2000a, Factors affecting the flow among domestic, synanthropic and sylvatic cycles of *Trichinella*, *Vet. Parasitol.* **93**:241–262.

Pozio, E., 2000b, Is horsemeat trichinellosis and emerging disease in EU? *Parasitol. Today* **16**:266.

Pozio, E., 2001a, New patterns of *Trichinella* infection, *Vet. Parasitol.* **98**:133–148.

Pozio, E., 2001b, Taxonomy of Trichinella and the epidemiology of infection in the Southeast Asia and Australian regions, *Southeast Asian J. Trop. Med. Public Health* **32**(Suppl 2):129–132.

Pozio, E., La Rosa, G., Rossi, P., and Fico, R., 1989, Survival of *Trichinella* muscle larvae in frozen wolf tissue in Italy, *J. Parasitol.* **75**:472–473.

Pozio, E., La Roza, G., and Amati, M., 1994a, Factors influencing the resistance of *Trichinella* muscle larvae to freezing, In Campbell, W. C., Pozio, E., and Bruschi, F. (eds), *Trichinellosis* Instituto Superiore di Sanità Press, Rome, Italy, pp. 173–178.

Pozio, E., Verster, A., Braack, L., DeMeneghi, K., and La Rosa, G., 1994b, Trichinellosis south of Sahara, In Campbell, W. C., Pozio, E., and Bruschi, F. (eds), *Trichinellosis*, Instituto Superiore di Sanità Press, Rome, Italy, pp. 527–532.

Pozio, E., La Rosa, G., Serrano, F. J., Barrat, J., and Rossi, P., 1996, Environmental and human influence on the ecology of *Trichinella spiralis* and *Trichinella brivoti* in Western Europe, *Parasitology* **113**:527–533.

Pozio, E., Owen, I. L., La Rosa, B., Sacchi, L., Rossi, P., and Corona, S., 1999, *Trichinella papuae* n.sp. (Nematoda), a new nonencapsulating species from domestic and sylvatic swine of Papua New Guinea, *Int. J. Parasitol.* **29**:1825–1839.

Pozio, E., Sacchini, D., Sacchi, L., Tamburrini, A., and Alberici, F., 2001, Failure of mebendazole in the treatment of humans with *Trichinella spiralis* infection at the stage of encapsulating larvae, *Clin. Inf. Dis.* **32**:638–642.

Pozio, E., Foggin, C. M., Marucci, G., La Rosa, G., Sacchi, L., Corona, S., Rossi, P., and Mukaratirwa, S., 2002, *Trichinella zimbabwensis* n.sp. (Nematoda), a new nonencapsulated species from crocodiles (*Crocodylus niloticus*) in Zimbabwe also infecting mammals, *Int. J. Parasitol.* **32**:1787–1799.

Pozio, E., Marucci, G., Casulli, A., Sacchi, L., Mukaratirwa, S., Foggin, C. M., and La Rosa, G., 2004, *Trichinella papuae* and *Trichinella zimbabwensis* induce infection in experimentally infected varans, caimans, pythons, and turtles. *Parasitology* **128**:333–342.

Prociv, P., and Luke, R. A., 1995, Evidence for larval hypobiosis in Australian strains of *Ancylostoma duodenale*, *Trans. R. Soc. Trop. Med. Hyg.* **89**:379.

Rockiene A., and Rocka, V. S., 1997, Seroprevalence studies on human trichinellosis in Lithuania, In Ortega-Pierres, Gamble, H. R., Van Knapen, F., and Wakelin, D. (eds), *Trichinellosis ICT9*, , Centro de Investigacion y Estudios Avanzados del Instituto Politecnico Nacional, Mexico, D.F. Mexico, pp. 503–506.

Rodriquez, R., Agostini, A. A., Porto, S. M., Olivaes, A. J., Branco, S. L., Genro, J. P., Laitano, A. C., Maurer, R. L., and Graeff-Teixeira, C., 2002, Dogs may be a reservoir host for *Angiostrongylus costaricensis*, *Rev. Inst. Med. Trop. Sao Paulo* **44**:55–56.

Rojas-Molina, N., Pedraza-Sanchez, S., Torres-Bibiano, B., Meza-Martinez, H., and Escobar-Gutierrez, A., 1999, Gnatostomosis, and emerging foodborne zoonotic disease in Acapulco, Mexico, *Emerg. Infect. Dis.* **5**:264–266.

Rojekittikhun, W., Waikagul, H., and Chaiyasith, T., 2002, Fish as the natural second intermediate host of *Gnathostoma spinigerum*, *Southeast Asian J. Trop. Med. Public Health* **33** (Suppl 3):63–69.

Rosen, L., Loison, G., Laignet, J., and Wallace, G. D., 1967, Studies on eosinophilic meningitis. 3. Epidemiologic and clinical observations on Pacific islands and the possible etiologica role of *Angiostrongylus cantonensis*, *Am. J. Epidemiol.* **85**:17–44.

Ruiz-Maldonado, R., and Mosqueda-Cabrera, M. A., 1999, Human gnathostomiasis (nodular migratory eosonophilic panniculitis). *Int. J. Dermatol.* **38**:52–57.

Rusnak, J. M., and Lucey, D. R., 1993, Clinical Gnathostomiasis: Case Report and Review of the English-Language Literature. *Clin. Infect. Dis.* **16**:33–50.

Sakanari J. A., and Mckerrow, J. H., 1989, Anisakiasis, *Clin. Microbiol. Rev.* **2**:278–284.

Schantz, P. M., and McAuley, J., 1991, Current status of food-borne parasitic zoonoses in the United States, *Southeast Asian J. Trop. Med. Public Health* **22**:65–71.

Shields, B. A., Bird, P., Liss, W. J., Groves, K. L., Olson, R., and Rossignol, P. A., 2002, The nematode *Anisakis simplex* in American shad (*Alosa sapidissima*) in Oregon rivers, *J. Parasitol.* **88**:1033–1035.

Slom, T. J., Cortese, M. M., Gerber, S. I., Jones, R. C., Holtz, T. H., Lopez, A. S., Zambrano, C. H., Sufit, R. L., Sakolvaree, Y., Chaicumpa, W., Herwaldt, B. L., and Johnson, S., 2002, An outbreak of eosinophilic meningitis caused by *Angiostrongylus cantonensis* in travelers returning from the Caribbean, *N. Engl. J. Med.* **346**: 668–675.

Sokolova, I. B., 1979, The effect of temperature on the viability of different species of *Trichinella*, *Voprosy Prirodnoi Ochagovosti Boleznei* **10**:185–187.

Smith, H. J., 1987, Vaccination of rats and pigs against *Trichinella spiralis* using the subspecies, *T. spiralis nativa. Can. J. Vet. Res.* **51**:370–372.

Smith, J. W., 1983, *Anisakis simplex* (Rudolphi, 1809, det. Krabbe, 1878) (Nematoda: Ascaridoidea): Morphometry of larvae from euphausiids and fish, and a review of the life-history and ecology, *J. Helminthol.* **57**:205–224.

Smith, J. W., and Wooten, R., 1975, Experimental studies on the migration of *Anisakis sp.* larvae into the flesh of herring *Clupea harrengus*, *Int. J. Parasitol.* **5**:133–136.

Steele, J. H., 2000, Food irradiation: A public health opportunity, *Int. J. Infect. Dis.* **4**:62–66.

Stoll, N. R., 1999, This wormy world. 1947, *J. Parasitol.* **85**:392–396.

Taira, K., Saeed, I., Permin, A., and Kapel, C. M., 2004, Zoonotic risk of *Toxocara canis* infection through consumption of pig or poultry viscera, *Vet. Parasitol.* **121**:115–124.

Taniguchi, Y., Hashimoto, K., Ichikawa, S., Shimizu, M., Ando, K., and Kotani, Y., 1991, Human Gnathostomiasis, *J. Cutan. Pathol.* **18**:112–115.

Ubelaker, J. E., and Hall, N. M., 1979, First report of *Angiostrongylus costaricensis* Morera and Céspedes 1971, in the United States, *Am. J. Parasitol.* **65**:307.

van Knapen, F., 2000, Control of trichinellosis by inspection and farm management practices, *Vet. Parasitol.* **93**:385–392.

van Thiel, P. H., Kuipers, F. C., and Roskam, R. T. H., 1960, A nematode parasite to herring causing acute abdominal syndromes in man, *Trop. Geogr. Med.* **2**:97–113.

Virchow, R., 1860, Recherches sur le développement du *Trichina spiralis*, *C.R. Acad. Sci.* **49**:660–662.

Waisberg, J., Corsi, C. E., Regelo, M. V., Vieira, V. T. T., Bromberg, S. H., dos Dantos, P. A., and Monteiro, R., 1999, Jejunal perforation caused by abdominal angiostrongyliasis, *Rev. Inst. Med. Trop. S. Paulo* **41**:325–328.

Waite, C. H., and Gorrie, R., 1935, *Gongylonema* in the role of a human parasite, *J. Am. Med. Assoc.* **105**:23–24.

Wakelin, D., and Denham, D. A., 1983, The immune response, In Campbell, W. C. (ed), *Trichinella and Trichinosis*, Plemum Press, New York, pp. 265–308.

Wilson, M. E., Lorente, C. A., Allen, J. E., and Eberhard, M. L., 2001, *Gongylonema* infection of the mouth in a resident of Cambridge, Massachussets, *Clin. Infect. Dis.* **32**:1378–1380.

Worley, D. E., Zarlenga, D. S., and Seesee, F. M., 1990, Freezing resistance of a *Trichinella spiralis nativa* isolate from a gray wolf, *Canis lupus*, in Montana, with observations on genetic and biological characteristics of the biotype, *J. Helminthol. Soc. Wash.* **57**:57–60.

Wu, S. S., French, S. W., and Turner, J. A., 1997, Eosinophilic ileitis with perforation caused by *Angiostrongylus (Parastrongylus) costaricensis*. A case study and review, *Arch. Pathol. Lab. Med.* **121**:989–991.

Yao, C., Prestwood, A. K., and McGraw, R. A., 1997, *Trichinella spiralis* (T1) and *Trichinella* T5: A comparison using animal infectivity and molecular biology techniques, *J. Parasitol.* **83**:88–95.

Zenker, F. A., 1860, Uber di Trichenen-Krankheiten des Menschen, *Virchow Arch. Path. Anat.* **18**:561–572.

CHAPTER 7

Foodborne Trematodes

Ann M. Adams

7.1 PREFACE

Digenetic trematodes comprise one of the most common groups of parasitic worms. They have a complex life cycle involving both sexual and asexual reproduction. The parasites require at least two hosts, the first of which is usually a mollusc. Over a 100 species of digenetic trematodes have been recorded from human hosts, but many cases may be spurious or accidental (Crompton, 1999). The association between humans and trematodes is longstanding, with reports of *Schistosoma* eggs in Egyptian mummies and of *Clonorchis* eggs in a 2000-year-old corpse from the Chu Dynasty in China.

In the present chapter, trematodes from six genera will be discussed (Table 7.1), including their geographic distribution, life cycles, epidemiology, and clinical aspects of disease. These trematodes account for approximately 40 million cases of human infections worldwide. Other mammalian hosts may also be infected, often acting as reservoir hosts for the parasites. Infection of the final host occurs from the consumption of foods contaminated with infectious larvae (metacercariae). The implicated sources of infection are usually freshwater or anadromous fishes, freshwater crustaceans, or aquatic vegetation, depending on the species of trematode involved.

Many of the species of food-borne trematodes are endemic in developing nations and have significant impact on public health. Developed countries, including the United States, also have foci of trematode infections. Some species are naturally present in the United States, such as *Nanophyetus salmincola*, *Fasciola hepatica*, and *Paragonimus kellicotti*. Others are introduced by the importation of contaminated foods or infected intermediate hosts, the development of new food habits, and the immigration of peoples from endemic countries.

The control of the trematode populations and prevention of their respective diseases are dependent on understanding the transmission and development of the parasites. Disruption of the life cycles is necessary for the control and eradication of disease, which can be attained through the treatment of infected populations, improved sanitation and agricultural practices, or the alteration of food habits.

7.2 *PARAGONIMUS* SPP.

7.2.1 Introduction

Trematodes of the genus *Paragonimus* invade various organs, but are found primarily within the lungs of the definitive hosts. WHO (1995) estimated that over 20 million people are infected worldwide. Over 40 species of *Paragonimus* have

Table 7.1. Comparison of foodborne trematodes.

Species	Estimate of human cases[a] (millions)	Usual infection site for adult parasite	Size range, adults (mm)	Size range, eggs (μm)	First intermediate hosts	Second intermediate hosts
Paragoninus spp.	20.68	Lungs	7–20 × 4–8	80–120 × 45–65	Snails	Freshwater crabs
Clonorchis sinensis	7.01	Bile Ducts	8–25 × 1.5–5	26–35 × 12–19	Snails	Freshwater fishes
Opisthorchis spp.	10.33	Bile Ducts	7–12 × 2–3	19–30 × 12–17	Snails	Freshwater fishes
Fasciola hepatica	2.39	Bile Ducts	20–40 × 8–13	130–150 × 63–90	Snails	Aquatic vegetation[b]
Fasciolopsis buski	0.21	Small Intestine	20–75 × 8–20	115–158 × 63–90	Snails	Aquatic vegetation[b]
Nanophyetus salmincola	0.019	Small Intestine	0.8–2.5 × 0.3–0.5	64–97 × 34–55	Snails	Freshwater and anadromous fishes

[a] WHO, 1995.
[b] Cercariae shed their tails and encyst as metacercariae on aquatic vegetation.

been described although not all may be valid. At least 15 species have been reported from humans, of which eight are considered important (Cross, 2001; Sinniah, 1997). *P. westermani* is the most studied since it has the broadest geographical distribution and accounts for most cases of pulmonary paragonimiasis. This species has been reported from China, Japan, Korea, Thailand, Taiwan, the Philippines, Indonesia, India, Nepal, and Manchuria. In addition to *P. westermani*, other species of medical importance in China are *P. skrjabini, P. hueitungensis*, and *P. heterotremus*. The latter is also distributed in southeast Asia, including Thailand. *P. miyazakii* is endemic in Japan in addition to *P. westermani*. Two species are endemic in Africa, *P. africanus* and *P. uterobilateralis*. In Latin America, including Peru and Ecuador, *P. mexicanus* is of greatest concern for human infections, although other species have been reported (e.g., *P. amazonicus* and *P. inca*. Note that *P. peruvianus, P. ecuadoriensis*, and *P. caliensis* are considered synonyms of *P. mexicanus*) (WHO, 1995). The North American species, *P. kellicotti*, is primarily known from wild and domestic carnivores, especially cats, dogs, pigs, minks, fishers, foxes, and muskrats (Schell, 1985; Yokogawa, 1965). However, human infections derived domestically have been reported (Castilla *et al.*, 2003; DeFrain and Hooker, 2002; Mariano *et al.*, 1986; Pachucki *et al.*, 1984; Procop *et al.*, 2000) and Weina and England (1990) suggest that the trematode may be more prevalent.

7.2.2 Life Cycle

The adult trematode of *P. westermani* is thick and oval shaped; 7–20 mm long, 4–8 mm wide, and 2.5–5 mm thick (Cross, 2001; Sinniah, 1997). The two suckers are equal in size and the tegument is spinose. The ovary is lobate and anterior of the two deeply lobed testes. The uterus lies opposite of the ovary. The cecae are unbranched and extend to the posterior end of the body. In addition to humans, other definitive hosts of this species include dogs, pigs, cats, tigers, leopards, panthers, and other wild cats (Yokogawa, 1965). Depending on the species of *Paragonimus*, weasels, badgers, monkeys, rats, and other carnivorous mammals in addition to those previously mentioned may also act as definitive hosts. Generally, the adults live in the respiratory tract of their hosts, although extrapulmonary infections can occur and are discussed below.

Two worms are usually present in a cyst although the trematodes are hermaphroditic. Eggs are passed unembryonated from the adults, through the cyst and into the bronchi. They are then either coughed up in thick sputum or are swallowed and passed out in the feces. The eggs are thick-shelled, operculate, and generally oval in shape (Fig. 7.1). If the eggs reach water, they will embryonate and hatch in 2 to 3 weeks. The resulting miracidia swim in search of appropriate snail intermediate hosts, of the families Hydrobiidae, Pleurocercidae, and Thiaridae (Cross, 2001). Numerous species of snails act as hosts for *Paragonimus* spp. in China and southeast Asia, demonstrating little host specificity. In North America however, only the snail *Pomatiopsis lapidaria* is known as the first intermediate host of *P. kellicotti* (WHO 1995; Yokogawa, 1965).

Within the molluscan host, the parasite develops as a sporocyst, then through two generations of rediae, prior to forming cercariae. These cercariae have small, knobby tails with spines (Fig. 7.2) and are poor swimmers. Therefore, after leaving

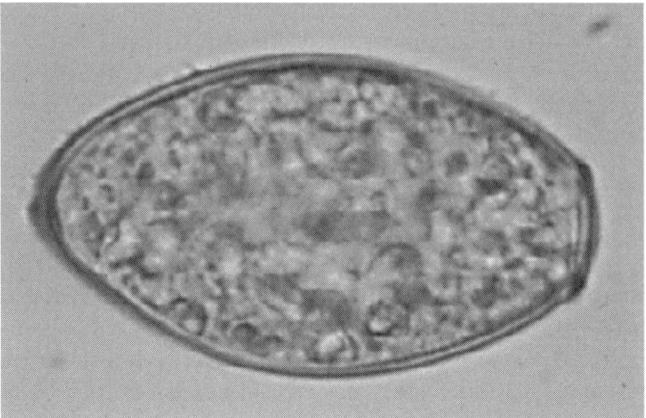

Figure 7.1. Egg of *Paragonimus westermani*.

the snails, the cercariae crawl along the sediment and rocks in search of crustacea to act as second intermediate hosts. Inside these hosts, the trematodes encyst in the viscera, gills, and muscles and form metacercariae. Numerous species of freshwater crabs and crayfish have been reported as second intermediate hosts of *Paragonimus* spp., again demonstrating little host specificity. Several species of crayfish of the genus *Cambarus*, including *C. propinquus*, are hosts for metacercariae of *P. kellicotti* in North America (Yokogawa, 1965). In Latin America, species of freshwater crabs from three genera (*Hypolobocerca*, *Potamocarcinum*, and *Pseudothelphusa*) are recognized as second intermediate hosts of *P. mexicanus* (WHO, 1995). In addition to penetration and encystment, experimental evidence indicates that crustaceans may also become infected from consuming infected snails (Noble, 1963).

When the infected crustacean is eaten raw or undercooked, the trematodes excyst in the intestine and pass through the intestinal wall. The worms penetrate the diaphragm and pleura, and encyst in the lungs. The trematodes reach maturity and

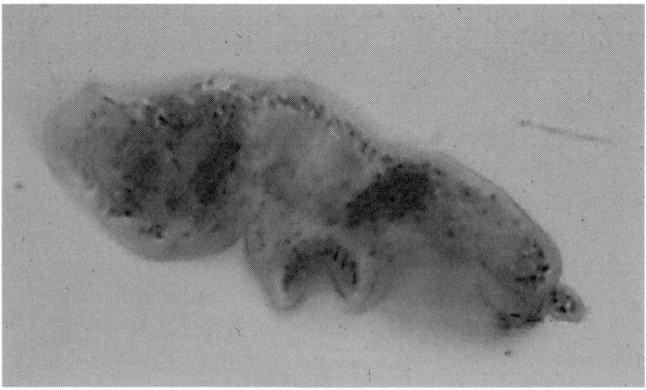

Figure 7.2. Cercaria of *Paragonimus westermani* showing small, knob-like tail.

begin to lay eggs in 5 to 6 weeks after infection. Although infections may persist up to 20 years, most adult worms die in about 6 years (Sinniah, 1997). If the host is not suitable for the trematodes to develop into adults, the worms will encyst in the tissues of that host and remain until consumed by an appropriate host (Sinniah, 1997). For example, people in Japan became infected after eating raw meat from a wild boar. Immature worms of *Paragonimus* were found in the muscle of the boar (Miyasaki and Hirose, 1976).

7.2.3 Epidemiology

Paragonimiasis is more common in children, with a peak prevalence in 10–14 year olds (Kum and Nchinda, 1982; WHO 1995). The prevalence also tends to be higher among males in comparison to females, but the difference is not always statistically significant (Kum and Nchinda, 1982; Moyou-Somo *et al.*, 2003). The exception to this observation concerns the use of raw crabs by females in Africa to aid fertility, resulting in higher infection rates. However, Kum and Nchinda (1982) conducted a survey which indicated this practice has decreased, with only 4% of the respondents considering this practice to be effective.

In general, humans become infected from the ingestion of infected crustaceans that are either raw or undercooked. Some traditional preparations of crustaceans by marinating, pickling, or salting may give the appearance of cooked flesh, but the metacercariae maintain their infectivity. In Korea, crab marinated in soy sauce is a major source of paragonimiasis, as are crabs soaked in wine (drunken crabs) in China (Cross, 2001). Throughout Asia, cultures have various dishes consisting of raw crab, shrimp, and crayfish which provide transmission of the trematode.

Improper cooking may account for many infections in cultures that infrequently consume raw crustaceans. Kum and Nchinda (1982) reported that only 12% surveyed in Cameroon admitted eating raw crabs; however, the local delicacy consisted of a preparation of crabs and plantain baked in hot ashes. The period of baking was not always sufficient to inactivate metacercariae. The authors also noted that children would roast crabs directly in the fire, but not long enough, since the metacercariae present were still viable. In a later study by Moyou-Somo *et al.* (2003), also in Cameroon, all children examined reported preparing crustaceans by roasting, boiling, or frying (in oil). However, duration of cooking was not determined by internal temperature or consistency of flesh, but rather by a color change of the crab shell. Sachs and Cumberlidge (1990) related that roasting of crabs by children was also common in Liberia, but that the claws and legs of crabs were often removed prior to roasting and children chewed these raw appendages while food preparation was underway. They proposed that this activity was the predominant route of infection for children and adolescents.

Cross-contamination of cooked foods with raw materials or utensils is another route of transmission for paragonimiasis. In Japan, crab soup is prepared by removing the shells and legs and chopping the bodies with a knife on a chopping block. The crab is strained through a bamboo basket and cooked 10 to 20 minutes with vegetables or noodles. A study of the preparation of the soup demonstrated metacercariae on the knife, on the cook's hands, on the chopping block and table, and on the bamboo basket (Yokogawa, 1965). Metacercariae can survive for weeks outside of the animal host, so contaminated utensils can be a serious source of infection (Cross, 2001).

In addition, juices from crustaceans may be used in food preparation or in traditional medicines. In Korea, raw juice from crabs were used to treat fever and diarrhea. Similarly, in Japan, juice from *Eriocheir japonicus* and *Potamon dehani* were used to treat fever or to make an ointment for urticaria (Yokogawa, 1965). Medicinal use of juice from crabs occurs also in South America, with the supernatant from ground crabs of the genus *Hypolobocera* used to treat children (WHO, 1995).

7.2.4 Clinical Signs, Diagnosis, and Treatment

The severity of symptoms is often determined by the location of the trematodes, the degree of infection, and the progression of the worms through the body of the host. Early in the infection, the human host may be asymptomatic; however, when the worms migrate from the intestine into the abdomen, diarrhea and abdominal pain may be experienced. Migration to the lungs can elicit an allergic response, with fever, chills, chest pain, urticaria, and eosinophilia (Cross, 2001). After maturity in the lungs, eggs may appear in the sputum without the presence of symptoms. Incidental lesions may form in the lungs with a granulomatous reaction that eventually results in fibrotic encapsulation of the adult trematodes. The most common clinical signs of infection are persistent cough, especially in the morning, and the production of gelatinous, odorous, rust-colored sputum. Other symptoms include fatigue, myalgia, fever, and dyspnea (WHO, 1995; Yokogawa 1965, 1969). Chronic infections can result in hemoptysis, pleural effusion, persistent rales, clubbed fingers, and pneumothorax (Cross, 2001). Patients with only chronic cough may be misdiagnosed with bronchitis, bronchial asthma, or bronchiectasis. Similarly, the dark sputum, heavy cough, rales, and hemoptysis of pulmonary paragonimiasis can cause confusion with tuberculosis. Chest radiography and CT scans of these infections may add to the possibility of misdiagnosis by mimicking the pleural and parenchymal lesions, and solitary nodular lesions of tuberculosis or other diseases (Mukae *et al.*, 2001). Kum and Nchinda (1982) recommended that chest X-rays be used only to evaluate the extent of damage to the lungs from the infections rather than for diagnosis. The literature is replete with reports of patients with paragonimiasis first being treated for tuberculosis (Pezzella *et al.*, 1981; Weina and England, 1990; Yokogawa, 1965). In Ecuador, 13% of the patients being treated for tuberculosis were actually infected with *Paragonimus* (WHO, 1995). Since treatment for the parasite was estimated to be 100 times less costly, the misdiagnoses had an economic impact as well as medical.

The lungs are the primary site for infections by *Paragonimus*, but the worms can wander and encyst throughout the body of the mammalian host. Reported extrapulmonary locations for encystment include the brain, spinal cord, liver, eyes, reproductive organs, subcutaneous tissues, diaphragm, pancreas, pericardium, and lymph nodes (Cross, 2001; Yokogawa, 1965, 1969). The spinal cord and brain are common sites for encystment outside of the lungs. Spinal involvement can result in paraplegia, monoplegia, limb weakness, and sensory deficiencies. Symptoms of cerebral paragonimiasis, usually manifested about 10 months after the appearance of pulmonary signs, are headache, fever, vomiting, seizures, and visual disturbances. Cerebral hemorrhage may occur, especially in children under 15 years of age (Sinniah, 1997). Oh and Jordan (1967) evaluated the intellectual capabilities of patients with cerebral paragonimiasis in Korea and reported that 90% of afflicted children under 15 years became mentally retarded. Children also frequently experience involvement of the liver. A study in China found 51% of infected children presenting with hepatic involvement (WHO, 1995).

Although *P. westermani* may occasionally demonstrate extrapulmonary infections in the abdomen or in subcutaneous tissues, these manifestations are usually caused by *P. skrjabini*, *P. heterotremus*, and *P. mexicanus* because humans are not the most suitable hosts for these parasites (Rim *et al.*, 1994). As a result, migratory subcutaneous lesions may form in the chest, abdominal wall, and extremities.

Microscopic examination of the sputum or stool for the characteristic eggs of *Paragonimus* is the usual approach for the clinical diagnosis of the infection. However, eggs may be intermittently discharged by the patient, decreasing the sensitivity of the procedure (Kong *et al.*, 1998). In light infections, sputum might not be produced or the patient may habitually swallow it, thereby eliminating the analysis of the sputum as a means of diagnosis. Pezzella *et al.* (1981) reported the treatment of 11 patients (15 to 39 years of age) in Korea with pulmonary paragonimiasis, of which only two had eggs in their sputum. In an earlier study, eggs were detected in sputum 72% of the time and in stools, 63% (Yokogawa, 1965). In those cases (13%) of which eggs were found in the stool specimens but not in the sputa, the patients were primarily children or the elderly.

Serological testing, in addition to the microscopical procedures, are of value for confirmation of other tests and for diagnosis in chronic infections of which eggs are difficult to isolate. The interdermal test using the veronal buffered saline (VBS) extract of adults of *P. westermani* is recommended as a screening method (Cross, 2001; Yokogawa, 1965). The test may be used in surveys or to differentiate paragonimiasis from tuberculosis, tumors, or other nonparasitic conditions. This method will provide positive results for up to 20 years after complete recovery; therefore, follow-up testing with complement fixation tests, ELISA, or immunoblot is recommended (Cross, 2001). Earlier work with these sero-diagnostic techniques indicated cross-reactivity with antigens from other trematodes, including *Clonorchis*, *Schistosoma*, and *Fasciola* (Hillyer and Serrano, 1983; Yokogawa, 1965). Advances in these procedures have increased the sensitivity and specificity, such that the cross-reactivity inherent in the earlier procedures have been decreased or eliminated. For example, ELISA methods have been developed using cysteine protease antigens which have high specificity and can differentiate between paragonimiasis and fascioliasis (Ikeda, 1998; Ikeda *et al.*, 1996). Similarly, immunoblot for paragonimiasis has been developed and refined such that sensitivity is estimated at 96% and specificity at 99% (Slemenda *et al.*, 1988). In a study of 40 separate cases of paragonimiasis caused by three species of *Paragonimus*, sera from patients with other trematodes or cestodes infections, with lung cancer, or from healthy subjects failed to react in the immunoblot (Kong *et al.*, 1998).

Several regimens are available for treatment. The drug of choice is praziquantel administered at 25 mg/kg, three times a day for 1 to 2 days (Medical Letter Inc, 1984). Dosage is the same for children as for adults. Bithionol is also efficacious for all forms of paragonimiasis (Yokogawa, 1969). The drug is given at a dose of 30–50 mg/kg every other day for 10 to 15 days (Sinniah, 1997). Niclofolan is also effective in a single oral dose of 2 mg/kg of body weight (Kum and Nchinda, 1982). When treating cerebral paragonimiasis, corticosteroids should be given to ameliorate the localized immune response to the dead parasites (Cross, 2001).

Surgical treatment may be necessary for severe pleural paragonimiasis; not to specifically remove the worms, but rather to alleviate some of the damage incurred during the infection (Pezzella et al., 1981). In these cases, patients were treated pharmologically 10 to 14 days before surgery. Surgical treatment of cerebral paragonimiasis could eliminate seizures with the removal of encapsulated abscesses, often including a trematode (Yokogawa, 1965). Only 30% of those patients operated upon were considered improved or cured after 2 years and 21% still experienced seizures after surgery.

7.3 *CLONORCHIS SINENSIS*

7.3.1 Introduction

Clonorchis sinensis, also known as the Chinese Liver Fluke, is widely distributed in China, Hong Kong, Taiwan, Japan, Korea, Vietnam (northern), and the far east of the Russian Federation. Stoll (1947) estimated that less than 19 million people were infected with *C. sinensis*, and that the disease was confined to Asia. Although some sources still support that estimate (e.g., Sun, 1997), the generally accepted estimate is approximately 7 million at present (Crompton, 1999; Rim et al., 1994; WHO, 1995).

The trematode has been present in China for over 2000 years, as demonstrated by the discovery of eggs in an excavated corpse from the Chu Dynasty of 206 B.C. (WHO, 1995). The parasite remains as a major public health issue in China with 4 million people infected, although efforts are underway to lower the prevalence of infection. A control program was implemented from 1983 to 1989 in the Sichuan Province, with a resulting drop from 21 to 24% prevalence to less than 1%. Similarly, 10.6% of the population were infected in the Henan Province in 1973, but only 0.7% in 1983 (WHO, 1995). Efforts in Korea have also resulted in a decrease in prevalence. A survey in the 1950s found an incidence of 11% of the population. From 1971 to 1992, clonorchiasis decreased from 4.6 to 2.2%, and although the trematode hasn't been eradicated, the mean fecal egg counts have decreased. The number of people currently infected is approximately one million (Cross, 2001; WHO, 1995). Japan has been very successful in the control of the trematode, such that infections are no longer detected in children and only sporadic cases are diagnosed in adults. In 1971, 780 cases were diagnosed; in 1976, 26 cases; in 1986, one case; and no cases in 1991 from an examination of one million stool specimens (WHO, 1995).

Migrations of people from endemic areas may present medical personnel with unexpected foci of infections. For example, clonorchiasis has been reported from Chinese immigrant communities in Canada and the United States. Among 150 immigrants in New York City, the prevalence of infection was 26%; 15.5% of Chinese immigrants examined in Montreal were infected (Sun, 1980). All were immigrants

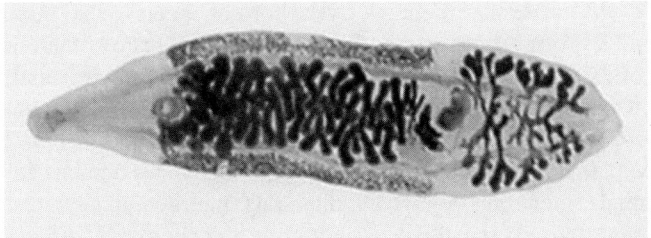

Figure 7.3. Adult of *Clonorchis sinensis*. Note the distinctly branched testes in the posterior portion of the body. Photograph by Charles Sterling, University of Arizona.

from Hong Kong. During 1979 to 1981, 13.4% of Chinese residents of Hong Kong applying for emigration to Canada had clonorchiasis (WHO, 1995).

7.3.2 Life Cycle

The adult trematode (Fig. 7.3) is small, 8–25 mm long, 1.5–5 mm wide, with a smooth cuticle (Chen *et al.*, 1994; Sun, 1997). The size of the parasite varies according to the species and size of the host, intensity of infection, the age of the trematode, and its location within the bile ducts (Chen *et al.*, 1994; Komiya, 1966). Adults have a single rounded ovary and two extensively branched testes posteriad of the ovary and in the lower third of the body. Although some place the parasite in the genus *Opisthorchis*, most parasitologists support the retention of the original taxonomic designation based on the distinctive structure of the testes. Eggs are small, ovoid, operculated with a prominent shoulder, and have a terminal knob at the abopercular end (Fig. 7.4). They measure 26–35 μm long and 12–19 μm at their widest point (Cross, 2001; Sun, 1997). Differentiation between eggs of *C. sinensis* and those of *Opisthorchis* is difficult because the dimensions and general appearance are similar, however, Ash

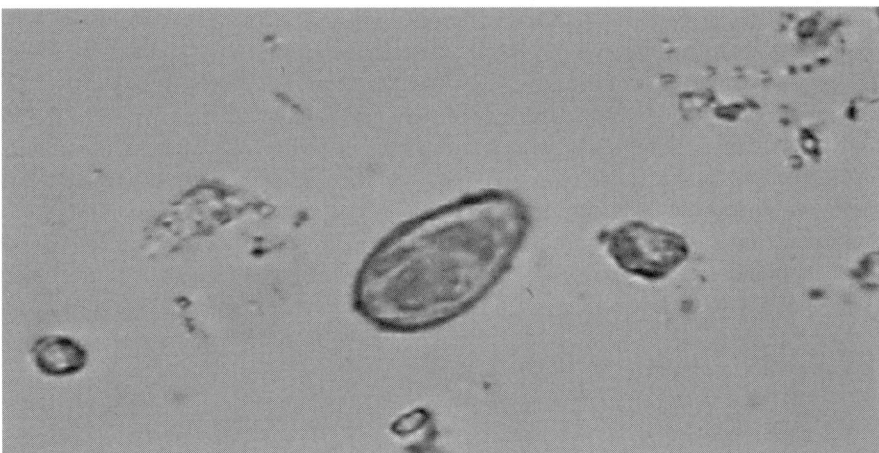

Figure 7.4. Egg of *Clonorchis sinensis*, with distinctive shoulders on egg for the operculum.

and Orihel (1997) state that the eggs of the former species have more prominent shoulders and well seated opercula. Eggs of *Clonorchis* are sometimes confused with those of *Heterophyes heterophyes*, *Metagonimus yokogawi*, and *Haplorchis taichui*, but these heterophyid eggs usually have inconspicuous opercula that appear to be flush with the surface of the shell (Ash and Orihel, 1997; Cross, 2001).

Adult trematodes reside in the intrahepatic bile ducts of humans and other fish-eating mammals, including dogs, cats, pigs, rats, tigers, mink, camels, and civets. Fish-eating birds may also be infected, such as the night heron, *Nycticorax nycticorax* (Komiya, 1966). Cats, dogs, pigs, and rats are probably the most important reservoir hosts in endemic areas. Cats are particularly susceptible. For example, in Japan from 1955 to 1961, the prevalence of *C. sinensis* in cats ranged from 45% to 87% (Komiya, 1966). Although the prevalence of infection in humans is generally lower than that in the reservoir hosts, the intensity of infection in humans is usually greater. As a result, infected humans drive the dynamics of transmission of clonorchiasis rather than reservoir hosts (WHO, 1995).

Adult trematodes produce eggs which are embryonated (each contain a ciliated miracidium) when passed into the bile ducts and excreted with the feces of the host. The eggs do not hatch when deposited in fresh water, rather they must be eaten by one of nine species of operculate snails for further development. (Rim *et al.*, 1994; WHO, 1995). *Parafossarulus manchouricus*, *Bithynia fuchsiana*, and *Alocinma longicornis* are the primary molluscan hosts for *C. sinensis* and are commonly found in ponds used for aquaculture (Cross, 2001). After ingestion by the snail, the miracidium hatches from the egg and develops into a sporocyst. Redia develop later and then cercariae form. After leaving the snail, the cercariae hang upside-down in the water and sink gradually to the bottom. They then vigorously rise in the water to resume their previous position. Cercariae do not actively search for the next host, rather, they react to disturbances in the water with sporadic movements. When a cercaria makes contact with a fish, it attaches with its suckers and penetrates within 6 to 15 minutes (Komiya, 1966). Within hours of penetration, the trematode encysts and forms a metacercaria. Metacercariae become infective to final hosts in 3 to 4 weeks. *C. sinensis* infects 113 species of freshwater fish belonging to 13 families; 95 species are members of the family Cyprinidae (Rim *et al.*, 1994; WHO, 1995). In the Fukien Province of China, some species of crayfish have been implicated in the transmission of clonorchiasis, including experimental infections of guinea pigs with metacercariae from crayfish (Komiya, 1966). When an infected fish is ingested, the trematodes excyst in the gastrointestinal tract and migrate through the common bile duct to the intrahepatic bile ducts. Migration through the ducts take 1 to 2 days and the trematodes mature into adults in about 30 days. Eggs thus appear in the stool about 4 weeks postinfection. In untreated infections, some trematodes can live up to 30 years (Cross, 2001; Sun, 1980). Unlike *Fasciola hepatica*, *C. sinensis* does not invade hepatic tissues.

7.3.3 Epidemiology

Generally, clonorchiasis is a disease of adults, with the incidence of infection increasing with age. The greatest prevalences of infection in Japan were found in those 30–50 years of age; whereas in some endemic areas of China, children under

15 years of age were also found to be infected (Komiya, 1966). In the latter circumstances, children tended to eat insufficiently cooked fishes and raw crayfishes. Early work in Vietnam indicated that in the heavily endemic region of the Red River delta, 40% of the adults and 8% of the children were infected (Komiya, 1966). Children are less likely to be infected than adults in Korea (Cross, 2001).

Depending on cultural practices, gender differences may also be present in regard to clonorchiasis. No difference was found in infection rates between males and females in Japan, but males were more often infected in China and Taiwan. In Korea, a social custom persists during which men gather to drink rice wine and eat raw fish, resulting in a higher prevalence of clonorchiasis in men 41–50 years of age (Cross, 2001; Komia, 1966). Women rarely participate in this custom.

Food habits are a primary component in the transmission of *C. sinensis*. As noted above, the improper cooking of fish or the consumption of raw fish and crayfish play important roles in the incidence of the trematode in endemic areas. In China, Taiwan, and Hong Kong, people eat a preparation called congee consisting of thin slices of raw fish over watery rice. Although neither the fish nor the snails are present in Hong Kong, this preparation from fish imported from the Chinese mainland has been implicated (Cross, 2001). Similarly, residents of Hawaii and the west coast of the United States who have been diagnosed with clonorchiasis but haven't been to the Orient probably acquired the parasite from eating imported fish containing viable metacercariae (Ash and Orihel, 1997).

Aquacultural practices in endemic areas directly impact the prevalence and persistence of the trematode. Freshwater fish, especially cyprinids, are commonly raised in ponds in Asia. These ponds also provide excellent habitat for snails which act as the first intermediate host for the parasite. In China and elsewhere, latrines were often placed over the ponds to allow immediate introduction of fecal material to the system (Komiya, 1966). Although efforts have been taken to restrict the use of human and animal wastes as fertilizer in aquaculture, the inadvertent contamination of ponds with human or animal excreta continues (WHO, 1995).

7.3.4 Clinical Signs, Diagnosis, and Treatment

In endemic areas, only about a third of those with clonorchiasis present with symptoms. Factors in the presentation of symptoms appear to be the numbers of worms and the length of infection. Moderate infections consist of 100–1000 trematodes, but infections of up to 21,000 have been reported (Komiya, 1966). The mean intensity of infections in endemic areas is 20 to 200 worms. Repeated exposure most likely results in an increase in intensity because antibody response by the host is not protective (Sun, 1980; Sun and Gibson, 1969).

Acute infections are usually asymptomatic and rarely reported. If food with a large number of metacercariae is ingested, symptoms of acute clonorchiasis can occur 10 to 26 days afterwards (Cross, 2001). Symptoms include fever, diarrhea, epigastric pain, indigestion, anorexia, eosinophilia, hepatosplenomegaly, and leucocytosis. Patients with heavy infections may also experience weakness, weight loss, feelings of abdominal fullness, anemia, and edema. Chronic infections may present with jaundice, portal hypertension, ascites, and upper gastrointestinal bleeding. Children with repeated or heavy infections with *C. sinensis* have been reported to suffer dwarfism with retardation of sexual development (WHO, 1995). In China, a study of 85 children with clonorchiasis found 40% experienced retarded growth (Sun, 1997).

The pathology of clonorchiasis is believed to be caused by the mechanical irritation from the suckers of the adults in the bile ducts and by toxic metabolic substances produced by the parasite. The presence of the worms in the ducts elicit proliferation of the epithelium, an increase in goblet cells, and an increase in the secretion of mucus. The cellular proliferation and infiltration of leucocytes result in the thickening of the ductal walls and the lumen becomes dilated, with possible fibrosis. The liver parenchyma and hepatic cell function are unaffected by *C. sinensis*.

An important complication of clonorchiasis is recurrent pyogenic cholangitis caused by secondary bacterial infections. Symptoms include repeated febrile attacks, abdominal pain, and jaundice. The high proportion of mucus in the bile and the presence of the eggs and adult trematodes form a favorable environment for bacteria, primarily *Escherichia coli*. Occasionally a patient with *C. sinensis* and cholangitis caused by *Salmonella* organisms can act as a typhoid carrier (Sun, 1980). Cholecystitis may also be caused by *C. sinensis*. The eggs of the trematode act as nuclei for the formation of stones in the bile ducts and the gall bladder (WHO, 1995). Although adults may be found in the gall bladder, they cannot survive in bile and do not directly cause cholecystitis (Sun, 1997); however, calculi may form around dead worms. Stones in the gall bladder and bile ducts were found in 70% of infected patients autopsied in Hong Kong (Cross, 2001). If adults enter the pancreatic ducts, dilation and hyperplasia of the ductal epithelium occur. Symptoms of acute pancreatitis may present, usually 1 to 3 h after a meal.

Cholangiocarcinoma is strongly associated with clonorchiasis, in that the cancer is more prevalent in areas where the parasite is endemic than in nonendemic regions such as Europe and North America (Chen et al., 1994; Patel, 2002). Similarly, examination of the ratio of cholangiocarcinoma to hepatocellular carcinoma in areas considered to have high rates of primary liver cancer provides further support. In Hong Kong, which has a high incidence of clonorchiasis, the ratio is 1:5; in the Pusan area of South Korea, 1:4. In contrast, ratios in areas in which the trematode is absent are: Java, 1:56; Africa, overall, 1:38; and Johannesburg, South Africa, 1:20 (Chen et al., 1994; Flavell, 1981). The development of cholangiocarcinoma requires long term and persistent infections by the liver fluke, although the presence of the trematode does not appear to be the sole factor in the development of the malignancy. Rather, the parasite most likely alters the biliary habitat, which becomes more susceptible to carcinogenic stimuli (Chen et al., 1994; Flavell 1981; Parkin et al., 1993; WHO, 1995).

Diagnosis is usually by direct microscopic examination of stool for the presence of eggs. A more sensitive test is the recovery of eggs from duodenal drainage, but patients tend to be reluctant. Single examinations of stool specimens have a high false-negative rate. Although repeated testing increases sensitivity, obtaining multiple stool specimens is difficult (Chen et al., 1994). As mentioned above, differentiation of eggs of *C. sinensis* from those of *Opisthorchis* spp., *Metagonimus* spp., or other heterophyid trematodes is difficult and requires intense study of specimens. Information such as food habits and past travel, in addition to the morphology of the eggs, may be needed to arrive at a diagnosis. For example, Canadian patients presented with abdominal pain, anorexia, and abnormalities in their liver function after eating raw fish. Further study indicated they had consumed the flesh of the white sucker, *Catostomus commersoni*, a fish common in North America, and that they were infected with *Metorchis conjunctus* (Ash and Orihel, 1997). Definitive diagnosis can be made from the identification of the worms passed by the patient after treatment.

Identification from histological specimens can also be done. In tissue sections, *C. sinensis* usually appears as 1 or 2 sections of an adult worm in the lumen of a bile duct, whereas *Opisthorchis* spp. is usually smaller and appears unfolded (Sun, 1997). Sections with *F. hepatica* show the branched digestive tract and cuticular spines of the trematode.

In early stages of study and treatment, the populations in endemic areas often willingly submitted stool specimens for testing. As *C. sinensis* has come under control and has a lower prevalence, the submission of stool specimens has decreased (WHO, 1995). Coupled with the rates of false-negatives with microscopical methods, other means of diagnosis are needed. Serologic tests are commonly used in surveys, but are not used in routine diagnosis. In Korea, screening is done with an enzyme-linked immunosorbent assay (ELISA) using a crude antigen of *C. sinensis*, followed by stool examinations for sero-reactive patients. Those found to be shedding eggs are then treated (WHO, 1995). Further development of ELISA techniques is ongoing to improve the sensitivity and specificity of the technology. Some approaches provide specificity, such as the use of serum IgG4 antibody, but lack sensitivity (Hong *et al.*, 1999). Other work involving excretory-secretory antigens indicated that patients with active infections can be differentiated from past cases, but further development is necessary (Kim, 1998). An ELISA technique utilizing cystatin capture with multiple cysteine proteinases appears to be sensitive and more specific than ELISA using crude extracts for serodiagnosis (Kim *et al.*, 2001). Other serologic techniques include the indirect fluorescent antibody test (IFAT), indirect hemaglutination test (IHA), complement fixation test (CFT), and counter-immunoelectrophoresis (CIEP), all of which have varying sensitivities and specificities (Chen *et al.*, 1994).

Indirect evidence of parasitic infection of the bile ducts may be provided by radiological techniques (Cross, 2001; Sun, 1997). Ultrasonography is noninvasive and may detect the movements of the trematodes. Computed tomography has been used to diagnose clonorchiasis, particularly in endemic regions.

Praziquantel is the drug of choice for treatment of clonorchiasis, at a dose of 25 mg/kg of body weight, three times in one day (Liu, 1996; Medical Letter, Inc., 1984). Treatment for 2 days provides cure rates of 97.7 to 100%, and is suggested for moderate infections (Sun, 1997; WHO, 1995; Yangco *et al.*, 1987). Praziquantel is suitable for adults and children older than 4 years of age. Although differential diagnoses of clonorchiasis, opisthorchiasis, and infections by other heterophyids can be difficult, treatment with praziquantel is similar such that the infections can be effectively treated. For those patients with cholangitis, antibiotics should also be administered. Prognosis of light infections is good; chronic infections may result in hepatobiliary disease.

7.4 *OPISTHORCHIS VIVERRINI* AND *OPISTHORCHIS FELINEUS*

7.4.1 Introduction

Several species of *Opisthorchis* infect humans and cause disease similar to clonorchiasis. *O. viverrini* is distributed in southeast Asia, including Thailand, Laos, and Kampuchea. In northeastern Thailand, the prevalence of *O. viverrini* rose from 3.5 million people infected in 1965 to 5.4 million in 1981 (Sun, 1997). The most recent estimate is approximately 7 million people infected in Thailand and 1.7 in Laos (Rim *et al.*, 1994). *O. felineus* (synonym: *O. tenuicollis*) is endemic in southern, central, and eastern Europe, including Poland and Germany, in the Russian commonwealth, Turkey, and western Siberia. Stoll (1947) estimated 1 million people were infected with the latter species. Sun (1997) suggests that several million people worldwide are presently infected. Approximately 1.2 million are infected in Russia and 1.5 million in the former USSR (Rim *et al.*, 1994). Western Siberia comprises the largest endemic region for *O. felineus*, with prevalences ranging from 40 to 95% in the Tyumen and Tomsk districts (WHO, 1995). A third species, *O. guayaquilensis* (synonym: *Amphimerus pseudofelineus*) infects humans in an isolated area of Ecuador (Sun, 1997).

Similar to that seen with *C. sinensis*, migrations of people from endemic areas can result in cases of opisthorchiasis being diagnosed far from the original foci of infections. The prevalence of opisthorchiasis among refugees from Asia range from 5 to 20% (WHO, 1995). Laotian immigrants in the United States and France are frequently found to have opisthorchiasis (WHO, 1995; Woolf *et al.*, 1984). Migrant labor can be a contributory factor in the introduction and spread of disease, including opisthorchiasis, as indicated by the movement of workers among the oil fields of Siberia. Similarly, Thai workers in Kuwait, Taiwan, and China were found to be infected with *O. viverrini*, but no transmission to the local population has been documented (WHO, 1995).

7.4.2 Life Cycle

Adults of *Opisthorchis* spp. are generally smaller than those of *C. sinensis*. The former trematodes are 7–12 mm long and 2–3 mm wide, with lobate testes. Adults of *O. viverrini* differ from those of *O. felineus* in that the former species has more indentations in the testes, the ovary is in greater proximity to the testes, and the vitellaria are clustered (Cross, 2001). The eggs of the two species closely resemble those of *C. sinensis* in appearance but are smaller, ranging from 19–30 μm in length and 12–17 μm in width (Ash and Orihel, 1997; Sun, 1997). Eggs of the opisthorchid species may be difficult to differentiate from eggs of the heterophyid trematodes.

The life cycle and development of the larval stages of the *Opisthorchis* spp. are nearly identical to those of *C. sinensis*. The adults reside in the biliary ducts of humans or other fish-eating mammals. In addition to humans, cats and dogs are commonly infected with *O. viverrini* in Thailand and Laos and are important in maintaining transmission of the parasite. Of the 28 species of mammals in the former USSR determined to be involved in the transmission of *O. felineus*, cats, dogs, foxes, and muskrats are the most important reservoir hosts (WHO, 1995). The trematode eggs pass in the feces of the host and must reach freshwater for further development. Eggs do not hatch in the water, rather they are ingested by the appropriate snails prior to hatching. Molluscan hosts for *O. viverrini* are *Bithynia siamensis goniomphalus* (synonym: *B. s. siamensis*), *B. s. funiculata*, and *B. s. laevis*. Snails of the species *Codiella (Bithynia) inflata, C. troscheli,* and *C. leachi* act as first intermediate hosts for *O. felineus* (WHO, 1995). The miracidia develop into sporocysts which in turn produce rediae and then cercariae. Cercariae of *Opisthorchis* spp. find and penetrate fishes similar to those of *C. sinensis*. Carp or cyprinoid fish commonly serve as the second intermediate hosts. In Thailand, 15 species of small freshwater fishes are the primary sources of infection by *O. viverrini*, with the fish eaten raw, pickled, smoked, or fermented. Infections by *O. felineus* in Russia are acquired from 22 species of cyprinids consumed raw, frozen, pickled, or smoked (Rim *et al.*, 1994).

7.4.3 Epidemiology

The age distribution of opisthorchiasis mirrors that of clonorchiasis. Generally, the incidence of the disease increases with age, from childhood to adulthood. In endemic areas such as Tyumen in the Russian Federation, chronic infections are highly prevalent; infections have been reported in 1-year-old infants and nearly 100% of children by 10 years of age (WHO, 1995). Infants and children become infected from eating raw or incompletely cooked fish. The intensity of infection as reflected by fecal egg counts also increases with age and then plateaus for those over 20 years of age, however, the number of worms in infections usually peaks between 30 and 40 years of age (Cross, 2001; Haswell-Elkins *et al.*, 1991).

No gender differences are observed in cases of opisthorchiasis, indicating similarities in diets and food habits between males and females in those countries in which *Opisthorchis* spp. are endemic. In early studies, the incidence of hepatobiliary disease as a result of infection appeared to be approximately equal for the sexes (Elkins *et al.*, 1990; Haswell-Elkins *et al.*, 1991). Later studies in northeast Thailand indicated a high prevalence of cholangiocarcinoma among heavily infected males (Haswell-Elkins *et al.*, 1994).

Food habits and aquacultural practices directly contribute to the transmission and persistence of the disease. Infection occurs from consuming infected fish raw or incompletely cooked, incompletely frozen, salted, or smoked. An outbreak of opisthorchiasis in Europe was traced to fish eaten after only 1 day of salting (Cross, 2001). In many endemic areas, traditional preparations of raw fish are eaten, such as koi-pla in Thailand. In some cultures, rituals involving food preparation may require that fish remain raw.

Aquacultural practices are common throughout Asia. Similar to China, ponds in Thailand are fertilized with human and animal excreta, ensuring the introduction of trematode eggs to freshwater. As aquaculture increases, the incidence of the disease may also increase if poor sanitation persists. Irrigation can also facilitate transmission of the trematode as a result of improved snail habitat and increased populations of freshwater fishes. *O. viverrini* is more prevalent among residents of the Nam Pong Resources Development Project, an irrigation district in Thailand, than in the population in the surrounding nonirrigated areas (WHO, 1995).

7.4.4 Clinical Signs, Diagnosis, and Treatment

Although Rim et al. (1994) stated that occurrence of symptoms did not appear to be related to the fecal egg count, others report a correlation between the presence of symptoms and the intensity and duration of infection (Elkins et al., 1990; Upatham et al., 1984). About 10% of those infected present with symptoms, with a higher occurrence of symptoms in heavy infections. Most individuals with light to moderate infections will be asymptomatic. Symptoms of acute infections include fever, eosinophilia, myalgia, and lymphadenopathy. Patients with chronic opisthorchiasis may present with diarrhea, flatulence, fever, dyspepsia after meals, anorexia, right upper quadrant pain, jaundice, and hepatomegaly. Although the liver may become enlarged and painful, the trematodes usually have no measurable effect on liver function. In areas with high prevalences of infection, opisthorchiasis has been shown to significantly contribute to malnourishment of children. In southern Laos, almost 70% of children weigh less than 80% of normal for their age (WHO, 1995).

Patients with chronic infections may also suffer from cholecystitis, cholangitis, biliary stones, and cholangiocarcinoma. Occurrence of these conditions are more common with opisthorchiasis than with clonorchiasis. In particular, the association of *O. viverrini* with cholangiocarcinoma is stronger than with *C. sinensis* (Cross, 2001; Sun, 1997). Cholangiocarcinoma is the leading cancer in northeast Thailand, but the cancer is rare in other areas of Thailand where *O. viverrini* is uncommon (Vatanasapt et al., 1990). In the Khon Kaen province of northeast Thailand, the annual rate of cholangiocarcinoma was about 135 per 100,000 among males and 43 per 100,000 among females (Green et al., 1991). Although fewer studies have been completed involving *O. felineus*, a correlation between the trematode and cholangiocarcinoma has also been reported in the central part of the Tyumen region of the Russian Federation. The prevalence of *O. felineus* in the central area was 45% and the rate of the cancer was almost 50 per 100,000. In the area south of Tyumen, 0.5% of the population was infected and the average prevalence of cholangiocarcinoma was approximately 4 per 100,000 (WHO, 1995).

Diagnosis of opisthorchiasis, like clonorchiasis, is usually by direct microscopic examination of eggs in a fecal sample. The eggs of *Opisthorchis* spp. are slightly different in size and have less prominent shoulders than those of *C. sinensis*, otherwise the eggs appear indistinguishable. Diagnosis may require additional information such as travel to endemic areas by the patient or food habits. Care must also be taken to differentiate the opisthorchid eggs from those of heterophyid trematodes.

Efforts to develop other means of diagnosis reflect approaches similar to those for clonorchiasis. Serological techniques detect antibodies to the trematodes, but active infections cannot be differentiated from those past. Antibodies can persist long after opisthorchiasis has been resolved through antihelminthic treatment. Specificity could also be an issue for serological tests. An indirect immunofluorescent antibody (IFA) technique for serology indicated 81% of patients had antibodies to the adult trematodes or eggs, and that about 5% of patients had a possible cross-reaction between the parasite antigens and "self" antigens (Boonpucknavig et al., 1986). ELISA techniques and DNA probes have been continuously studied and pursued. A sandwich ELISA with a mixture of three IgG_1 monoclonal antibodies was used to detect as little as 0.05 ng of antigens of *O. viverrini*. A probe to detect DNA from the trematode eggs was specific in a dot blot hybridization procedure. Both techniques displayed good specificity and their sensitivity was similar to that of microscopy (Sirisinha *et al.*, 1991). The development of a polymerase chain reaction (PCR) technique to detect *O. viverrini* in experimentally infected hamsters was expanded to test for eggs in human feces. The PCR method was highly specific, with a sensitivity somewhat less than the microscopical techniques (Wongratanacheewin *et al.*, 2002). The difference in sensitivity was suggested to be the effect of the amount of sample analyzed in PCR (0.1 g) versus that in microscopy (1 g or 2 g). The PCR method may be suitable for screening large numbers of samples in epidemiological studies.

The drug of choice, in the treatment of opisthorchiasis, is praziquantel at 25 mg/kg of body weight, three times in one day (Lui, 1996). Mebendazole at 30 mg/kg/day for 3 to 4 weeks or albendazole at 400 mg twice a day for 3 to 7 days are also effective (Cross, 2001). Patients with cholangiocarcinoma were not treated with praziquantel to prevent possible worsening of obstruction by dead trematodes (Elkins *et al.*, 1990). However, surgery may be necessary to alleviate jaundice caused by obstructions (Cross, 2001). Patients with cholangitis should also be treated with antibiotics.

7.5 *NANOPHYETUS SALMINCOLA*

7.5.1 Introduction

Nanophyetus salmincola is an intestinal trematode of the family Nanophyetidae and is present in the Pacific Northwest of the United States and in eastern Siberia. Skriabin and Podiapolskaia (1931) described the trematode from humans in Siberia as *N. schikhobalowi*, based on smaller body size and eggs, and on the absence of a cirrus pouch. However, subsequent work has shown that the two species are synonymous and the Siberian parasite is considered a subspecies (Filimonova, 1966; Gebhardt *et al.*, 1966; Witenberg, 1932).

7.5.2 Life Cycle

The adult trematode is small, about 1–2.5 mm in length, and oval or pyriform in shape (Fig. 7.5). Definitive hosts are primarily piscivorous mammals, including dog, wolf, fox, bear, raccoon, spotted skunk, otter, and bobcat (Millemann and Knapp, 1970; Schlegel *et al.*, 1968). Piscivorous birds such as the Great Blue Heron, Hooded

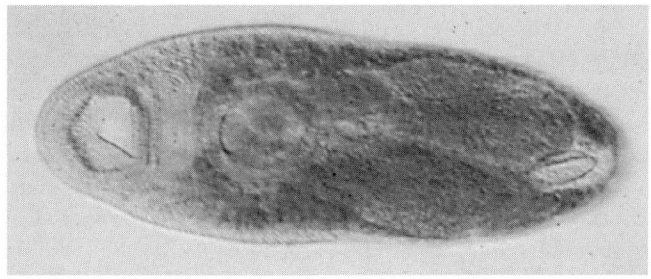

Figure 7.5. Adult of *Nanophyetus salmincola*. Photograph by Thomas R. Fritsche.

Merganser, and Belted Kingfisher were also found to be naturally infected, but other birds such as domestic mallards and chickens were refractive to experimental infection (Schlegel et al., 1968). In eastern Siberia, human infections are common among the native peoples, with up to 80% prevalence in some villages along the Amur River (WHO, 1995). Until a report of infections in Oregon (Eastburn et al., 1987), only the subspecies *N. s. schikhobalowi* was known to infect humans, although according to Gephardt et al. (1966) a parasitologist successfully infected himself with the North American subspecies by eating raw trout from Oregon. His infection was asymptomatic, but small numbers of eggs were observed in his stool 10 days after ingestion of the fish. Therefore, one can assume that people were infected with *N. salmincola* previously in the Pacific Northwest, but the infections were not diagnosed because of vague symptoms or none at all.

The life cycle of *N. salmincola* is complex, requiring two intermediate hosts. The eggs are passed from the definitive host and must be introduced to fresh water to hatch. The resulting miracidia are not attracted to snails, the first intermediate hosts, but have been observed to repeatedly bump into the snails and swim away (Bennington and Pratt, 1960). The trematode has high specificity for its snail hosts and its geographic distribution is limited by the presence of the snails in rivers and streams. In the Pacific Northwest, *Jugo plicifera* (*Oxytrema silicula* and *Goniobasis plicifera* are synonyms) acts as the molluscan host; whereas, in Siberia, *Semisulcospira cancellata* and *S. laevigata* fulfill that role (WHO, 1995). The miracidium penetrates the snail and develops into redia. The prevalence of rediae in snails depends on the time of the year, ranging from 9 to 52% in Oregon and 9 to 17% in Siberia (Millemann and Knapp, 1970).

Cercariae develop within the rediae and are then shed. The cercariae are somewhat elongate, measuring 0.31 to 0.47 mm long by 0.03 to 0.15 mm wide, with a small blunt tail. Although cercariae rhythmically contract, they do not actively search for the second intermediate host. When fish come in contact with the cercariae, the trematodes immediately begin to penetrate the skin, taking no more than 2 minutes (Bennington and Pratt, 1960). The cercariae penetrate quickly into the muscles and into the blood vessels and spread throughout the tissues of the piscine host where they encyst and form metacercariae (Fig. 7.6).

The trematode does not display the same degree of host specificity for the second intermediate host. Both salmonid and non-salmonid fishes are known to be naturally infected, as well as the Pacific giant salamander, *Dicamptodon ensatus* (Filimonova, 1963; Gebhardt et al., 1966; Millemann and Knapp, 1970; WHO, 1995). Millemann and Knapp (1970) summarizes the pathology of infection by *N. salmincola*. Experimental infections confirmed the pathogenicity of the trematode to fish when exposed to large numbers of cercariae in a limited time period, although different tolerances to infection were displayed by different species of fish. Salmon and trout can have high numbers of metacercariae. For example, up to 2000 trematodes in salmon from Siberia; 14,062 in a cutthroat trout and 2002 in a steelhead trout, both from Oregon. The prevalence and numbers of metacercariae in salmonids are of concern, since most human infections reported by Eastburn et al. (1987) and Fritsche et al. (1989) resulted from the ingestion of salmon or steelhead trout. The growth of infected fish was also retarded in comparison to control fish. Heavy infections in fish

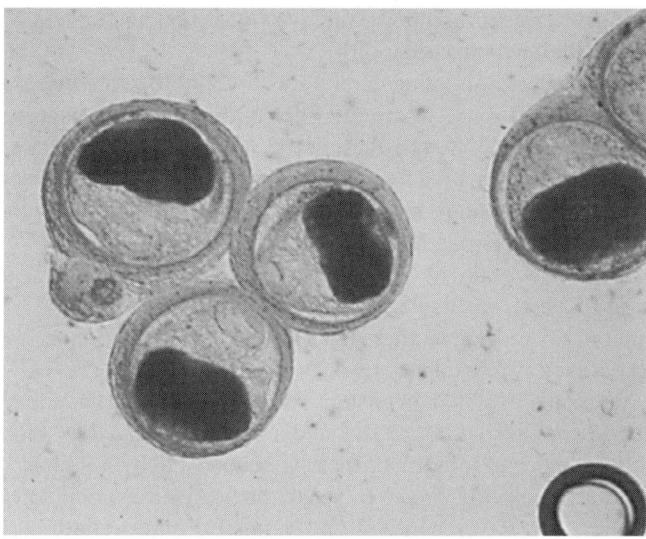

Figure 7.6. Metacercariae of *Nanophyetus salmincola*. Photograph by Thomas R. Fritsche.

with metacercariae can result in tail curvature, increased respiration, decrease in movement or erratic swimming, drifting, and loss of equilibrium.

In the Pacific Northwest, some hatcheries experience losses of heavily infected salmon when the fish begin to migrate to saltwater, probably due to the inability to adapt to the saline environment. Initially, the migration of salmon to the ocean was thought to eliminate the trematode, but this was probably a reflection of the loss of heavily infected fish. Weiseth *et al.* (1974) examined three species of ocean-caught salmon from along the coast of Oregon; king (*Oncorhynchus tshawytscha*), coho (*O. kisutch*), and pink (*O. gorbuscha*). King and coho salmon had overall infections of 31 and 53% respectively. Numbers of metacercariae were generally low, with only 5% of the king salmon and 18% of the coho having more than 24. The pink salmon were free of *N. salmincola*. Although some spawning of pink salmon occurs along the Pacific Northwest coast, the more abundant runs occur throughout the Alaskan coastal waters. In addition, after emergence from the gravel, the fry of pink salmon quickly migrate downstream to the ocean (Heard, 1991), thereby decreasing exposure to cercariae. These two factors most likely explain the lack of the metacercariae in the pink salmon (Weiseth *et al.*, 1974).

7.5.3 Salmon Poisoning Disease

N. s. salmincola is unique in that it acts as a vector for a rickettsial disease of canids referred to as "salmon poisoning disease." The Siberian subspecies is not known to transmit the rickettsia. The correlation between the consumption of fish, usually salmon, and the often fatal disease for dogs was well known to settlers and Indians of the Pacific Northwest (Philip, 1955; Simms *et al.*, 1931). Donham (1925) found small trematodes in the intestines of necropsied dogs and later reported the association of the disease with the cysts (metacercariae) in fish (Donham *et al.*, 1926). In the 1950s the etiological agent for salmon poisoning was identified and named *Neorickettsia helminthoeca* (Millemann and Knapp, 1970).

Dogs, foxes, coyotes, and other canids are susceptible to the disease. Raccoons may experience a slight temperature after consuming infected fish, but are otherwise without symptoms. However, Philip (1955) injected a dog with a suspension made from trematodes recovered from a raccoon, resulting in the characteristic fatal infection. Therefore, although raccoons are refractive to infection by the rickettsia, they can perpetuate the disease in the environment.

The course of infection by *N. helminthoeca* in canids is well documented (Millemann and Knapp, 1970; Philip, 1955; Simms *et al.*, 1931). The incubation period is generally 5 to 7 days after ingestion of the infected fish, followed by a sudden onset of fever. Anorexia usually occurs within 24 h of the fever. The fourth or fifth day of symptoms results in persistent vomiting with excessive thirst. Diarrhea ensues on the fifth to seventh day. The stools become dark with blood. Although the temperature may drop to normal, the dog becomes dehydrated and weak from loss of weight. Death usually occurs by the tenth day. Untreated infections are 90% fatal. *N. helminthoeca* is susceptible to a broad spectrum of antibiotics including sulfonamides, penicillin, tetracycline, and streptomycin (Millemann and Knapp, 1970; Philip, 1955). Dogs that recover, either spontaneously or through treatment, are immune to further infection by the rickettsia.

7.5.4 Clinical Signs, Diagnosis, and Treatment

Although we now know that both subspecies of *Nanophyetus* can infect humans, the presentation of symptoms is dose-dependent, requiring approximately 100 or more worms (WHO, 1995). The most common complaint is vague, nonlocalized abdominal pain or discomfort, often associated with diarrhea (Eastburn *et al.*, 1987; Fritsche *et al.*, 1989). Patients also reported bloating, nausea and vomiting, weight loss, fatigue, and decreased appetite. Symptoms are often described as resembling those of influenza, resulting in the general description of the disease as "fish flu." At least half of those infected also presented with peripheral blood eosinophilia ranging from 6 to 43% (Eastburn *et al.*, 1987). Untreated infections can persist for 8 months or more (Filimonova, 1965; Fritsche *et al.*, 1989).

With the presentation of the above symptoms, plus a history of eating raw or undercooked fish, particularly salmon or trout, a diagnosis of nanophyetiasis should be contemplated. Further consideration of the geographic distribution of the trematode in the Pacific Northwest of the United States or in eastern Siberia may also indicate infection by this parasite. However, the latter information should not preclude the consideration of nanophyetiasis in the differential diagnosis of patients. Infection by this trematode has been reported in New Orleans, Louisiana (US) and attributed to the consumption of raw salmon that had been shipped fresh from the Pacific Northwest (Adams and DeVlieger, 2001).

Diagnosis is made from the observation of eggs in stool specimens, but routine laboratory procedures may be insufficient to identify all cases. Eggs generally appear in the stool 1 week after ingestion of infected fish. However, adult worms contain few eggs, which probably results in the small number of eggs in stool specimens from

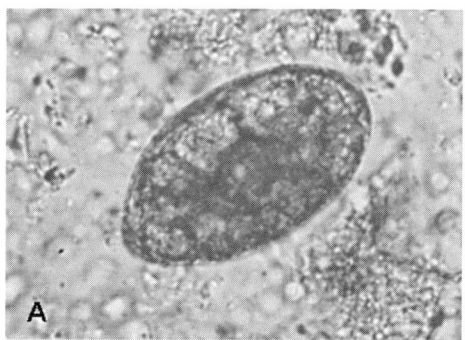 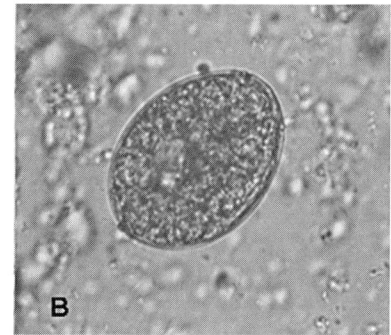

Figure 7.7. (a) Egg of *Nanophyetus salmincola*. Photograph by Thomas R. Fritsche. (b) Egg of *Diphyllobothrium* sp. Photograph by Robert L. Rausch.

patients with light infections. Eastburn *et al.* (1987) also indicated that the clinical preparation of specimens greatly impacts the detection of the trematode eggs. Most of their diagnoses were first made from trichrome stained preparations rather than from the typical formalin-ethyl acetate method. The reason for this discrepancy is unknown.

Because the eggs of *Nanophyetus* (Fig. 7.7 a) resemble those of tapeworms of the genus *Diphyllobothrium* (Fig. 7.7 b), misdiagnosis may occur (Adams and Rausch, 1997). This may be compounded in that several species of *Diphyllobothrium* are transmitted to humans through the consumption of salmonids. Recovery of segments from the cestodes or of whole worms would obviously clarify the etiological agent involved. Pharmacological treatment for both the trematode and the cestodes is similar (Table 7.2), except that nanophyetiasis usually requires a higher dosage for efficacy. This is in contrast to veterinary use of praziquantel to treat nanophyetiasis in canids such that dosages effective against cestodes (6.68–38.73 mg/kg) were adequate for the treatment of the trematode (Foreyt and Gorham, 1988). Note that treatment for the trematode in canids, is separate from that for the rickettsial infection.

Table 7.2. Comparison of Pharmacological Treatments for *Nanophyetus salmincola* and *Diphyllobothrium* spp.

Drug	Dosage for N. salmincola	Dosage for Diphyllobothrium spp.[a]
Bithional	50 mg/kg on alternate days ×2 doses[b]	30 mg/kg total in 1 or 2 doses
Niclosamide	2 g on alternate days ×3 doses[b]	2 g, single dose
Praziquantel	20 mg/kg ×3 daily for 1 day[c]	5–10 mg/kg for 1 day

[a] Adams and Rausch, 1997.
[b] Eastburn *et al.*, 1987.
[c] Fritsche *et al.*, 1989.

7.6 *FASCIOLA* SPP.

7.6.1 Introduction

Fasciola hepatica and *Fasciola gigantica* are both known to infect humans, although they are primarily parasites of ruminants. The former species has a worldwide distribution, but *F. gigantica* is found in Asia, Africa, and Hawaii. The latter trematode was probably introduced to Hawaii with the importation of water buffaloes from Asia (Alicata, 1964). Of the two species, *F. hepatica* more commonly infects humans. Stoll (1947) estimated that there were less than 100,000 cases of fascioliasis worldwide. Fifty years later, estimates are given of 2.4 million human infections in over 56 countries, primarily distributed in China, Egypt, Europe (especially France and Portugal), Iran, and South America (especially Bolivia, Ecuador, and Peru) (Crompton, 1999; Rim *et al.*, 1994; WHO, 1995). Although no consistent quantitative correlation has been shown, areas with high prevalences in domestic animals also tend to have high rates of human infections. For example, in the Altiplano region of Bolivia, infection rates of sheep and cattle range from 25 to 95%. In some villages, 65% of the people were found to pass eggs in their stool and 92% were serologically positive (WHO, 1995).

Fascioliasis is a serious disease in cattle and sheep throughout the world with an enormous economic impact. In 1969, Boray (1969) reported that one quarter of the sheep and cattle (40 million and 5 million, respectively) were grazing on pastures in which the infective metacercariae were potentially endemic. In Great Britain, 53% of the farms had livestock with fascioliasis (Froyd, 1975), with adult stock affected more than twice that of young animals. In the United States, 17% of cattle slaughtered in Montana were found to be infected (Knapp *et al.*, 1992). In a survey conducted by McKown and Ridley (1995), 33% of 278 veterinarians in Kansas reported diagnosing at least one case of liver fluke disease in their practice. Foreyt and Todd (1976) disclosed that 2.1% of beef livers examined in Kansas were condemned because of damage by the trematodes. Fascioliasis causes decreased wool production in sheep, decreased milk production and weight gain, increased numbers of condemned livers at slaughter, and decreased reproductive performance. In the United States alone, financial losses due to these trematodes were estimated at $30 million in 1973 (McKown and Ridley, 1995). Similarly, total losses due to fascioliasis in animals amounted to US $11 million in Peru (WHO, 1995).

7.6.2 Life Cycle

The adult trematodes are large, somewhat leaf-shaped, with a "cone-shaped" projection on the anterior. *F. hepatica* is generally 30 mm long and 13 mm wide (Fig. 7.8); *F. gigantica* may measure 73 mm long. The ventral sucker is larger than the oral sucker and located anteriad, near the base of the "cone." The intestinal ceca are highly branched, as are the two testes. The ovary is smaller, located near the ventral sucker. The operculate eggs of *F. hepatica* measure 130 to 150 µm by 63 to 90 µm (Fig. 7.9); eggs of *F. gigantica* are typically larger, 160 to 190 µm by 70 to 90 µm. Eggs of the former may be differentiated from those of *Fasciolopsis buski* by the roughened or irregular area at the abopercular end of the shell of *F. hepatica* (Ash and Orihel, 1997). Although humans may be infected, definitive

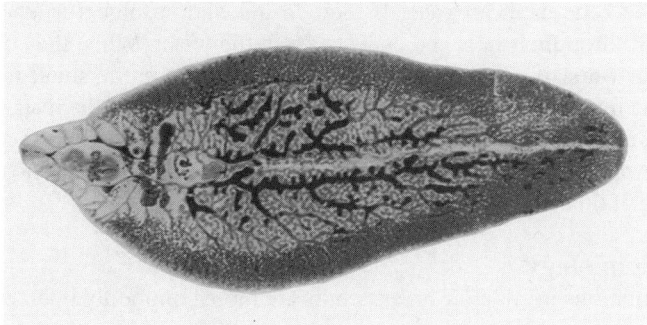

Figure 7.8. Adult of *Fasciola hepatica*, showing the distinct shape of the body and the "cone-shaped" projection on the anterior. Photograph by Charles Sterling, University of Arizona.

hosts are generally ruminants, including cattle, sheep, goats, and swine. Horses may also be hosts, but infections generally lack clinical symptoms and produce low egg counts (Alicata, 1964, Boray, 1969). Rabbits, mice, rats, and guinea pigs are all susceptible to infection. The rabbit has been recognized as a natural reservoir host.

Fasciola adults reside in the bile ducts of the liver. Operculated eggs are passed out in the feces of the host and must reach freshwater for continued development. When the egg hatches, a miracidium escapes and searches for a snail host. Suitable snails belong to several genera including *Lymnaea*, *Fossaria*, *Pseudosuccinea*, and *Austropeplea* (WHO, 1995). The miracidium penetrates the snail and develops into a sporocyst. Two generations of rediae ensue. Cercariae develop in the daughter rediae and later emerge from the snail and become free-swimming. Unlike other trematodes, these cercariae do not infect a second intermediate host. When they encounter vegetation or submerged bark, they shed their tails and encyst on the

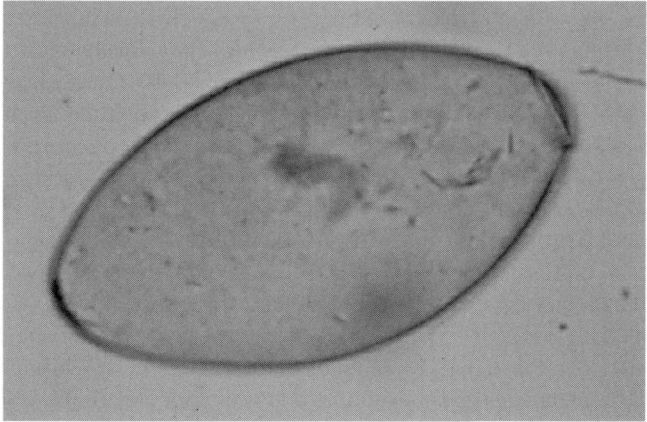

Figure 7.9. Egg of *Fasciola hepatica*.

plants and become metacercariae. If they do not find an object to encyst on, the cercariae will drop their tails and encyst free in the water. When the contaminated vegetation or water is consumed, the trematodes excyst in the small intestine and penetrate the intestinal wall. They travel around the viscera to the liver, where they burrow in and continue to move and feed for a couple months before entering the bile ducts. Approximately 4 months (range 3–18 months) after ingestion, the worm has matured and begun passing eggs (WHO, 1995).

7.6.3 Epidemiology

Within populations in endemic areas, adults are more commonly infected than children. Similarly, women have a higher prevalence of fascioliasis than men (Binkley and Sinniah, 1997). Women more often pick and gather vegetation and may consume more salads and raw vegetation than men.

In general, people become infected from eating raw vegetation contaminated with *Fasciola* metacercariae. Watercress (e.g., *Nasturtium officinale*) is the most common source of infections for people, although mint, lettuce, parsley, and wild watercress may also be contaminated with metacercariae (Rim *et al.*, 1994; WHO, 1995). Food habits often play a role in the transmission of parasites. For example, southern Europeans frequently consume watercress. When they migrated to Algeria, an outbreak of fascioliasis occurred among the Europeans, even though no infections of native Algerians were reported. Unlike the Europeans, the Algerians did not eat watercress (WHO, 1995).

Agricultural practices contribute to the contamination of vegetation by the metacercariae. The use of animal excreta as fertilizer helps distribute the eggs in the environment and increases the chances that the parasites may encounter freshwater and snails. The use of effluent from slaughterhouses or livestock pens as fertilizer for watercress plots increases the concentration and distribution of metacercariae on the plants. Emphasis on "organic" or natural healthy foods has increased the collection and consumption of wild watercress and other vegetation, introducing the infection to higher economic classes in some countries.

Agricultural practices may help maintain or increase fascioliasis among livestock. For example, many farmers allow their livestock to wander freely to graze and to drink from streams and ponds. This provides the parasite eggs ready access to freshwater in order to develop and hatch and gives the livestock direct access to vegetation that may be contaminated with metacercariae. Many farmers also cut forage from wet or swampy areas to feed cattle; in essence feeding the infecting metacercariae to their livestock (Alicata, 1964).

7.6.4 Clinical Signs, Diagnosis, and Treatment

The onset of symptoms usually occur 4 to 6 weeks after ingestion of the metacercariae, but can vary depending on the number of parasites ingested and the host immune response (WHO, 1995). Light infections may be asymptomatic. Patients often complain of fever, sweating, abdominal pain, dizziness, cough, bronchial asthma, fatigue, general malaise, loss of weight, and loss of appetite (Binkley and Sinniah, 1997; WHO, 1996). Patients may have gastrointestinal complaints including nausea and vomiting. Allergic reactions such as urticaria and pruritis may also occur.

The liver may be tender and enlarged, with jaundice. During the acute infection, children may have severe clinical manifestations such as right upper quadrant or general abdominal pain, fever, and anemia (WHO, 1995). These infections in children can be fatal.

The chronic phase of the infection begins when the trematodes reach the bile ducts. Often, this stage of the infection is asymptomatic. After arrival in the bile ducts, the trematodes mature into adults, and may cause irritation and inflammation of the ducts. The resulting symptoms may include dyspepsia, diarrhea, jaundice, biliary colic, cholecystitis, and cholelithiasis. If juvenile worms do not migrate properly into the liver, they may cause ectopic fascioliasis in other organs, including the intestines, lungs, heart, brain, and skin (the most common extrahepatic site). Nodules may form in the skin around the abdomen, sometimes reaching 6 cm across (Binkley and Sinniah, 1997).

In an endemic region, fascioliasis is suggested if a patient presents with fever, hepatomegaly, and eosinophilia (Binkley and Sinniah, 1997). Patients have been misdiagnosed as having visceral larval migrans or toxoplamosis when the fever is intermittent (WHO, 1995). Diagnosis is made either by observation of eggs in a fecal sample or from sero-immunological tests. However, diagnosis dependent on the former may be problematic for several reasons. Passage of eggs may be inconsistent, resulting in negative exams. In a study involving patients with a history of consuming watercress, 79% tested positive by serology but only 4% had positive stool specimens during their hospital stay (Chen and Mott, 1990). Symptoms of infection during the acute phase arise within 4 to 6 weeks (see above), but worms do not mature and pass eggs until approximately 4 months after ingestion of metacercariae. Therefore, the clinical findings cannot be supported by parasitological exams until months later. Patients with fascioliasis may be subjected to exploratory surgery to establish the cause of their disease because of the difficulties in diagnosis (WHO, 1995). In contrast, people who consume infected livers of cattle, sheep, or water buffalo may be incorrectly diagnosed as having fascioliasis when eggs are passed in their stools (Ash and Orihel, 1997). To prevent the diagnosis of spurious fascioliasis, the patient should be placed on a liver-free diet for at least 3 days and then retested. If eggs are still observed in the feces, the diagnosis of fascioliasis can be supported (Binkley and Sinniah, 1997).

Given the lag between the onset of symptoms and the production of parasite eggs, the use of serologic/immunodiagnostic methods are invaluable, particularly during the acute phase. Immunodetection of coproantigens of *Fasciola* rather than the observation of eggs in stool specimens is a viable alternative for diagnosis. Youssef *et al.* (1991) used partially purified *F. gigantica* worm antigens in counterimmunoelectrophoresis to test stool extracts. They found that the test was capable of detecting both early and chronic infections and that false-positives from spurious fascioliasis or cross-reactions with other parasitic infections were avoided. Espino *et al.* (1998) developed a sandwich enzyme-linked immunosorbent assay (ELISA) to detect coproantigens of *F. hepatica*; which was especially useful if patients with prepatent infections (e.g., not excreting eggs) did not have circulating antigens in their sera. If circulating antigens are present, various forms of ELISA are available for diagnosis, including the sandwich ELISA (Espino *et al.*, 1990), the Falcon assay

Table 7.3. Pharmacological Treatments for *Fasciola hepatica* and *F. gigantica*. Dosages are the same for adults and children.

Bithional	30–50 mg/kg on alternate days × 10–15 doses[a]
Praziquantel	25 mg/kg × 3 daily for 1 day[a] or 25 mg/kg × 3 daily for 3 to 7 days[b]
Triclabendazole[c]	10 mg/kg, single dose after overnight fast[d] or 5 mg/kg × 2 postprandially on same day with 6–8 h interval[e] (total dose 10 mg/kg)

[a] Medical Letter Inc., 1984.
[b] Binkley and Sinniah, 1997.
[c] Not approved for use in the United States or Canada.
[d] Apt *et al.* (1995).
[e] WHO, 1995.

screening test (FAST-) ELISA (Hillyer *et al.*, 1992), and Micro-ELISA (Carnevale *et al.*, 2001). Those that use excretory-secretory antigens appear to be more sensitive and specific. In particular, those tests that use worm cysteine proteinases such as cathepsin L1 (O'Neill *et al.*, 1998) or the Fas1 and Fas2 antigens isolated by Cordova *et al.* (1997) have excellent sensitivity and specificity. In addition, these antigens are excreted by the worms at all stages of development. If the trematodes themselves (or portions thereof) have been recovered, differentiation between *F. hepatica* and *F. gigantica* is possible with isoelectric focusing of soluble proteins from the adult worms (Lee and Zimmerman, 1993).

Praziquantel is the drug of choice for most trematode infections. Some consider the drug to also be effective against fascioliasis, however, the WHO (1995) and others (Binkley and Sinniah, 1997; Lui and Weller, 1996; Wessely *et al.*, 1988) recommend the use of other pharmaceuticals (Table 7.3). The current drug of choice is bithionol, but this is expected to be replaced by triclabendazole after appropriate testing. The latter is a benzimidazole antihelminthic and is effective against the adults in the bile ducts and the immature worms which migrate through the hepatic parenchyma (Wessely *et al.*, 1988; WHO, 1995). Usually, a single oral dose is effective, but a second dose may be administered for those infections which persist. Apt *et al.* (1995) treated 24 patients, with 19 testing negative after 2 months. Three of the remaining five were treated again with positive resolution. The drug was well tolerated and effective. Triclabendazole is also effective in the treatment of cattle, reducing the worm burden and improving weight gain (Fuhui *et al.*, 1989). The drug was considered to be safer and more efficacious than other drugs against fascioliasis, but has not been approved by the US Food and Drug Administration for use in the United States.

7.7 *FASCIOLOPSIS BUSKI*

7.7.1 Introduction

Fasciolopsis buski is another fasciolid trematode infectious to humans. Although it is similar in many ways to *Fasciola hepatica*, it also has some striking differences. Rather than residing in the bile ducts and liver, adults of *F. buski* are located in the small intestine. The geographic distribution of the latter species is generally restricted to Asia, but particularly in India, Thailand, China, Indonesia, Taiwan, east Pakistan, Laos, Vietnam, Cambodia, and Bangladesh (Crompton, 1999; WHO, 1995). Other countries, such as Israel and Hong Kong, have been concerned about the possible establishment of the parasite through the importation of foods and animals, and the arrival of infected immigrants from endemic countries (Sinniah and Binkley, 1997). The concern is legitimate. Prior to the 1940s, reports of fasciolopsiasis in Bangladesh were at best sporadic. However, after substantial immigrations from India, more cases were diagnosed until Gilman *et al.* (1982) considered the parasite to be endemic.

Earlier, Stoll (1947) estimated 10 million infections by this trematode, 100 times more than he reported for fascioliasis. Interestingly, Crompton (1999) reported somewhat the opposite situation, with only 210,000 human cases of fasciolopsiasis versus 2.1 million infections by *Fasciola*, a 10-fold difference. His estimates are taken from WHO (1995), which provided numbers of fasciolopsiasis from only China (204,000 cases) and Thailand (10,000). No data are provided for the other endemic countries. The prevalence of infection can vary greatly within the endemic areas, with fasciolopsiasis ranging up to 70% in some villages of Bangladesh, India, and China (Gilman *et al.*, 1982; WHO, 1995).

7.7.2 Life Cycle

The adult trematode is large, 20 to 75 mm long and up to 20 mm wide. Unlike the adult of *Fasciola*, *F. buski* does not have a conical projection on the anterior of the body and does not have branched cecae. The eggs are operculated and large (Fig. 7.10), 115 to 158 µm by 63 to 90 µm (Ash and Orihel, 1997; Hadidjaja *et al.*, 1982). The eggs may be difficult to differentiate from *F. hepatica*. The life cycle of *F. buski* is also very similar to that of *Fasciola*. The eggs pass out in the feces of the host and must reach freshwater. Miracidia hatch from the eggs and penetrate snails of the genera *Hippeutis*, *Segmentina*, *Gyraulus*, and *Polypylis* (Sinniah and Binkley, 1997; WHO, 1995). Development within the snails mirrors that of *F. hepatica*. Cercariae emerge from the snails and encyst on underwater vegetation. Infection of the mammalian host occurs when these plants are eaten raw or when people

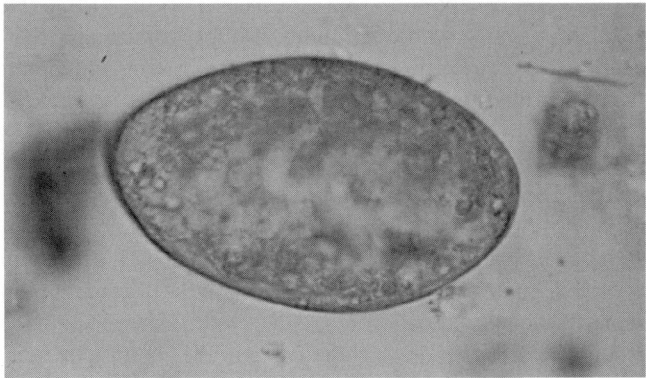

Figure 7.10. Egg of *Fasciolopsis buski*.

use their teeth to crack or peel the edible portions. In addition, a small number of cercariae, up to 4%, may encyst free in the water rather than attach to vegetation. Ingestion of the contaminated water may also result in fasciolopsiasis. In a study of case histories, 13% of people and 40% of pigs were found to be infected by this mechanism in China (Sinniah and Brinkley, 1997). The metacercariae excyst in the small intestine and develop into adults without migrating within the host. Once mature, each trematode may produce about 25,000 eggs daily and live for about 6 months in humans (Sinniah and Binkley, 1997). In addition to humans, pigs are commonly infected, although dogs and water buffalo may also be hosts for the parasite.

7.7.3 Epidemiology

In endemic regions, the trematode has a higher prevalence and intensity in children, particularly those between 2 and 10 years of age (Gilman *et al.*, 1982). Children often consume vegetation such as water lotus while playing in water. Socioeconomic status may also be a factor in the prevalence of infection. In Bangladesh, water plants are cheap and readily available and thus, are a common food source for the poor (Gilman *et al.*, 1982). Plants that are commonly implicated in the transmission of fasciolopsiasis include water chestnut, water caltrop, lotus, bamboo, water hyacinth, water mimosa, and water spinach (Sinniah and Binkley, 1997; WHO, 1995). Water caltrop (*Trapa natans* or *T. bicornis*) is a favorite snack, particularly for children in rural areas, and is the most important plant vector of *F. buski* in east Asia (WHO, 1995).

Similar to *F. hepatica*, agricultural practices directly contribute to the persistence of *F. buski* in the environment and the transmission of the parasite to humans and domestic animals. The use of untreated manure as fertilizer for the cultivation of plants, including water plants in ponds, perpetuate the life cycle of the trematode. In China, the drainage systems from pig farms are often physically linked to ponds where the various edible plants mentioned above are raised (WHO, 1995). In Thailand and elsewhere, ponds used for cultivation are often fertilized with human feces.

Human fasciolopsiasis is generally correlated to a high prevalence of *F. buski* in pigs (WHO, 1995). An exception to this observation occurs in Bangladesh, a predominantly Muslim country, where the numbers of pigs are small. In those villages in which fasciolopsiasis were reported, no pigs were present, and therefore humans were responsible for the introduction and transmission of the trematode (Gilman *et al.*, 1982).

7.7.4 Clinical Signs, Diagnosis, and Treatment

Most infections with *F. buski* are light and asymptomatic. However, in heavy infections, the patient may first present with epigastric pain and diarrhea; with vomiting, nausea, and possibly anorexia ensuing. Edema, particularly of the face may occur in heavy infections, especially in children (Cook, 1986). Toxemia and allergic symptoms may occur. Localized inflammation occurs at the site of attachment which in turn causes excess mucus secretion. Some ulceration can occur as well as the occasional formation of abscesses. Although heavy infections may sometimes result in the partial obstruction of the intestine, no clear evidence exists of general malabsorption (Cook, 1986; WHO 1995). Intense infections, defined as greater than 300 worms, are associated with malnutrition and increased mortality (Gilman et al., 1982). Toxemia occurs from the absorption of metabolites from the trematodes, resulting in sensitization and possibly death of the patient. Mortality of children from profound intoxication has been reported in China, India, and Thailand (WHO, 1995).

Diagnosis is primarily by detection of eggs in feces. Rarely, a patient may simplify diagnosis by vomiting intact adult worms (Hadidjaja et al., 1982). Peripheral eosinophilia and a slight macrocytic anemia may be present (Cook, 1986; WHO, 1995). However, no immunologic method for diagnosis is presently available. Care must be taken in diagnosis since the epigastric pain can simulate a peptic ulcer. In both acute and chronic fasciolopsiasis, the diarrhea does not respond to antibiotics. If the infection is thought to be caused by bacteria or viruses, the misdiagnosis can be costly because of repeated treatments (WHO, 1995).

Unlike *F. hepatica*, fasciolopsiasis responds well to treatment with praziquantel. The dose is similar to that of other trematodes: 25 mg/kg, three times in 1 day (Lui and Weller, 1996; Medical Letter Inc., 1984). Niclosamide, tetraclorethylene, and hexylresorcinol are also known to be effective (Cook, 1986).

7.8 PREVENTION AND CONTROL

The prevention and control of infections caused by these foodborne trematodes can be achieved primarily by the disruption of the life cycles of the parasites. Since the cycles are complex, involving multiple hosts and developmental stages, disruption of development and transmission can occur at multiple sites within the cycles.

The elimination of adults and any resulting eggs can be pursued through treatment of populations at risk. A control program may be implemented in an endemic area which relies heavily on drug treatment. WHO (1995) recommends that a program for community-based treatment provides annual drug therapy for the parasite for up to 3 years. In an endemic area of Thailand, 90% of people treated for opisthorchiasis were negative for eggs 2 years after a single treatment. Although infected, the remaining patients had light infections and were essentially asymptomatic. As mentioned previously, control programs for *C. sinensis* in Japan, Korea, and China have been successful at reducing the incidence of the parasite. Additionally, the broad efficacy of praziquantel, the drug of choice for many parasitic infections, allows the targeting of multiple species of parasites, even when the specific identity of a helminth cannot be determined. The general safety of the drug and the accepted use for pediatric patients as well as for adults underscore the value of this treatment for helminthiases.

Generally, patients do not become refractive to reinfection by these foodborne trematodes. Therefore, once treated for the parasites, changes in food preparation and food habits are required to keep patients from acquiring new infections. In addition to drug therapy, the community should receive education as to the life cycle of the parasite and dietary changes that could prevent transmission. Mothers in particular should receive training on the importance of children consuming only fully cooked or adequately processed foods. However, changes in traditional preparation and consumption of foods are difficult and reinfection is not uncommon. For example, in Bangladesh aquatic plants are particularly attractive to and consumed extensively by children (Gilman et al., 1982). Therefore, educational programs to prevent or reduce infections by *F. buski* in children were considered unlikely to be effective.

Infections by foodborne trematodes can be prevented if the intermediate hosts are not consumed raw, undercooked, or improperly (or incompletely) pickled, salted, dried, or smoked. Fully cooking fishes and crabs would neutralize any metacercariae present. This process is time-temperature dependent, e.g., cooking time decreases with an increase in temperature. For example, metacercariae of *O. viverrini* survives for 5 h at 50°C, but only 5 minutes at 80°C (WHO, 1995). Immersion of aquatic vegetation in boiling water for a few minutes kills metacercariae of *F. buski*.

Freezing is also very effective at neutralizing the metacercariae. The texture of the flesh may change after freezing and therefore, many people are reluctant to eat raw previously-frozen fish. Freezing is also time-temperature dependent, with colder temperatures requiring less time. Metacercariae of *O. felineus* in fish are killed after 32 h at −28°C; 14 at −35°C; and 7 at −40°C (WHO, 1995). The type of freezer, thickness of the fish, and the form of the product (e.g., whole, headed and gutted, fillets) can affect the time needed to attain the required temperature and thereby increase the time in the freezer.

Food preparation requiring pickling, salting, drying, or smoking may be insufficient to kill the metacercariae. During pickling, the acidity of the solution and the thickness of the fish may be factors in the continued viability of the larvae. Generally, the higher the acidity the more effective the control measure. However, pickling solutions are usually about 4% acid; similar to vinegar. Metacercariae of *O. viverrini* can survive only 1–1.5 h when placed free in the solution (WHO, 1995), but would require much longer if still situated within the flesh of the fish. Conditions are similar for salting in that the effectiveness of the salt is decreased (1) with a decrease in the concentration, and (2) with an increase in the thickness of the flesh. Consumption of fish salted for only 1 day resulted in an outbreak of opisthorchiasis in Europe (Cross, 2001). The smoking process may include a substantial increase in temperature. If the fish is cooked during the smoking process (also known as hot smoking), the parasites and other pathogens will be killed. However, if the temperature of the product is not raised sufficiently during smoking (e.g., cold smoking), the product is still raw and the parasites remain viable and infectious (Adams and DeVlieger, 2001). Safety of these food preparations can be enhanced considerably by freezing the fish prior to pickling, salting, drying, or smoking (Adams et al., 1997). The previously-mentioned change in texture caused by freezing is usually not noticeable after further processing.

Agricultural practices, particularly those involving sanitation, have a direct impact on the completion of the life cycles of these trematodes. The use of human feces as fertilizer, also known as night soil, assists in the distribution of parasites and their infective stages. In China, latrines were often placed above ponds to enrich the system for fish-rearing. In so doing, eggs of *C. sinensis* were efficiently introduced to the necessary habitat of snails and fishes. Runoff from areas fertilized with human and animal excreta can inadvertently contaminate nearby ponds and waterways with the parasites and other pathogens. Alternatives include composting or sterilizing the excreta prior to use as fertilizer and the use of other compounds for enrichment, such as rice bran. The diversion of runoff and the use of cleaner water sources for fish-rearing ponds also reduces or prevents contamination.

Another approach is the reduction of infected intermediate and reservoir hosts. Regular antihelminthic treatment for farm and companion animals can reduce the available pool of trematode eggs in endemic areas. Restricting access for animals to streams and ponds may also prevent the consumption of metacercariae and the direct introduction of eggs from animal feces into freshwater. Livestock will readily graze on aquatic vegetation, exposing them to metacercariae of *F. hepatica*. The control of snails is difficult. The use of molluscicides is often ineffective since snails can climb up on vegetation to avoid the chemicals and the products can become diluted. Molluscicides are also harmful to fish and the water environment (Cross, 2001).

The increase in aquaculture can inadvertently lead to larger trematode populations. An increase in ponds also creates additional snail habitat. The expansion in fish-farming increases the number of potential fish hosts. In 1990, 46% of the freshwater fish produced in Thailand came from aquaculture. In 1984, approximately 49,000 tons of freshwater fish were produced by Thai aquaculture; increasing to almost 99,000 in 1991 (WHO, 1995). Expanded aquaculture may also result in the fishes being sold over a larger area or exported to nonendemic areas. The parasite may then become established in areas with suitable intermediate hosts and insufficient sanitation. Under these circumstances, the diagnosis and treatment of resulting helminth infections may be difficult because medical personnel would have little experience with these parasitic diseases.

REFERENCES

Adams, A. M., and Rausch, R. L., 1997, Diphyllobothriasis, In Connor, D. H., Chandler, F. W., Schwartz, D. A., Manz, H. J., and Lack, E. E. (eds), *Pathology of Infectious Diseases*, Vol. 2, Appleton and Lange, Stamford, CT, pp. 1377–1389.

Adams, A. M., and DeVlieger, D. D., 2001, Seafood parasites: Prevention, inspection, and HACCP, In Hui, Y. H., Sattar, S. A., Murrell, K. D., Nip, W. -K., and Stanfield, P. S. (eds), *Foodborne Disease Handbook*, Vol. 2, 2nd edn., Marcel Dekker, Inc., New York, pp. 407–423.

Adams, A. M., Murrell, K. D., and Cross, J. H., 1997, Parasites of fish and risks to public health, *Rev. Sci. Tech. Off. Int. Epiz.* **16**:652–660.

Alicata, J. E., 1964, *Parasitic Infections of Man and Animals in Hawaii*, University of Hawaii, Honolulu, Hawaii.

Apt, W., Aguilera, X., Vega, F., Miranda, C., Zulantay, I., Perez, C., Gabor, M., and Apt, P., 1995, Treatment of human chronic fascioliasis with triclabenzadole: Drug efficacy and serologic response, *Am. J. Trop. Med. Hyg.* **52**:532–535.

Ash, L. R., and Orihel, T. C., 1997, *Atlas of Human Parasitology*, 4th edn. American Society of Clinical Pathologists Press, Chicago.

Bennington, E., and Pratt, I., 1960, The life history of the salmon-poisoning fluke, *Nanophyetus salmincola* (Chapin), *J. Parasitol.* **46**:91–100.

Binkley, C. E., and Sinniah, B., 1997, Fascioliasis, In Connor, D. H., Chandler, F. W., Schwartz, D. A., Manz, H. J., and Lack, E. E. (eds), *Pathology of Infectious Diseases*, Vol 2, Appleton and Lange, Stamford, CT, pp. 1419–1425.

Boonpucknavig, S., Kurathong, S., and Thamavit, W., 1986, Detection of antibodies in sera from patients with opisthorchiasis, *Clin. Lab. Immunol.* **19**:135–137.

Boray, J. C., 1969, Experimental fascioliasis in Australia, In Dawes, B. (ed), *Advances in Parasitology*, Vol 7, Academic Press, New York, pp. 95–210.

Carnevale, S., Rodriguez, M., Santillán, G., Labbé, J. H., Cabrera, M. G., Bellegarde, E. J., Velásquez, J. N., Trgovcic, J. E., and Guarnera, E. A., 2001, Immunodiagnosis of human fascioliasis by an enzyme-linked immunosorbent assay (ELISA) and a Micro-ELISA, *Clin. Diag. Lab. Immunol.* **8**:174–177.

Castilla, E. A., Jessen, R., Scheck, D. N., and Procop, G. W., 2003, Cavitary mass lesion and recurrent pneumothoraces due to *Paragonimus kellicotti* infection: North American paragonimiasis, *Am. J. Surg. Pathol.* **27**:1157–1160.

Chen, M. G., and Mott, K. E., 1990, Progress in assessment of morbidity due to *Fasciola hepatica*: A review of recent literature, *Trop. Dis. Bull.* **87**:R1–R38.

Chen, M. G., Lu, Y., Hua, X., and Mott, K. E., 1994, Progress in assessment of morbidity due to *Clonorchis sinensis* infection: A review of recent literature, *Trop. Dis. Bull.* **91**:R7–R65.

Cook, G. C., 1986, The clinical significance of gastrointestinal helminths—A review, *Trans. Roy. Soc. Trop. Med. Hyg.* **80**:675–685.

Cordova, M., Herrera, P., Nopo, L., Bellatin, J., Naquira, C., Guerra, H., and Espinoza, J. R., 1997, *Fasciola hepatica* cysteine proteinases: Immunodominant antigens in human fascioliasis, *Am. J. Trop. Med. Hyg.* **57**:660–666.

Crompton, D. W. T., 1999, How much human helminthiasis is there in the world? *J. Parasitol.* **85**:397–403.

Cross, J. H., 2001, Fish- and invertebrate-borne helminths, In Hui, Y. H., Sattar, S. A., Murrell, K. D., Nip, W.-K., and Stanfield, P. S. (eds), *Foodborne Disease Handbook*, 2nd ed. Marcel Dekker, Inc., New York, pp. 249–288.

DeFrain, M., and Hooker, R., 2002, North American paragonimiasis: Case report of a severe clinical infection, *Chest* **121**:1368–1372.

Donham, C. R., 1925, So-called salmon poisoning of dogs. Preliminary report, *J. Am. Vet. Med. Assoc.* **66**:637–739.

Donham, C. R., Simms, B. T., and Miller, F. W., 1926, So-called salmon poisoning in dogs, *J. Am. Vet. Med. Assoc.* **68**:701–715.

Eastburn, R. L., Fritsche, T. R., and Terhune, C. A., Jr., 1987, Human intestinal infection with *Nanophyetus salmincola* from salmonid fishes, *Am. J. Trop. Med. Hyg.* **36**:586–591.

Elkins, D. B., Haswell-Elkins, M. R., Mairiang, E., Mairiang, P., Sithithaworn, P., Kaewkes, S., Bhudhisawasdi, V., and Uttaravichien, T., 1990, A high frequency of hepatobiliary disease and suspected cholangiocarcinoma associated with heavy *Opisthorchis viverrini* infection in a small community in north-east Thailand, *Trans. R. Soc. Trop. Med. Hyg.* **84**:715–719.

Espino, A. M., Mareet, R., and Finlay, C. M., 1990, Detection of circulating excretory secretory antigens in human fascioliasis by sandwich enzyme-linked immunosorbent assay, *J. Clin. Microbiol.* **28**:2637–2640.

Espino, A. M., Diaz, A., Pérez, A., and Finlay, C. M., 1998, Dynamics of antigenemia and coproantigens during a human *Fasciola hepatica* outbreak, *J. Clin. Microbiol.* **36**:2723–2726.

Filimonova, L. V., 1963, Biologicheskii tsikl trematody *Nanophyetus schikhobalowi*, *Tr. Gel'mint. Lab.* **13**:347–357.

Filimonova, L. V., 1965, Eksperimental'noe izuchenie biologii *Nanophyetus schikhobalowi* Skrjabin et Podjapolskaja, 1931 (Trematoda, Nanophyetidae), *Tr. Gel'mint. Lab.* **15**:172–184.

Filimonova, L. V., 1966, Rasprostranenie nanofietoza na territorii sovetskogo dal'nego vostoka, *Tr. Gel'mint. Lab.* **17**:240–244.

Flavell, D. J., 1981, Liver-fluke infection as an aetiological factor in bile-duct carcinoma of man, *Trans. R. Soc. Trop. Med. Hyg.* **75**:814–824.

Foreyt, W. J., and Todd, A. C., 1976, Liver flukes in cattle, *Vet. Med. Small Anim. Clin.* **71**:816–822.

Foreyt, W. J., and Gorham, J. R., 1988, Evaluation of praziquantel against induced *Nanophyetus salmincola* infections in coyotes and dogs, *Am. J. Vet. Res.* **49**:563–565.

Fritsche, T. R., Eastburn, R. L., Wiggins, L. H., and Terhune, C. A., Jr., 1989, Praziquantel for treatment of human *Nanophyetus salmincola* (*Troglotrema salmincola*) infection, *J. Infect. Dis.* **5**:896–899.

Froyd, G., 1975, Liver fluke in Great Britain: A survey of affected livers, *Vet. Rec.* **97**:492–495.

Fuhui, S., Bangfa, L., Chengui, Q., Ming, L., Mingbao, F., Jiliang, M., Wei, S., Siwen, W., and Xueliang, J., 1989, The efficacy of triclabendazole (fasinex®) against immature and adult *Fasciola hepatica* in experimentally infected cattle, *Vet. Parasitol.* **33**:117–124.

Gebhardt, G. A., Millemann, R. E., Knapp, S. E., and Nyberg, P. A., 1966, "Salmon poisoning" disease. II. Second intermediate host susceptibility studies, *J. Parasitol.* **52**:54–59.

Gilman, R. H., Mondal, G., Maksud, M., Alam, K., Rutherford, E., Gilman, J. B., and Khan, M. U., 1982, Endemic focus of *Fasciolopsis buski* infection in Bangladesh, *Am. J. Trop. Med. Hyg.* **31**:796–802.

Green, A., Uttaravichien, T., Bhudhisawasdi, V., Chartbanchachai, W., Elkins, D. B., Marieng, E. O., Pairqjkul, C., Dhiensiri, T., Kanteekaew, N., and Haswell-Elkins, M. R., 1991, Cholangiocarcinoma in north east Thailand. A hospital-based study, *Trop. Geogr. Med.* **43**:193–198.

Hadidjaja, P., Dahri, H. M., Roesin, R., Margano, S. S., Djalins, J., and Hanafiah, M., 1982, First autochthonous case of *Fasciolopsis buski* infection in Indonesia, *Am. J. Trop. Med. Hyg.* **31**:1065.

Haswell-Elkins, M. R., Elkins, D. B., Sithithaworn, P., Treesarawat, P., and Kaewkes, S., 1991, Distribution patterns of *Opisthorchis viverrini* within a human community, *Parasitology* **103**(Pt 1):97–101.

Haswell-Elkins, M. R., Mairiang, E., Mairiang, P., Chaiyakum, J., Chamadol, N., Loapaiboon, V., Sithithaworn, P., and Elkins, D. B., 1994, Cross-sectional study of *Opisthorchis viverrini* infection and cholangiocarcinoma in communities within a high-risk area in northeast Thailand, *Int. J. Cancer* **59**:505–509.

Heard, W. R., 1991, Life history of pink salmon (*Oncorhynchus gorbuscha*), In Groot, C., and Margolis, L. (eds), *Pacific Salmon Life Histories*, University of British Columbia Press, Vancouver, BC, pp. 119–230.

Hillyer, G. V., and Serrano, A. E., 1983, The antigens of *Paragonimus westermani*, *Schistosoma mansoni*, and *Fasciola hepatica* adult worms, Evidence for the presence of cross-reactive antigens and for cross-protection to *Schistosoma mansoni* infection using antigens of *Paragonimus westermani*, *Am. J. Trop. Med. Hyg.* **32**:350–358.

Hillyer, G. V., Soler de Galanes, M., Rodriguez-Perez, J., Bjorland, J., Silva de Lagrava, M., Ramirez Guzman, S., and Bryan, R. T., 1992, Use of the Falcon[TM] assay screening test–enzyme-linked immunosorbent assay (FAST-ELISA) and the enzyme-linked immunoelectrotransfer blot (EITB) to determine the prevalence of human fascioliasis in the Bolivian Altiplano, *Am. J. Trop. Med. Hyg.* **46**:603–609.

Hong, S. T., Lee, M., Sung, N. J., Cho, S. R., Chai, J. Y., and Lee, S. H., 1999, Usefulness of IgG4 subclass antibodies for diagnosis of human clonorchiasis, *Korean J. Parasitol.*, **37**:243–248.

Ikeda, T., 1998, Cystatin capture enzyme-linked immunosorbent assay for immunodiagnosis of human paragonimiasis and fascioliasis, *A. J. Trop. Med. Hyg.* **59**:286–290.

Ikeda, T., Oikawa, Y., and Nishiyama, T., 1996, Enzyme-linked immunosorbent assay using cysteine proteinase antigens for immunodiagnosis of human paragonimiasis, *Am. J. Trop. Med. Hyg.* **55**:434–437.

Kim, S. I., 1998, A *Clonorchis sinensis*-specific antigen that detects active human clonorchiasis, *Korean J. Parasitol.* **36**:37–45.

Kim, T. Y., Kang, S.-Y., Park, S. H., Sukontason, K., Sukontason, K., and Hong, S.-J., 2001, Cystatin capture enzyme-linked immunosorbent assay for serodiagnosis of human clonorchiasis and profile of captured antigenic protein of *Clonorchis sinensis*, *Clin. Diag. Lab. Immunol.* **8**:1076–1080.

Knapp, S. E., Dunkel, A. M., Han, K., and Zimmerman, L. A., 1992, Epizootiology of fascioliasis in Montana, *Vet. Parasitol.* **42**:241–246.

Komiya, Y., 1966, *Clonorchis* and clonorchiasis, In Dawes, B. (ed) *Advances in Parasitology*, Vol 4, Academic Press, New York, pp. 53–106.

Kong, Y., Ito, A., Yang, H.-J., Chung, Y.-B., Kasuya, S., Kobayashi, M., Liu, Y.-H., and Cho, S.-Y., 1998, Immunoglobin G (IgG) subclass and IgE responses in human paragonimiases caused by three different species, *Clin. Diag. Lab. Immunol.* **5**:474–478.

Kum, P. N., and Nchinda, T. C., 1982, Pulmonary paragonimiasis in Cameroon, *Trans. Roy. Soc. Trop. Med. Hyg.* **76**:768–772.

Lee, C. G., and Zimmerman, G. L., 1993, Banding patterns of *Fasciola hepatica* and *Fasciola gigantica* (Trematoda) by isoelectric focusing, *J. Parasitol.* **79**:120–123.

Lui, L.X, and Weller, P. F., 1996, Antiparasitic drugs, *New Engl. J. Med.* **334**:1178–1184.

Mariano, E. G., Borja, S. R., and Vruno, M. J., 1986, A human infection with *Paragonimus kellicotti* (lung fluke) in the United States, *Am. J. Clin. Pathol.* **86**:685–687.

McKown, R. D., and Ridley, R. K., 1995, Distribution of fasciolosis in Kansas, with results of experimental snail susceptibility studies, *Vet. Parasitol.* **56**:281–291.

Medical Letter, Inc., 1984, Drugs for parasitic infections, *Med. Letter on Drugs and Therapeutics* **26**:27–34.

Millemann, R. E., and Knapp, S. E., 1970, Biology of *Nanophyetus salmincola* and "Salmon poisoning" disease, In Dawes, B. (ed), *Advances in Parasitology*, Vol. 8, Academic Press, New York, pp. 1–41.

Miyasaki, I., and Hirose, H., 1976, Immature lung flukes first found in the muscle of the wild boar in Japan, *J. Parasitol.* **62**:836–837.

Moyou-Somo, R., Kefie-Arrey, C., Dreyfuss, G., and Dumas, M., 2003, An epidemiological study of pleuropulmonary paragonimiasis among pupils in the peri-urban zone of Kumba town, Meme Division, Cameroon, *BMC Public Health* **3**:40–44.

Mukae, H., Taniguchi, H., Matsumoto, N., Iiboshi, H., Ashitani, J., Matsukura, S., and Nawa, Y., 2001, Clinicoradiologic features of pleuropulmonary *Paragonimus westermani* on Kyusyu Island, Japan, *Chest* **120**:514–520.

Noble, G. A., 1963, Experimental infection of crabs with *Paragonimus*, *J. Parasitol.* **49**:352.

Oh, S. J., and Jordan, E. J., 1967, Findings of intelligence quotient in cerebral paragonimiasis, *Jpn. J. Parasitol.* **16**:436–440.

O'Neill, S. M., Parkinson, M., Strauss, W., Angles, R., and Dalton, J. P., 1998, Immunodiagnosis of *Fasciola hepatica* infection (fascioliasis) in a human population in the Bolivian

Altiplano using purified cathepsin L cysteine proteinase, *Am. J. Trop. Med. Hyg.* **58**:417–423.

Pachucki, C. T., Cort, W. W., and Yokogawa, M., 1984, American paragonimiasis treated with praziquantel, *N. Engl. J. Med.* **311**:582–583.

Parkin, D. M., Ohshima, H., Srivatanakul, P., and Vatanasapt, V., 1993, Cholangiocarcinoma: Epidemiology, mechanisms of carcinogenesis and prevention, *Cancer Epidemiol. Biomarkers Prev.* **2**:537–544.

Patel, T., 2002, Worldwide trends in mortality from biliary tract malignancies, *BMC Cancer*, **2**:10–16.

Pezzella, A. T., Yu, H. S., and Kim, J. E., 1981, Surgical aspects of pulmonary paragonimiasis, *Cardiovascular Dis., Bull. Texas Heart Inst.* **8**:187–194.

Philip, C. B., 1955, There's always something new under the "Parasitological" sun (The unique story of helminth-borne salmon poisoning disease), *J. Parasitol.* **41**:125–148.

Procop, G. W., Marty, A. M., Scheck, D. N., Mease, D. R., and Maw, G. M., 2000, North American paragonimiasis. A case report, *Acta Cytol.* **44**:75–80.

Rim, H.-J., Farag, H. F., Sornmani, S., and Cross, J. H., 1994, Food-borne trematodes: Ignored or emerging? *Parasitol. Today* **10**:207–209.

Sachs, R., and Cumberlidge, N., 1990, Distribution of metacercariae in freshwater crabs in relation to *Paragonimus* infection of children in Liberia, West Africa, *Ann. Trop. Med. Parasitol.* **84**:277–280.

Schell, S. S., 1985, *Trematodes of North America, North of Mexico*, University Press of Idaho, Moscow, Idaho.

Schlegel, M. W., Knapp, S. E., and Millemann, R. E., 1968, "Salmon poisoning" disease. V. Definitive hosts of the trematode vector, *Nanophyetus salmincola*, *J. Parasitol.* **54**:770–774.

Simms, B. T., Donham, C. R., and Shaw, J. N., 1931, Salmon poisoning, *Am. J. Hyg.* **13**:363–391.

Sinniah, B., 1997, Paragonimiasis, In Connor, D. H., Chandler, F. W., Schwartz, D. A., Manz, H. J., and Lack, E. E. (eds), *Pathology of Infectious Diseases*, Vol. 2, Appleton and Lange, Stamford, CT, pp. 1527–1530.

Sinniah, B., and Binkley, C. E., 1997, Fasciolopsiasis–Infection by *Fasciolopsis buski*, In Connor, D. H., Chandler, F. W., Schwartz, D. A., Manz, H. J., and Lack, E. E. (eds), *Pathology of Infectious Diseases*, Vol. 2, Appleton and Lange, Stamford, CT, pp. 1427–1430.

Sirisinha, S., Chawengkirttikul, R., Sermswan, R., Amornpant, S., Mongkolsuk, S., and Panyim, S., 1991, Detection of *Opithorchis viverrini* by monoclonal antibody-based ELISA and DNA hybridization, *Am. J. Trop. Med. Hyg.* **44**:140–145.

Skrjabin, K. J., and Podjapolskaja, W. P., 1931, *Nanophyetus schikhobawi* n. sp., ein neuer Trematode aus dem Darm des Menschen, *Zbl. Bakt. I. Orig.* **119**:294–297.

Slemenda, S. B., Maddison, S. E., Jong, E. C., and Moore, D. D., 1988, Diagnosis of paragonimiasis by immunoblot, *Am. J. Trop. Med. Hyg.* **39**:469–471.

Stoll, N. R., 1947, This wormy world, *J. Parasitol.* **33**:1–18.

Sun, T., 1980, Clonorchiasis: A report of four cases and discussion of unusual manifestations, *Am. J. Trop. Med. Hyg.* **29**:1223–1227.

Sun, T., 1997, Clonorchiasis and opisthorchiasis, In Connor, D. H., Chandler, F. W., Schwartz, D. A., Manz, H. J., and Lack, E. E. (eds), *Pathology of Infectious Diseases*, Vol. 2, Appleton and Lange, Stamford, CT, pp.1351–1360.

Sun, T., and Gibson, J. B., 1969, Antigens of *Clonorchis sinensis* in experimental and human infections. An analysis by gel-diffusion technique, *Am. J. Trop. Med. Hyg.* **18**:241–252.

Upatham, E. S., Viyanant, V., Kurathong, S., Rojborwonwitaya, J., Brockelman, W. Y., Ardsungnoen, S., Lee, P., and Vajrasthira, S., 1984, Relationship between prevalence and intensity of *Opisthorchis viverrini* infection, and clinical symptoms and signs in a rural community in northeast Thailand, *Bull. World Health Org.* **62**:451–461.

Vatanasapt, V., Uttaravichien, T., Mairiang, E., Pairqjkul, C., Chartbanchachai, V., and Haswell-Elkins, M. R., 1990, Northeast Thailand: A region with a high incidence of cholangiocarcinoma, *Lancet* **335**:116–117.

Weina, P. J., and England, D. M., 1990, The American lung fluke, *Paragonimus kellicotti*, in a cat model, *J. Parasitol.* **76**:568–572.

Weiseth, P. R., Farrell, R. K., and Johnston, S. D., 1974, Prevalence of *Nanophyetus salmincola* in ocean-caught salmon, *J. Am. Vet. Med. Assoc.* **165**:849–850.

Wessely, K., Reischig, H. L., Heinerman, M., and Stempka, R., 1988, Human fascioliasis treated with triclabendazole (Fasinex®) for the first time, *Trans. R. Soc. Trop. Med. Hyg.* **82**:743–744.

WHO, 1995, Control of foodborne trematode infections, *WHO Tech. Rep. Ser.* **849**:1–157.

Witenberg, G., 1932, On the anatomy and systematic position of the causative agent of so-called salmon poisoning, *J. Parasitol.* **18**:258–263.

Wongratanacheewin, S., Pumidonming, W., Sermswan, R. W., Pipitgool, V., and Maleewong, W., 2002, Detection of *Opisthorchis viverrini* in human stool specimens by PCR, *J. Clin. Microbiol.* **40**:3879–3880.

Woolf, A., Green, J., Levine, J. A., Estevez, E. G., Weatherly, N., Rosenberg, E., and Frothingham, T., 1984, A clinical study of Laotian refugees infected with *Clonorchis sinensis* or *Opisthorchis viverrini*. *Am. J. Trop. Med. Hyg.* **33**:1279–1280.

Yangco, B. G., De Lerma, C., Lyman, G. H., and Price, D. L., 1987, Clinical study evaluating efficacy of praziquantel in clonorchiasis, *Antimicrobial Agents and Chemotherapy*, **31**:135–138.

Yokogawa, M., 1965, Paragonimus and paragonimiasis, In Dawes, B. (ed), *Advances in Parasitology*, Vol 3, Academic Press, New York, pp. 99–158.

Yokogawa, M., 1969, Paragonimus and paragonimiasis, In Dawes, B. (ed), *Advances in Parasitology*, Vol 7, Academic Press, New York, pp. 375–387.

Youssef, F. G., Mansour, N. S., and Aziz, A. G., 1991, Early diagnosis of human fascioliasis by the detection of copro-antigens using counterimmunoelectrophoresis, *Trans. R. Soc. Trop. Med. Hyg.* **85**:383–384.

CHAPTER 8

Cestodes

Natalie Bowman, Joseph Donroe, and Robert Gilman

8.1 PREFACE

Cestodes, or tapeworms, belong to the class Cestoidea of the phylum Platyhelminthes. Members of this family vary greatly in size and behavior; however, they share the same basic body plan. They attach to the intestinal wall of definitive hosts by the scolex, or head. The scolex is followed by the neck, behind which grow the body segments, or proglottids. The proglottids together form the strobila, or body, of the worm; the number of proglottids in the adult worm depends on the species. Proglottids have both longitudinal and transverse muscles and are motile. Each proglottid also has both male and female reproductive organs, but mating usually occurs with adjacent segments rather than by self-fertilization. The oldest proglottids are farthest from the scolex and contain the tapeworm's eggs. Cestodes have no digestive or circulatory system and must absorb nutrients from the lumen of the host's small intestine through microvilli. These cover the surface of each proglottid and excrete waste through a pair of excretory tubules. Tapeworms do have a rudimentary nervous system consisting of ganglia in the scolex and nerves in the proglottids.

8.2 TAENIA

Three *Taenia* species, members of the Cyclophyllidea order, claim human beings as their definitive hosts: *T. saginata*, *T. solium*, and *T. asiatica*. Humans acquire intestinal infection with the worms by ingestion of undercooked meat—pork in the cases of *T. solium* and *T. asiatica*, and beef in the case of *T. saginata*—that contains encysted larvae. *T. solium* is presumably the only parasite of the three to cause significant human pathology in the form of human cysticercosis, although the pathogenicity of *T. asiatica* in humans has not been fully characterized. *T. saginata* and *T. asiatica* are morphologically very similar and closely related genetically. They cause limited pathology in humans, but their economic impact on the livestock market is significant.

Cestodes of the order Cyclophyllidea typically require two hosts in order to complete their life cycles (Fig. 8.1). The life cycles of *Taenia* species are similar, and differences will be highlighted in the "Biology" section in the discussions of the individual species. In general, Taenia eggs are ingested by an intermediate host, allowing larva to mature to the metacestaode stage of development. The metacestode stage is the encysted, infective larva that is referred to as a cysticercus. Maturation to the adult tapeworm continues once the cyticercus is ingested by a definitive host, at which point egg production and release into the environment permits the life cycle to continue.

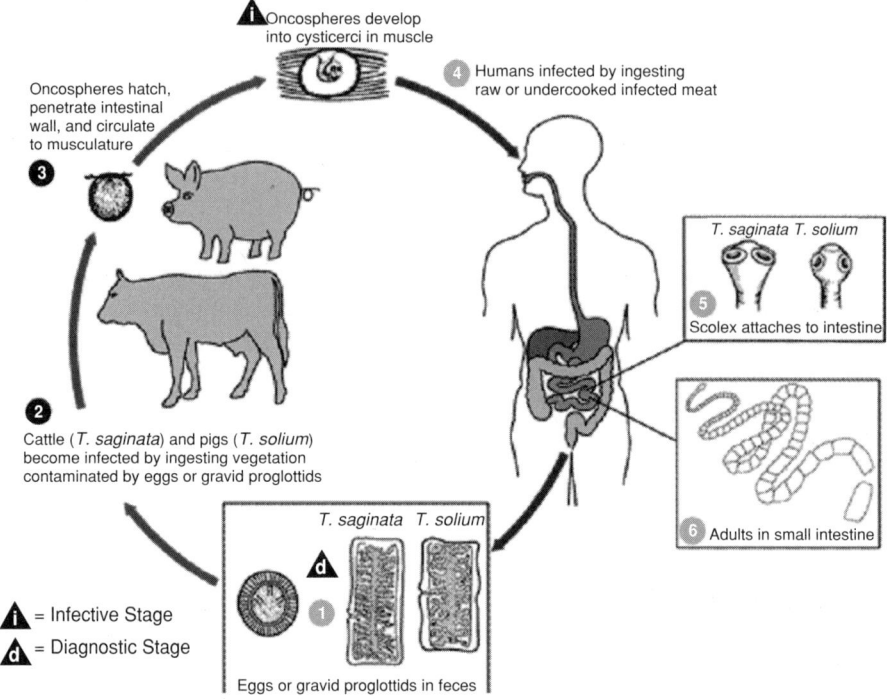

Figure 8.1. Life cycle of *Taenia* species. http://www.dpd.cdc.gov/dpdx

The adult tapeworms of all three *Taenia* species consist of a scolex characterized by four suckers, a short neck, and a strobila of varying length. Morphologic differences detectable by microscopy are the most reliable method of distinguishing between the three species. Injection of India ink into the uterus of the gravid proglottid allows for visualization of uterine branches and provides a means of species identification. *T. saginata* has 15–20 uterine branches per side, *T. asiatica* has 12–30, and *T. solium* has 7–13. Other notable morphologic species differences detectable by microscopy include the presence (*T. saginata*, *T. asiatica*) or absence (*T. solium*) of the vaginal sphincter muscle and the unilobed (*T. solium*) or bilobed (*T. saginata*, *T. asiatica*) ovary (Table 8.1). Polymerase chain reaction with restriction enzyme analysis (PCR-REA) can be used to distinguish the species (Mayta et al., 2000). Eggs of all three species are identical and therefore do not permit species identification by microscopy. They are spherical, typically 31–43 μm in diameter, with a thick brown shell easily recognized by its radially striated appearance (Fig. 8.2). Within each egg is an embryonated larva, or oncosphere, with six hooks.

8.2.1 *Taenia solium*

8.2.1.1 Biology

The adult *T. solium* tapeworm lives in the human small intestine, attaching to the gut wall by its scolex; causing mild inflammation at the attachment site. The *T. solium* adult has a scolex with four suckers and a rostellum armed with a double row of hooklets followed by a narrow neck. A mature tapeworm can reach a length of

Table 8.1. Morphological differences between *Taenia solium*, *Taenia saginata*, and *Taenia asiatica*.

	1 T. solium	*2 T. saginata*	*3 T. asiatica*
Intermediate host	Pig	Cattle	Pig
Human pathology	Taeniasis; cysticercosis	Taeniasis	Taeniasis; unknown capacity for cysticercosis
Rostrellum	Double row of hooklets	unarmed	Rudimentary, wart-like formation
Strobila	≤1000 proglottids	>1000 proglottids	<1000 proglottids
Uterine branches	7–13 per side	15–20 per side	12–30 per side
Vaginal sphincter muscle	Absent	Present	Present
Ovary	Unilobed	Bilobed	Bilobed
Metacestode cysts	0.5–1.5 cm	7–10 mm	2 mm

2–4 m, containing up to 1000 hermaphroditic proglottids, and can live for several years in the upper jejunum, absorbing food through its tegument. Maturation to the reproductive adult stage occurs in the upper jejunum and requires 10–12 weeks. Egg release begins approximately 2 months postinfection and occurs via the passage of 3–5 whole proglottids at a time in the human feces. Each proglottid can contain 50,000 eggs, which can survive for months in the external environment. Once an

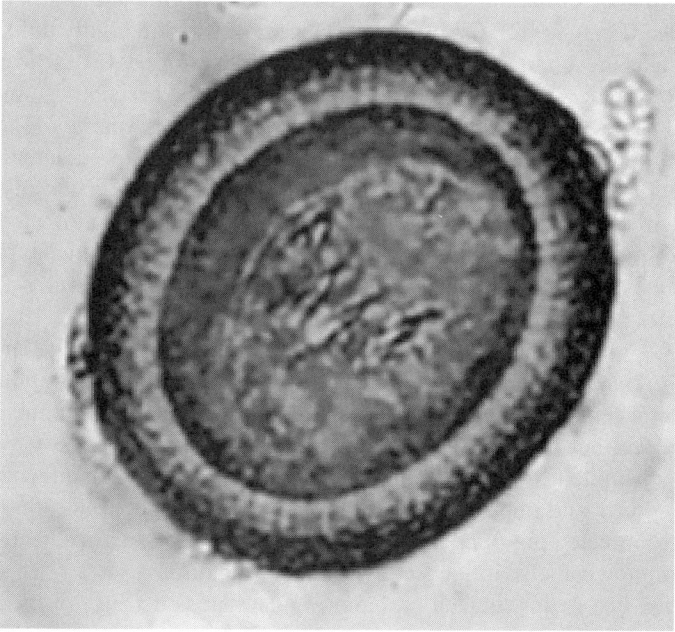

Figure 8.2. Taeniid egg with embryonated oncosphere with six hooklets within a characteristic radially striated brown shell. http://www.dpd.cdc.gov/dpdx

intermediate host ingests the eggs, they hatch in the host's gastrointestinal tract, releasing oncospheres that penetrate the intestinal wall and migrate to other tissues through the bloodstream. The larvae encyst in various tissues, particularly striated muscle in the neck, tongue, and trunk of pigs, and the central nervous system and skeletal muscle of humans. The encysted metacestode (cysticercus) matures over 60–90 days reaching a size of approximately 1 cm (Fig. 8.1). Normally, the encysted organism consists of only a scolex and can exist for many years in the host tissue. The cysticercus suppresses the host's immune response; thus, there is little or no inflammation while the cyst remains alive. Once the cyst begins to degenerate, however, there is a localized cellular immune response that may result in edema and scarring.

8.2.1.2 Epidemiology

Taenia solium has a worldwide distribution, but is most common in South America, Africa, India, and Southeast Asia. Prevalence is highest in rural areas where humans and pigs live in close proximity and good hygiene and sanitation are lacking. In many non-Muslim developing nations, where pork is a common food source, *T. solium* is the most common helminthic infection (Carangelo *et al.*, 2001; Garcia-Noval *et al.*, 2001). *T. solium* is no longer endemic in industrialized nations, but the prevalence there has been increasing in recent years due to immigration; identification increased due to improved diagnostic techniques. Although neurocysticercosis is much more common in developing nations, there are more than one thousand cases in the United States each year, mainly in immigrants from endemic areas (García *et al.*, 1999). In a recent study of emergency room patients with seizures, neurocysticercosis was diagnosed in 10% of the patients in Los Angeles, California and in 6% of those in New Mexico (Craig 2002). Outbreaks have even occurred in highly unlikely populations such as in a community of orthodox Jews in New York City who contracted *T. solium* cysticercosis from infected employees who had recently immigrated to the United States from tapeworm-endemic areas in Latin America. The infected employees were shedding eggs, which were then ingested in contaminated food (Schantz *et al.*, 1992).

In endemic regions, large portions of the population may be infected. In these areas up to 6% of people can have intestinal worms at a given time and serve as a source of infectious eggs that cause cysticercosis (Allan 1996a; Cruz *et al.*, 1989). Between 10 and 25% of people may be seropositive, and of these, 10–18% have abnormal CT findings suggestive of cysticercosis (Cruz *et al.*, 1999; Garcia-Noval *et al.*, 2001; Sánchez *et al.*, 1999; Schantz *et al.*, 1994). In general, prevalence is higher among females. There is an association between socioeconomic status and seroprevalence, but because of the transmission of *T. solium* eggs, the disease is certainly not limited to poor populations. Other risk factors for infection include age, open sewers and improper disposal of human feces, poor education, inability to recognize infected pigs, and inadequate inspection of pork. Local techniques for food preparation can also influence the spread of the disease.

T. solium is a major societal problem in many areas because of the morbidity associated with neurocysticercosis, which is of public health and economic concern. A large proportion of the patients admitted to neurology wards in developing countries suffer from neurocysticercosis (Garcia, 2003). In areas where *T. solium* infection is endemic, neurocysticercosis is a major cause of epilepsy, and in many areas, is the leading cause of epilepsy. The prevalence of epilepsy in these countries is 3–6 times higher than in industrialized nations: 1.7% in Ecuador, 2.1% in Colombia, and 3.7% in Nigeria (Commission on Tropical Diseases, 1994; Jimenez, 1991; Nicoletti *et al.*, 2002; Osuntokun and Schoenberg, 1982; Placenceia, 1992). Seropositive people have a two- to threefold greater risk of developing epilepsy compared to seronegative people (Sánchez *et al.*, 1999; Schantz *et al.*, 1994). In Peru, 5% of seropositive people have a seizure disorder, 19,000–31,000 in total, and 23,500–39,000 Peruvians have some form of symptomatic neurocysticercosis. When similar analysis is applied to all of Latin America, it is estimated that 400,000 people have symptomatic neurocysticercosis. One-quarter to one-half of adult seizure disorders in tapeworm-endemic areas can be attributed to *T. solium* neurocysticercosis (Bern, 1999). Because neurocysticercosis tends to affect older children and young adults, it has a disproportionate economic effect due to lost wages. Hospitalization costs are also very high, estimated at $15 million per year in Mexico alone (Flisser *et al.*, 1998). Loss of livestock also has a large economic impact on the pork industry in endemic nations, where 30–60% of pigs may be infected.

8.2.1.3 Transmission
The human is the only definitive host for *T. solium*. The intermediate host is usually the pig, but humans can also act as intermediate hosts. Other species such as dogs, cats, and sheep can occasionally be infected, but rarely develop significant disease. Infection in the intermediate host, known as cysticercosis, is manifested by encysted metacestode larvae in skeletal muscle or other tissue. Humans may develop numerous cysts in muscle or in the central nervous system, a disease known as neurocysticercosis. Infection by the oncosphere does not always result in cysticercosis because the host immune system is often capable of resolving light infections in the early stages. People do not need to have an intestinal worm to acquire cysticercosis since cysticercosis is caused by ingestion of eggs, but about 15% of patients with neurocysticercosis do harbor a tapeworm at the time of diagnosis (García *et al.*, 1999; Gilman *et al.*, 2000) and 25% report having had one at some point (Dixon and Lipscomb, 1961).

The cestode life cycle is propagated via foodborne and fecal-oral transmission. Taeniasis, human gastrointestinal infection by the adult worm, occurs when people eat undercooked pork that contains live cysticerci (Fig. 8.3). Cysticercosis in the intermediate host is a result of the ingestion of the cestode's eggs, usually in contaminated food. Pigs can acquire cysticercosis through exposure to raw sewage containing infected human waste. Autoinfection can occur in people when gravid proglottids of an intestinal worm rupture and hatching oncospheres are activated in the small intestine, rendering them capable of causing cysticercosis. The prevalence of antibodies to oncospheres suggests that this occurs more frequently than previously thought. Human-to-human spread can occur by fecal-oral transmission if poor hygiene is practiced when preparing food. Infections are often clustered in households, a phenomenon likely related to food preparation practices. *T. solium*, therefore, can become a problem anywhere that pigs are raised and fed in places

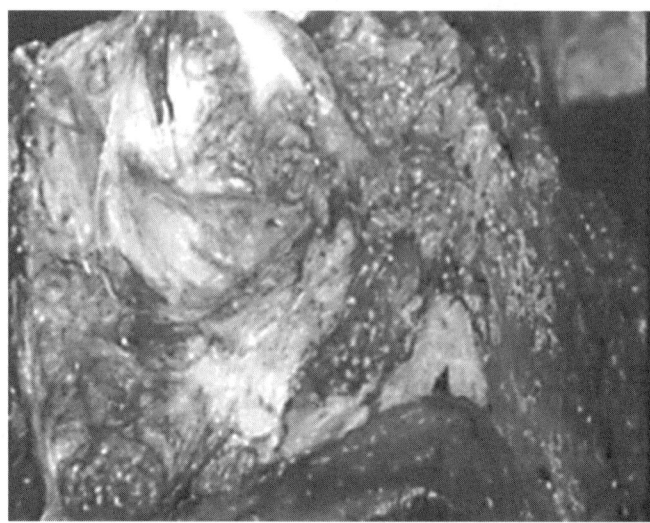

Figure 8.3. Pork flesh containing *T. solium* cysticerci.

contaminated with human feces, and where pork is not cooked thoroughly before eating. *T. solium* infection almost never occurs in industrialized countries, where nearly all pigs are raised commercially under heavy regulation, ensuring that there is no possibility of contact with human feces.

8.2.1.4 Clinical Aspects

Taeniasis, the infection of the human gut with the adult *T. solium*, is usually asymptomatic, but may occasionally cause epigastric pain, nausea, diarrhea, sensation of hunger, or weight loss. In contrast to the mobile proglottids of *T. saginata*, immobile *T. solium* proglottids can be visible in the person's stool, but usually pass unnoticed. *T. solium* cysticercosis can occur in almost any part of the body. In humans, cysticerci are most commonly found in skeletal muscles and in the brain. They are also found in the skin, subcutaneous tissue, eye, and heart. Cysticerci in the tissues are rarely symptomatic unless they encyst in the eye or in neural tissue, a condition known as neurocysticercosis. Cysticercosis can also cause subcutaneous nodules in the skin and muscular pseudohypertrophy when there is very high worm burden in skeletal muscles.

In the brain or other neural tissues, the cysticercus resembles a tumor and can cause many problems, including hydrocephaly, meningitis, and seizures. Cysts can be located in the brain parenchyma, especially at the grey-white junction, in the subarachnoid space, in the ventricles, and in the spinal cord. Cysts in the central nervous system are usually larger than cysts in other tissues. When cysticerci are found in the ventricular systems, they most commonly reside in the fourth ventricle, followed by the third ventricle. The racemose form of the disease consists of a large, frequently lobulated, translucent cyst without a scolex that is usually found in the ventricles or at the base of the brain (Fig. 8.4). This form of disease carries a high mortality rate due to hydrocephalus. Typical presentation of neurocysticercosis

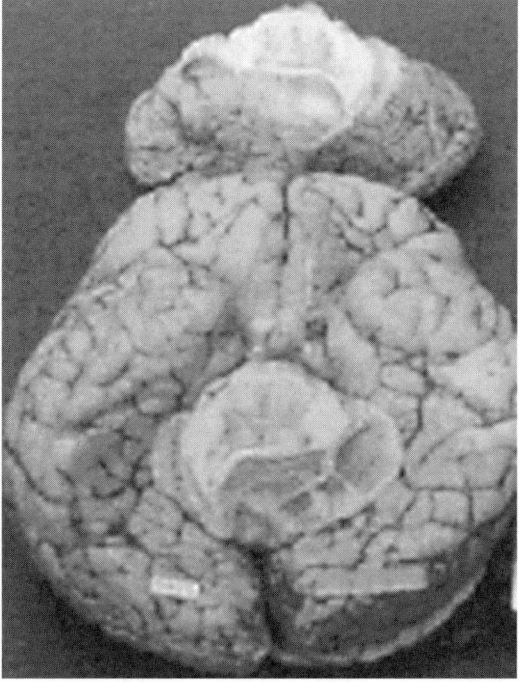

Figure 8.4. *Taenia solium* racemose cyst in the brain of a neurocysticercosis patient.

varies by geographical location: single brain lesions are most common in Asia while the presence of several viable cysts is more common in Latin America. Symptoms depend on the location of the cysticerci, the amount of inflammation or scarring, and the parasite load. The signs and symptoms of neurocysticercosis are usually due to inflammation, but the cyst can have a mass effect when located in the cerebrospinal fluid system or in the parenchyma when they are exceptionally large.

Parenchymal cysts are rarely fatal, but they can cause significant morbidity. Epilepsy is the most common symptom of neurocysticercosis. In fact, cysticercosis is the most common cause of epilepsy in the developing world. In endemic areas, new onset of seizures in adolescence or young or middle adulthood is suggestive of neurocysticercosis. Between 50 and 80% of people with parenchymal brain cysts have at least one seizure (Commission on Tropical Diseases, 1994; Chopra *et al.*, 1981; Del Brutto *et al.*, 1992). The resulting seizures can be partial or general, and EEG will often reveal focal slowing in the region of the lesion. Cysticerci have been reported to cause psychiatric symptoms as well, including one case report of binge-eating disorder (Fernandez-Aranda *et al.*, 2001). In contrast to parenchymal cysts, intraventricular cysts can frequently be fatal if they cause hydrocephalus. Twenty to thirty percent of patients develop increased intracranial pressure or hydrocephalus when CSF outflow is impeded (García *et al.*, 2003), causing headache, nausea or vomiting, dizziness, visual problems, ataxia, confusion, or papilloedema. Cysticerci in the subarachnoid space can cause stroke. Neurocysticercosis can also cause encephalitis (especially in younger adults or children and in females), meningitis, or arachnoiditis. About 1% of patients with spinal cord neurocysticercosis develop spinal cord compression (García et al., 2003).

In the eye, the cysticercus may invade the anterior chamber, the vitreous humor, or the retina, giving rise to visual problems. Only 1–3% of infected patients have ocular cysts, but, nevertheless, *T. solium* is the most common intraocular parasite. Subretinal cysticerci can cause retinal detachment and vision loss. Intraocular cysts can often be seen on fundoscopic examination. Rarely, cysts form in the optic nerve or extraocular muscles.

8.2.1.5 Diagnosis

Diagnosis of intestinal infection can be difficult because stool microscopy is not very sensitive and the eggs of *T. solium* are identical to those of other *Taenia* and *Echinococcus* species. Eggs from different species can, however, be distinguished with PCR, enzyme electrophoresis, or immunological tests. Diagnosis becomes easier if proglottids or the scolex are recovered. Use of a purgative before and after administration of treatment improves the chances of recovering the worm's scolex and strobila. The proglottids of *T. solium* and *T. saginata* can be differentiated by the anatomy of the female reproductive organs: *T. solium* has 5–10 uterine branches on each side, no vaginal sphincter muscle, and only one ovarian lobe while *T. saginata* has at least 15 uterine branches, a vaginal sphincter muscle, and two ovarian lobes. Sensitivity for the detection of eggs in stool may be improved by using concentration techniques. Sedimentation, as opposed to flotation, techniques for the isolation of eggs are required due to the density of *Taenia* eggs. Of note, it is important to fix eggs or proglottids before examination to prevent infection of laboratory personnel. There is also a coproantigen ELISA that is about 95% sensitive and 99% specific (Allan, 1990; Allan AJTMH, 1996).

Species differentiation is a difficult but important part of the diagnosis of taeniasis. Overlapping geographic distributions and the morphologic similarities make speciation a challenging endeavor. Nonetheless, it is important for clinical as well as investigational purposes. Clinically, *T. solium* infection must be ruled out in order to identify potential cases of human cysticercosis, which can be worsened by inappropriate antiparasitic therapy.

Diagnosis of *T. solium* cysticercosis is not straightforward and must take into account history, examination, epidemiology, radiological studies, and serology. Del Brutto, et al. established a detailed set of criteria to diagnose definite or probable neurocysticercosis. Their system includes a weighted list of absolute, major, minor, and epidemiological diagnostic criteria (Table 8.2). A definitive diagnosis of neurocysticercosis is established by the presence of one absolute criterion or the presence of two major, one minor, and one epidemiologic criterion. A probable diagnosis is established by one of three combinations of criteria: (1) one major plus two minor criteria; (2) one major plus one minor and one epidemiologic criterion; or (3) three minor plus one epidemiologic criterion (Del Brutto et al., 2001).

Examination of cerebrospinal fluid may show pleocystosis with a predominance of lymphocytes, neutrophils or eosinophils, increased protein, and normal or low

Table 8.2. Criteria for diagnosis of neurocysticercosis (Del Brutto, 2001).

Absolute Criteria	1. Histological demonstration of the parasite from biopsy of a brain or spinal cord lesion 2. Cystic lesions showing the scolex on CT or MRI 3. Direct visualization of subretinal parasites by fundoscopic examination
Major Criteria	1. Lesions highly suggestive of neurocysticercosis on neuroimaging studies 2. Positive serum EITB for the detection of anticysticercal antibodies 3. Resolution of intracranial cystic lesions after therapy with albendazole or praziquantel 4. Spontaneous resolution of small single enhancing lesions
Minor Criteria	1. Lesions compatible with neurocysticercosis on neuroimaging studies 2. Clinical manifestations suggestive of neurocysticercosis 3. Positive CSF ELISA for detection of anticysticercal antibodies of cysticercal antigens 4. Cysticercosis outside the CNS
Epidemiological Criteria	1. Evidence of a household contact with *Taenia solium* infection 2. Individuals coming from or living in an area where cysticercosis is endemic 3. History of frequent travel to disease-endemic areas

glucose. In particular, eosinophils in the CSF are suggestive of neurocysticercosis, but they are found in only about 40% of cases.

Neuroimaging is one of the most important elements of diagnosis and is important for discovering the number, size, and location of cysts. CT is about 95% sensitive and specific for neurocysticercosis. Cysticerci typically appear as hypodense lesions with well-defined edges and with a hyperdense lesion inside. As the cyst degenerates and causes inflammation, the lesion enhances with contrast. Dead, calcified cysts usually do not enhance, but the presence of inflammation is predictive of relapse of symptoms. Because cyst fluid is isodense with CSF and cysts have thin walls, it is difficult to see cysticerci in the CSF. CT is better at identifying calcified lesions, but MRI is better to visualize cystic lesions, intraventricular cysts, or enhancing lesions. MRI is also more useful to stage lesions; although it cannot detect calcifications. Gradient refocused echo MRI is the best modality to visualize the scolex in a calcified lesion. Periventricular enhancement on MRI can distinguish neurocysticercosis from lymphoma or other infections. MRI and B-scan ultrasonography are superior to CT for diagnosis of ocular cysticercosis.

Immunological testing is also extremely useful for diagnosis of neurocysticercosis, though antibodies can disappear over time. Cysticercosis causes an increase in the serum levels of specific IgA, IgE, IgG, and IgM antibodies. ELISA for IgG is sensitive (88.5%) and specific (93.2%) (Odashima *et al.*, 2002), especially IgG4 (Huang *et al.*, 2002). Antigen detection by enzyme-linked immunoelectrotransfer blot assay (EITB) is even better, with 98 to 99.4% sensitivity and 100% specificity (García *et al.*, 2003; Gekeler *et al.*, 2002). This test is less sensitive if the patient has only one cyst. Because ELISA frequently cross-reacts with antibodies to *Hymenolepis nana* and *Echinococcus granulosus*, it is best used for screening only, and positive results should be confirmed by EITB. EITB is useful for screening and in clinical settings and is the current test of choice. Antibodies may persist for weeks after the cyst degenerates even with treatment.

8.2.1.6 Treatment

Niclosamide and praziquantel are the drugs used to treat intestinal *T. solium* infection. Both are effective in a single dose—2 g for niclosamide or 5–10 mg/kg for praziquantel. Praziquantel is cysticidal and can cause inflammation around dying cysts in patients with cysticercosis; this may trigger seizures or other symptoms. Praziquantel-induced seizures can be problematic when the drug is used in mass-treatment campaigns. Because niclosamide is not absorbed from the gastrointestinal tract, it does not cause an inflammatory response and is considered the drug of choice for intestinal taeniasis.

Treatment of cysticercosis, especially neurocysticercosis, can be difficult and must be individualized. Cysticercosis outside of the central nervous system is usually asymptomatic and does not require therapy. Symptomatic treatment of seizures and increased intracranial pressure in neurocysticercosis is crucial. Anticonvulsants usually control seizures. They can be discontinued after 2 years if the patient remains seizure-free and intracranial paranchymal lesions resolve without calcifications. In patients with hydrocephalus, antiparasitic treatment should be avoided until the increased intracranial pressure has resolved, to avoid worsening of the patient's condition due to drug-induced inflammation. Hydrocephalus due to neurocysticercosis should be treated as with other diseases, using techniques such as ventriculoperitoneal shunt or removal of the cysticercus by craniotomy or ventriculoscopy if necessary. Arachnoiditis and associated vasculitis usually respond to steroids. Ocular cysticerci or spinal medullary lesions usually require surgery because inflammation in these areas can cause irreversible damage.

There is controversy about treatment of parenchymal brain cysticerci because as the cysts die in response to medication, surrounding inflammation can cause transient worsening of symptoms. Certain groups of patients have definite benefit from antiparasitic therapy, like those with giant cysts, subarachnoid cysts, or growing cysts. One argument against the use of antiparasitic therapy for other types of neurocysticercosis is that most cysts will resolve on their own; however, there is some evidence that treatment results in less scarring as lesions heal (García *et al.*, 2004; Padma *et al.*, 1995). García *et al.*, performed a placebo-controlled trial that showed treatment with 800 mg of albendazole daily, with adjuvant steroid therapy, reduced the frequency of generalized seizures. In the first month after treatment, the albendazole group had more seizures than the placebo group, but this relationship reversed at 2 months and at 30 months, when treated patients had significantly fewer generalized seizures. Intracranial lesions also resolved more quickly in the albendazole group (García *et al.*, 2004). Both albendazole (15 mg/kg/day for 8 days) and praziquantel (50–60 mg/kg/day for 15 days) kill encysted worms, though albendazole appears to penetrate the brain more effectively. Concurrent administration of glucocorticoids and, if needed, anticonvulsants, can control the increased inflammation. Steroids, however, increase first-pass hepatic metabolism of praziquantel, so cimetidine should also be coadministered to inhibit hepatic enzymes and maintain therapeutic concentrations of praziquantel. All patients with seizures should receive antiepileptic drugs.

8.2.1.7 Control

As with other food-borne diseases, good hygiene and sanitation are crucial measures to control the spread of *T. solium*. Only humans with intestinal infections and pigs with cysticercosis transmit the parasite, so control measures should be targeted at these stages. Proper management of human waste disposal, such as the construction of latrines, decreases human-pig and human-human transmission. Improved sanitation in pig husbandry and the pork industry is also important. Regulation and effective screening in slaughterhouses, proper housing and care of pigs on farms or in peridomestic areas, and preparation of pork by salt pickling, freezing at $-10°C$ for at least 9 days or cooking above $65°C$, decreases transmission from pigs to humans. Implementation of these measures can be difficult in developing countries, where the pork industry is often not well-regulated and many pigs are raised and slaughtered in private settings outside slaughterhouses. Because pigs are cheaper than cows, easy to feed with garbage, and easy to sell or eat, people in developing countries often raise them in their homes and sell them, without having them inspected for cysticercosis. In Peru, for example, more than half of pigs entering the pork market do so illegally, and many are infected with *T. solium*. Diseased pork is much cheaper, so pork sellers may disguise infected pork or mix it with uninfected meat (Cysticercosis Working Group in Peru, 1993). Public education on the need to isolate pigs from human feces is necessary to change habits in these areas. Because of economic pressures on poor rural populations, however, economic incentives like compensation or introduction of other agricultural products must complement education (Gonzales *et al.*, 2003).

Medical interventions also play an important role in prevention of transmission. Treatment of human intestinal infections with niclosamide or praziquantel reduces cysticercosis. In some areas, mass treatment of the population with praziquantel reduced the prevalence of intestinal infection by 53%, seizures by 70%, and the prevalence of antibodies in both pigs and humans (Sarti *et al.*, 2000). Pigs can be treated as well with a single dose of oxfendazole 30 mg /kg (Gonzales *et al.*, 1997).

Difficulty with large-scale propagation of cestodes *in vivo* or *in vitro* thwarted vaccine development efforts prior to the introduction of recombinant DNA technology. However, advances in molecular science have led to the successful development of recombinant vaccines for *Taenia* cestodes in their intermediate hosts. A vaccine has been developed for use in pigs; it remains to be seen if it will be economically feasible for widespread use in endemic areas.

8.2.2 *Taenia saginata*

8.2.2.1 Biology

Taenia saginata has a scolex characterized by four suckers and a unique, retracted, unarmed rostellum. The size of the mature proglottids ranges from 2.1–4.5 mm and gravid proglottids range from 0.3–2.2 cm. Proglottids are longer than they are wide and have a large central genital pore, and each hermaphroditic proglottid is capable of producing thousands of eggs per day. On average *T. saginata* grows to a length of

5–10 m and can have more than 1000 proglottids. The intact proglottids are mobile and can occasionally be seen moving in the stool.

The life cycle of *T. saginata* occurs in cattle and human beings, although llamas, buffalo, and giraffes occasionally can act as intermediate hosts (Fig. 8.1). Cattle are infected when they ingest eggs on local vegetation that has been contaminated by human feces. Cattle develop cysticerci in skeletal muscles, which are then ingested by humans in undercooked beef. The life cycle of *T. saginata* is essentially identical to that of *T. solium*, with two notable exceptions: the cow, not the pig, is the intermediate host of *T. saginata*, and the human, the definitive host, never acts as an intermediate host of *T. saginata*.

8.2.2.2 Epidemiology

Taenia saginata is a ubiquitous parasite, and human taeniasis occurs in all countries where raw or undercooked beef is consumed. The World Health Organization estimates that over 60 million people are infected with taeniasis worldwide (WHO report, 1992). The *T. saginata* tapeworm is most prevalent in Sub-Saharan Africa and the Middle East; other regions with high prevalence of *T. saginata* (defined as greater than 10% of the population) include Central Asia, the Near East, and Central and Eastern Africa. Areas of low prevalence (defined as less than 1% of the population) include Europe, the United States, Southeast Asia, and Central and South America.

Bovine cysticercosis resulting from *T. saginata* infection is a global problem occurring in cattle rearing regions of the world and resulting in significant financial loss. Bovine cysticercosis renders beef unmarketable, and is globally responsible for over 2 billion dollars in yearly economic losses (Hoberg, 2002).

8.2.2.3 Transmission

Taenia saginata transmission and propagation is closely tied to both food consumption and sanitary habits. Human taeniasis results from the consumption of contaminated beef that has not been frozen or thoroughly cooked. Cows subsequently become infected by ingesting eggs excreted by humans. Once eggs are released into the environment, they can remain viable for months to years until an appropriate intermediate host ingests them and the life cycle resumes. Perpetuation of bovine and human disease in an agricultural setting can occur in several ways (Hoberg, 2002), including direct transmission to cattle through fecal contamination of pastureland by agricultural workers, application of untreated human sewage onto pastureland, and indirect contamination of the cattle food or water supply.

8.2.2.4 Clinical Aspects and Diagnosis

Human taeniasis caused by *T. saginata* is typically asymptomatic. Patients may become aware of infection upon the passage of proglottids, or even several feet of strobila, in the stool. Proglottids are often motile, thus causing discomfort with discharge. A small percentage of patients complain of colicky abdominal pain, nausea, changes in appetite, weakness, weight loss, constipation or diarrhea, pruritis ani, and general malaise. Abdominal discomfort and nausea are the most common complaints and are often relieved with the ingestion of food. In infants, increased irritability may be the only sign of infection. Complications from *T. saginata* infection are rare and are a result of the motile nature of the proglottids. Migrating proglottids rarely cause biliary duct, pancreatic duct, or appendiceal obstruction.

Clinically, taeniasis may be associated with an elevated serum IgE and a mild eosinophilia of 5–15% in a minority of patients. Definitive diagnosis of taeniasis, however, is made by direct visualization of eggs or proglottids in the stool or by cellophane tape swab of the perianal region. *T. solium* releases eggs or proglottids into the stool erratically, in contrast to *T. saginata*, which releases eggs or proglottids on a daily basis. Large sections of *T. saginata* strobila can break off in a day, without subsequent release of eggs or gravid proglottids for several days thereafter; therefore, collection of multiple stool samples is recommended.

8.2.2.5 Treatment

A single 5–10 mg/kg dose of praziquantel is highly effective for cestode infections. Alternatively, a single dose of niclosamide is also effective. Dosing of niclosamide is 2 g for adults, 1.5 g for children greater than 34 kg, and 1 g for children 11–34 kg. Both praziquantel and niclosamide are class B drugs, although treatment can and should be delayed until after pregnancy unless clinically indicated. Since praziquantel is released in breast milk and safety in children under 4 years of age has not been investigated, it is recommended that nursing mothers should not breastfeed for 72 h after treatment.

8.2.2.6 Control

There is a *T. saginata* recombinant vaccine developed based upon the identification of homologues to host protective antigens of *T. ovis*. The vaccination of cattle using a combination of the *T. saginata* proteins tsa9 and tsa18 has resulted in 94–99% protection against parasite infection (Lightowlers *et al.*, 2003). Research to develop a more practical and affordable vaccine is ongoing.

In the meantime, disease prevention depends upon interrupting the parasite life cycle through public health efforts directed at raising awareness of disease transmission mechanisms and at improving sanitary practices. Cysticerci in beef can be inactivated by cooking meat at least 56°C or freezing meat at -10°C for at least 9 days. Reliable meat inspection to remove infected meat from the market and proper disposal of human feces to interrupt transmission to cattle are also important preventive measures.

8.2.3 *Taenia asiatica*

Taenia asiatica is a relatively recently described species of *Taenia* that closely resembles *T. saginata* both morphologically and genetically. Debate is ongoing as to whether or not the "Asian *Taenia*" should be considered a subspecies of *T. saginata*, and it is still referred to as *T. saginata-asiatica* in some current literature. Genetic studies of ribosomal genes and the mitochondrial cytochrome C oxidase I (COI) gene identified a close relationship between *T. saginata* and *T. asiatica*, supporting the subspecies theory (Flisser *et al.*, 2004). However, comparative morphologic studies and life cycle differences lend support to the idea that the "Asian *Taenia*" may be closely related to *T. saginata*, but is an individual species nonetheless (Galan-Puchades and Fuentes, 2000).

8.2.3.1 Biology

The life cycle of *T. asiatica* is almost identical to that of *T. solium* (Fig. 8.1) with the exception that humans probably cannot act as intermediate hosts for *T. asiatica*. The pig is the primary intermediate host, but other intermediate hosts include cattle, goat, monkey, and wild boar. While *T. asiatica* cysticercosis can occur in many organs, the larva seems to have a special tropism for the liver. When the definitive human host ingests the cysticercus in undercooked pork, maturation to adulthood occurs in the small intestine.

The scolex of *T. asiatica* has four suckers and a rostellum armed with rudimentary hooklets referred to as wart-like formations (Flisser *et al.*, 2004). The strobila typically is composed of fewer than 1000 hermaphroditic proglottid segments. Like *T. saginata*, a vaginal sphincter muscle is present in mature proglottids, and gravid proglottids have a central uterus with 12–30 uterine branches per side. However, *T. asiatica* has more than 57 uterine twigs per side, with a larger ratio of uterine twig: Uterine branches than other *Taenia* species. The prominent protuberance on posterior aspect of *T. asiatica* gravid proglottids is also unique. The metacestode cysts contain rudimentary hooklets and typically measure 2 mm, compared with the 7–10 mm cysts of *T. saginata*, and 0.5–1.5 cm cysts of *T. solium*.

8.2.3.2 Epidemiology

Thus far, *T. asiatica* has been identified in Taiwan, Indonesia, Thailand, Korea, China, Malaysia, Vietnam, and the Philippines. Prevalence of the tapeworm can be as high as 21% (Galan-Puchades and Fuentes, 2000) in some endemic regions, particularly where people eat undercooked pork with viscera. The distribution of *T. asiatica* may not be limited to Asia, as the same epidemiological conditions that led to its discovery in Taiwan have been noted in other parts of the non-Asian world as well.

8.2.3.3 Transmission

Taenia asiatica was identified using basic epidemiology combined with knowledge of the food consumption habits in the effected population. Investigators in Taiwan noted that the prevalence of *T. saginata* did not correlate well with the food consumption habits of the local people. Pork is a staple part of the diet in Southeast Asia. In fact, it is a common practice in many rural parts to house pigs and other animals under the floor of the house, creating an ideal environment for *Taenia* spp. transmission. Consumption of pig meat and viscera was far more common than that of beef, yet *T. solium* had a surprising low prevalence compared to *T. saginata*. This observation led investigators to hypothesize that perhaps there was another *Taenia* parasite present that closely resembled *T. saginata*, and *T. asiatica* was subsequently classified.

8.2.3.4 Clinical Aspects and Diagnosis

The debate over whether or not *T. asiatica* is its own species or a subspecies of *T. saginata* has importance beyond the merely academic. It has yet to be definitively determined whether or not *T. asiatica* is capable of causing human cysticercosis. Evidence suggesting that it does not cause human cysticercosis includes its genetic and morphologic similarities with *T. saginata*. In addition, in regions of the world where the prevalence of *T. asiatica* is highest, namely regions of Taiwan and Samosir Island (Indonesia), there are very few cases of human cysticercosis. Evidence suggesting that it could cause human cysticercosis includes the fact that, similar to *T. solium*, pigs are the primary intermediate hosts of *T. asiatica*. Additionally, *T. asiatica* cysticerci demonstrate liver tropism in porcine models, suggesting cases of human *T. asiatica* cysticercosis may not present like the traditional neurocysticercosis cases of *T. solium*. The potential for *T. asiatica* to cause human cysticercosis needs further investigation (Galan-Puchades and Fuentes, 2000).

Definitive diagnosis of *T. asiatica* taeniasis is by microscopic examination of stool. Since the eggs of all *Taenia* species are identical, only by noting morphologic differences in proglottids or scolex, if expelled, is speciation possible. If resources permit, observations can be confirmed by PCR analysis of DNA. Coproantigen testing can also be performed on stool samples, but this methodology is only specific to the level of genus.

8.2.3.5 Treatment
Effective treatment for *T. asiatica* taeniasis includes single doses of praziquantel or niclosamide, using the same dosage for taeniasis as for other *Taenia* species (see *T. solium*, 1.2.1.6 Treatment).

8.2.3.6 Control
Preventive measures involve improving hygiene and education regarding the tapeworm lifecycle. Meat inspection for cysticerci, keeping pigs indoors and without contact with human feces, and avoiding the consumption of undercooked pork or pig viscera are other important measures of prevention. Although yet to be demonstrated, it is hoped that the recombinant vaccine for *T. saginata* will be equally successful for *T. asiatica*, given their genetic and morphologic similarities.

8.3 DIPHYLLOBOTHRIUM

8.3.1 *Diphyllobothrium Latum*

8.3.1.1 Biology
Diphyllobothrium latum, also known as the fish tapeworm or broad tapeworm, is the longest of the human tapeworms, reaching lengths of 10–20 mm/cm, with 3000–4000 proglottids. Pseudophyllidea typically have a scolex with two sucking grooves, as opposed to the four true suckers of cestodes of the Cyclophilidea order. The sucking grooves, or bothria, are located along opposite sides of the scolex. Proglottids are wider than they are long, typically measuring 2 to 4 mm by 10 to 12 mm. The gravid proglottid has a central, coiled uterus with a distinctive rosette-like appearance. The central ventral genital pore is the site of egg expulsion. Eggs are yellow, oval, and distinctly operculated at one end. On the end opposite the operculum is a small knob like feature. Eggs measure 58 to 76 μm by 40 to 51 μm, and are passed in the stool unembryonated (Fig. 8.5).

Diphyllobothrium latum belongs to the order Pseudophyllidea, and its life cycle requires three hosts. The life cycle of *D. latum* (Fig. 8.6) requires two intermediate

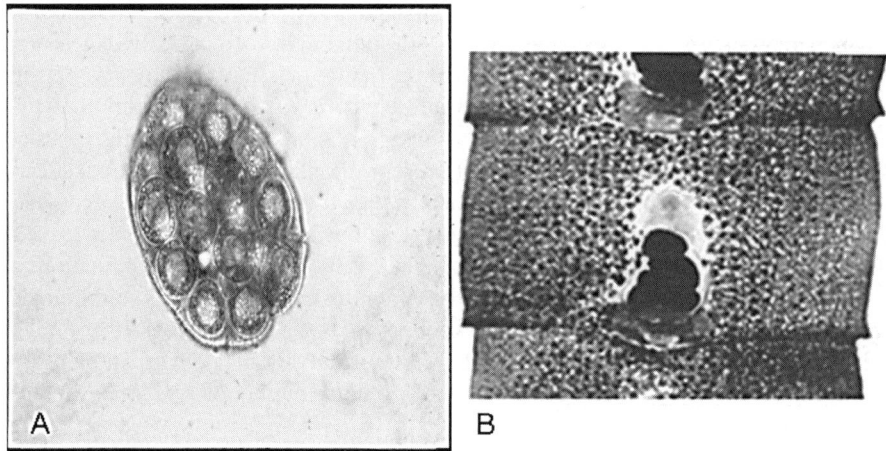

Figure 8.5. (a) *Diphyllobothrium* egg packet. (b) Proglottid with central coiled uterus. http://www.dpd.cdc.gov/dpdx

marine hosts before adulthood is reached in the definitive mammalian host. Immature eggs released with the feces must promptly be deposited into fresh water, where they mature and eventually release a freely swimming, ciliated, six hooked embryo, called a coracidium, from the opened operculum. The coracidia are ingested by crustaceans such as copepods or water fleas, where the first larval, or procercoid, stage of development occurs. The procercoid larva develops in the crustacean tissue and maturation is arrested until it is ingested once again by a larger fish. In the muscle tissue of the larger fish, the procercoid larva matures into a plerocercoid cyst, or sparganum. The sparganum grows to 6 mm and can be transmitted up the food chain to progressively larger fish, until finally it is ingested by a definitive mammalian host and can mature to adulthood. Interestingly, as progressively larger fish consume the sparganum it does not mature; but rather re-encysts in the fish muscle tissue. Once consumed by the definitive host, the sparganum matures to adulthood in the small intestine. The tapeworm attaches to the gut mucosa utilizing its bothria, and attains adulthood in 3–6 weeks. Definitive hosts can include, but are not limited to, humans, dogs, cats, foxes, bears, and pigs.

8.3.1.2 Epidemiology
Diphyllobothrium latum has a worldwide distribution and is most common in regions where undercooked or raw fish are consumed. Areas of high endemnicity (>2%) include lake and delta areas of Siberia, parts of Europe, particularly Scandinavian countries, North America, Japan, Peru, and Chile. *D. latum* primarily infects older children and adults, and those who eat or prepare raw fish for home or for commercial distribution are at greatest risk. The prevalence in the United States is estimated to be less than 0.5% (Masci, 2004), though outbreaks have been associated with increased availability of fresh salmon and sushi.

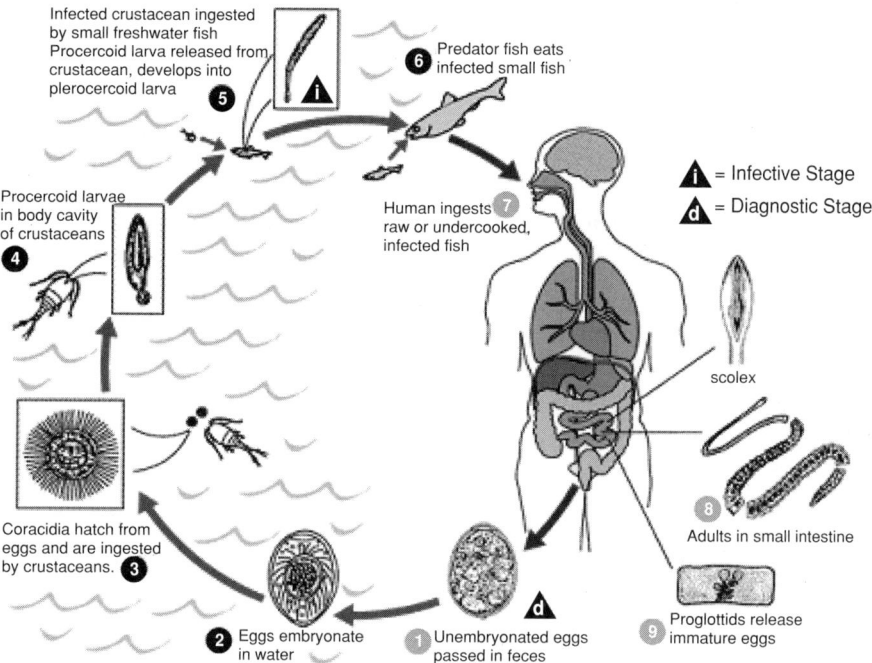

Figure 8.6. *Diphyllobothrium latum* life cycle. http://www.dpd.cdc.gov/dpdx

8.3.1.3 Transmission

Human infection is most commonly associated with the consumption of salmon, whitefish, rainbow trout, pike, walleye, perch, turbot, and ruff, particularly in areas where raw, marinated, or undercooked fish are eaten.

8.3.1.4 Clinical Aspects and Diagnosis

The majority of *D. latum* infections are asymptomatic, although infection can be associated with diarrhea, abdominal distension, flatulence, abdominal cramping, and general malaise in a minority of patients. Rare complications include intestinal obstruction due to high worm burden and megaloblastic anemia due to impaired vitamin B_{12} absorption in up to 2% of those infected (King, 2000). The adult tapeworm affects vitamin B_{12} absorption through two mechanisms. First, the tapeworm is a scavenger of vitamin B_{12}, which is important for worm growth and development. Second, *D. latum* can uncouple the vitamin B_{12}-intrinsic factor complex, thus preventing ileal B_{12} absorption. Historically, anemia had been described with heavy worm burden (the presence of multiple worms) and prolonged infection in the Scandinavian population. It is currently very rare to find and may only result in genetically predisposed individuals.

Diagnosis is made by visualizing eggs in the stool. They are usually so numerous that concentration techniques are unnecessary. Their characteristic operculum is evident on microscopy. Egg discharge can be intermittent, and diagnosis may be missed with only one stool sample. Scolex and proglottids are unlikely to be passed in with the feces, although case reports have described parts of the strobila emitted in vomitus (Richards, 2003).

8.3.1.5 Treatment
Single doses of 5–10 mg/kg of praziquantel or 2 g (adult)/50 mg/kg (children) of niclosamide are highly effective. Stool should be reexamined in 6 weeks to determine disease persistence. Cobalamin injections are recommended for severe anemia resulting from B_{12} deficiency.

8.3.1.6 Control
Altering human eating habits is the only effective means of interrupting the *D. latum* life cycle. Regulating proper disposal of human sewage is not sufficient because egg deposition into fresh water occurs from many of the animal sources that act as definitive hosts. Brief cooking of fish at 56°C for 5 minutes or freezing at −18°C for 24 h will effectively kill the sparganum. Importantly, sparganum remain viable in dried or smoked fish.

8.3.2 *Diphyllobothrium pacificum*

8.3.2.1 Life Cycle
Causal agents of human diphyllobothriasis include *D. latum* and *D. pacificum*. The life cycle of *D. pacificum* is similar to that described for *D. latum*, with the exception that the second intermediate, or paratenic, hosts are salt water species rather than fresh water species of fish. The typical definitive hosts for *D. pacificum* are sea lions, with humans infrequently becoming accidental definitive hosts upon ingestion of raw or under-prepared marine fish infected with *D. pacificum* plerocercoid larvae.

8.3.2.2 Epidemiology
Diphyllobothrium pacificum was first identified in South America. It is considered endemic to Peru, with the prevalence reaching 2% on the southern coast, particularly in underdeveloped urban areas. To date, it is the only known cause of human diphyllobothriasis in Peru (Medina Flores *et al.*, 2002). Sixteen commercially important species of fish have been identified as paratenic hosts (Cabrera, 2001) including Jurel (*Trachurus symmetricus murphyi*), Bonito (*Sarda chiliensis*), Coco (*Paralonchurus peruanus*), Perico (*Coryphaena hippurus*), and Lisa (*Mugil cephalus*).

Cases of *D. pacificum* diphyllobothriasis in Chile have been associated with the El Niño phenomenon, presumably due to changes in water temperatures resulting in the southern displacement of fish native to Peruvian waters and the creation of conditions favorable for the overgrowth of copepods, the first intermediate host of *D. pacificum* (Sagua *et al.*, 2001).

8.3.2.3 Transmission
Human infection is related to the consumption of raw or undercooked fish, common practice in Peruvian coastal regions. *D. pacificum* infection has been associated with dishes such as cebiche (raw fish prepared with lemon juice and ají), tiradito (raw fish prepared with a lemon, garlic, and ají sauce), and chinguirito (raw fish prepared with lemons and red peppers).

8.3.2.4 Clinical Presentation and Diagnosis

Diphyllobothrium pacificum diphyllobothriasis can be asymptomatic or can present with abdominal pain, vomiting, or diarrhea. There is no association between *D. pacificum* infection and pernicious anemia (Medina Flores *et al.*, 2002). Diagnosis is made upon visualization of the expelled proglottids or eggs in the feces. Eggs typically measure 48 to 60μm long by 35 to 40μm wide. Infection can be successfully treated with niclosamide.

8.3.2.5 Control

Human *D. pacificum* diphyllobothriasis results from the consumption of raw or improperly prepared marine fish. Prevention measures include thoroughly cooking fish in endemic regions or adequately freezing fish prior to food preparation.

8.4 SPIROMETRA

8.4.1 *Spirometra mansoides*

8.4.1.1 Biology and Transmission

Spirometra mansoides is a common tapeworm of dogs, cats, and other carnivores. It is a member of the Pseudophyllidea order, and thus requires two intermediate hosts in order to complete its life cycle. Eggs are discharged into freshwater, and, similar to *D. latum*, hatch to release a free-swimming, ciliated coracidium. The first intermediate host, a copepod, ingests the coracidium where it develops into a procercoid larva. At this point, the life cycle of *S. mansoides* diverges from that of *D. latum*, because the former cannot have a fish as the second intermediate host. Rather, the second intermediate, or paratenic, host of *S. mansoides* may be a reptile, amphibian, bird, or other mammal. Within the paratenic host, the procercoid larva migrates through the intestinal mucosa and eventually into muscle or connective tissue where it encysts and develops into the plerocercoid larva. This process will continue until the paratenic host is consumed by a definitive host, usually a dog, cat, or raccoon. In the small intestine of the definitive host, the tapeworm develops to adulthood, reaching maturity in 10–30 days. Eggs are discharged from gravid proglottids and passed with the feces. Eggs are brown and approximately 57 μm by 39 μm. The proglottids have a pinkish hue and a coiled uterus when gravid.

8.4.1.2 Epidemiology

Spirometra mansoides is an uncommon parasite, but widespread in warm climates. Human cases of sparganosis are most common in Asia, though cases have been reported in North and South America, Europe, and Africa.

8.4.1.3 Transmission

Humans are an accidental paratenic host, and infection can occur in several ways. If infected copepods, such as cyclops (water fleas), are ingested through contaminated water, the procercoid will continue to mature to the plerocercoid stage in human tissue. Similarly, ingestion of undercooked, infected paratenic hosts will likewise result in human infection with the plerocercoid larvae. Transfer of plerocercoids can also occur through direct contact with the skin or organs of infected paratenic hosts. Skin or eye poultices sometimes used in ritualistic healing practices provide an unusual conduit for the migrating plerocercoid cyst to infect humans.

8.4.1.4 Clinical Aspects and Diagnosis

Plerocercoid encystment or reencystment in human tissue results in sparganosis. Larva may grow up to 30 cm and slowly migrate, resulting in cutaneous or visceral migrans, depending on the site of infection. Skin and intestinal mucosa are the two most common sites of plerocercoid encystment. Sparganosis may present as an erythematous, pruritic, subcutaneous swelling that slowly migrates through the skin. Migration into the eye can occur through direct contact with a poultice and leads to significant inflammatory pathology with pain, periorbital swelling, and conjuntivitis. Invasion of the CNS and bowel perforation have been reported.

8.4.1.5 Treatment

Treatment of sparganosis with antihelminthic agents has not proven to be beneficial. Surgical excision for localized sparganosis is the only effective therapy.

8.4.1.6 Control

Most cases of human sparganosis result from the consumption of water contaminated with infected copepods and the consumption of a paretenic host. Control can be achieved through proper water purification and food preparation techniques in endemic regions. Education regarding the dangers posed by application of poultices in healing practices should also be addressed.

8.5 ECHINOCOCCUS

The Echinococcus species are small tapeworms that usually reside in canine or feline definitive hosts and herbivore intermediate hosts. Humans are accidental intermediate hosts. There are several species of Echinococcus that can cause disease in humans, but by far the most important are *Echinococcus granulosus*, which causes cystic echinococcosis, and *Echinococcus multilocularis*, which causes alveolar echinococcosis. Although *Echinococcus* is not a truly food-borne human parasite, it is included here because of its importance as a human pathogen.

8.5.1 *Echinococcus granulosus*

8.5.1.1 Biology

Echinococcus granulosus is a small tapeworm in the Taenidiae family that infects humans in its larval form. The adult worm inhabits a variety of canine definitive hosts, with the domestic dog being the most important from the standpoint of human health. The intermediate hosts include sheep and other ungulates, camels, goats, and cattle. Infrequently, humans are intermediate hosts for the tapeworm.

A mature *E. granulosus* worm is 3–6 mm long and contains only three proglottids, one immature, one mature, and one gravid. Dogs are infected by consuming infected tissue, most commonly raw sheep viscera (Fig. 8.7). The tapeworm matures over 4–5 weeks in the dog's intestine and survives for 5 to 20 months. The gravid proglottid ruptures to release fertilized eggs. Upon ingestion by the intermediate

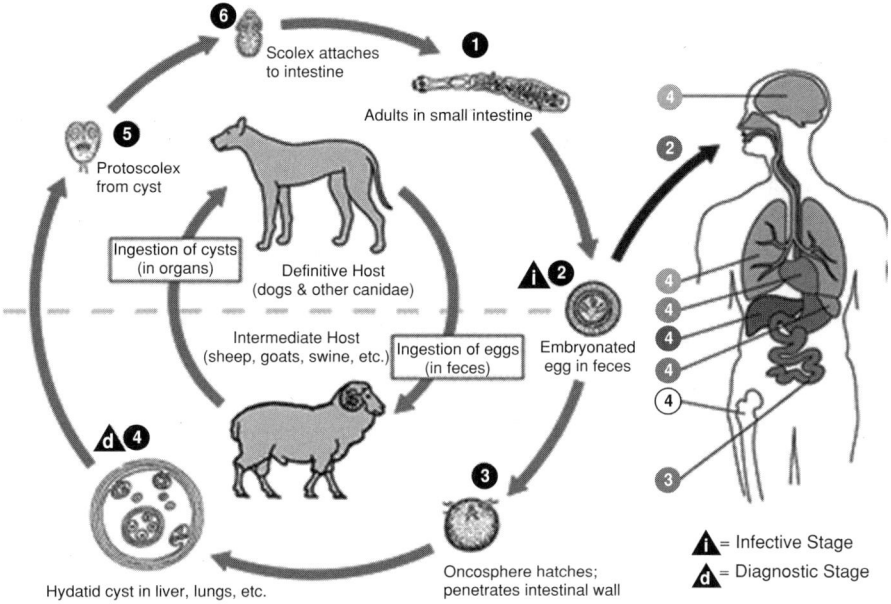

Figure 8.7. *Echinococcus granulosus* life cycle. http://www.dpd.cdc.gov/dpdx

host, the egg hatches and the released larva migrates through the intestinal wall and travels through the bloodstream to other sites, most notably the liver and the lung in humans, where the larva encysts.

8.5.1.2 Epidemiology
Echinococcus granulosus has a worldwide distribution and is the most common *Echinococcus* infection in humans. There are two forms of the disease, "European" and "Northern." The European form is globally distributed in domestic animals while the Northern form is restricted to the tundra and taiga of North America and Eurasia. There are several sylvatic cycles that may bring the parasite in contact with humans, including a cycle involving dingoes and marsupials in Australia (Thompson and McManus, 2001) and a cycle involving wolves or sled dogs and cervids in North America and Eurasia (Castrodale *et al.*, 1999). Sylvatic strains appear to have less severe clinical manifestations in humans than the more widespread domestic forms. There is considerable genetic variability between strains in different geographical regions, and often demonstrate host specificity. Regions with the highest prevalence are Eurasia, especially Russia, China, and various Mediterranean nations, northern and eastern Africa, Australia, and South America. Prevalence is highest where people raise sheep, especially when dogs are involved in the care of sheep.

Prevalence in humans varies widely. In some endemic areas such as rural Greece, prevalence can be as high as 29% in the general population (Sotiraki *et al.*, 2003). Hydatid disease appears to be more common in women than in men in many endemic areas. Risk factors for infection include occupation as a hunter and ownership of many sheep, and in children, infected family members, a father who slaughters sheep, and heavy exposure to dogs in the first year of life (Bai *et al.*, 2002; Larrieu, 2002). Drinking tap water (rather than from another source) appears protective, probably because tap water is unlikely to be contaminated with dog feces (Larrieu 2002).

8.5.1.3 Transmission

Dogs become infected with *E. granulosus* when they eat raw sheep viscera containing protoscolices, the larval worms. Worm burden can be very high; therefore, a single dog can disseminate a large quantity of eggs. The eggs may be carried by the wind and can survive for years in the environment. Transmission of *E. granulosus* to humans is by the fecal-oral route, predominantly through contact with surfaces, food, or water contaminated with dog feces. Humans do not develop intestinal infections from consuming contaminated sheep flesh, so transmission is predominantly associated with contact with infected dogs.

8.5.1.4 Clinical Aspects

Echinococcus granulosus infection usually occurs during childhood, and because the cyst grows slowly, is often asymptomatic for many years before it causes symptoms or is discovered incidentally on an imaging study. Up to 60% of infected persons may remain asymptomatic throughout life, but when symptomatic disease does occur, it is usually diagnosed during childhood or young adulthood. *E. granulosus* cysts are usually unilocular and filled with fluid, and they may have internal membranes. The outer acellular membrane protects the parasite from the host's immune response. The inner layer consists of a germinal membrane that produces new larvae, called protoscolices, daughter cysts, and brood capsules. Protoscolices and parasite hooklets accumulate in the cystic fluid to form a substance called hydatid sand. There may also be a layer of host granulomatous tissue surrounding the cyst. Symptoms are usually the result of the space-occupying effect and depend on the location of the cyst. More emergent symptoms may result with cyst rupture, when the release of cyst contents can cause fever, urticaria, pruritis, eosinophilia, and anaphylaxis. The cysts rarely metastasize or seed other body parts unless they rupture. Cystic echinococcosis appears to provoke a protective immune response that prevents reinfection.

Echinococcus granulosus can encyst in nearly any part of the body, but the liver is the site in more than half of patients. The average size of hepatic cysts is 12 cm. People usually present with abdominal pain or mass. Hepatic cysts can cause hepatobiliary problems by mass effect, cyst rupture, or communication with the biliary tree, causing jaundice, cholestasis, cholecystitis, cholangitis, or fistula formation. Hepatic cyst rupture can occur with blunt trauma, frequently in a motor vehicle accident, and may cause hepatobiliary symptoms, pain, or anaphylaxis. The lung is the second most common site of hydatid cyst formation, occurring in approximately 25% of patients and typically unilaterally. Pulmonary cysts may cause pain, fever, dyspnea, cough, and hemoptysis, and sputum may contain protoscolices or hooklets. Complications of pulmonary cysts include pleural or pericardial disease, pneumothorax, empyema, and fistula formation. Patients also may have cysts in both the liver and the lungs, a syndrome called hepatopulmonary hydatidosis. Bone cysts, most commonly in the spine or pelvic bones, are not common but can cause

Table 8.3. Classification of hepatic cystic echinococcosis lesions based on ultrasound examination (Pawlowski, I., Eckert, J., Vuitton, D., *et al.*, 2001).

Type	Active	Fertile	Cyst Wall	Remarks
CL*	Yes	No	Not visible	If cysts are due to cystic Echinococcus—early stage of development.
CE1	Yes	Yes	Visible	Unilocular, anechoic, or "snowflake sign."
CE2	Yes	Yes	Visible	Multiseptate and multivesicular, daughter cysts present.
CE3	Transitional	Yes	Visible	Anechoic content with detached laminated membrane ("waterlilly sign"). Decreased intracystic pressure. Cyst starting to degenerate.
CE4*	Inactive	No	Not visible	Heterogeneous hyperechoic or hypoechoic contents. No visible daughter cysts. "Ball of wool" sign due to degenerate membranes. Usually no viable protoscolices.
CE5*	Inactive	No	Calcified	Thick, variably calcified wall producing a cone-shaped shadow. Usually no viable protoscolices.

Cysts subclassified according to size: small <5 cm; medium 5–10 cm; large >10 cm.
*Further tests required to ascertain a diagnosis of cystic echinococcosis.

significant morbidity. Hydatid disease occurs rarely in other organs. Twenty to forty percent of patients have multiple cysts.

8.5.1.5 Diagnosis

Radiological imaging is the cornerstone of diagnosis of cystic echinococcosis in humans. Ultrasonography is excellent for screening and diagnosis because it is safe and noninvasive, provides immediate results, and can often be performed in the community. CT and MRI are also quite sensitive, but are only available in hospitals. CT may be better than ultrasonography for the diagnosis of extrahepatic hydatid disease (Gottstein and Reichen, 1996). The visualization of daughter cysts on ultrasound, CT, or MRI is pathognomonic for cystic echinococcosis; another characteristic finding is wall calcification. The World Health Organization has developed a classification system for the staging of hepatic cystic echinococcosis based on ultrasound findings (Table 8.3). Plain film chest X-ray is also useful for the diagnosis of pulmonary cysts or calcified hepatic cysts. On X-ray, pulmonary cysts appear as irregular rounded masses of uniform density. Endoscopic retrograde cholangiography can be used preoperatively to distinguish simple from complicated cysts, and endoscopic retrograde cholangiopancreatography can be used after surgery to look for biliary tree damage.

Serologic testing is also useful for diagnosis of cystic echinococcosis. Antigen and antibody tests exist, but both have problems with specificity. Serologic testing by enzyme-linked immunotransfer blot (EITB) is positive in 80% of patients with hepatic cysts, but in only 56% of patients with pulmonary cysts or cysts in multiple organs. Specificity approaches 100%, yet there are reports of cross-reactivity in patients with *T. solium* cysticercosis. An ELISA based on swine hydatid fluid had similar sensitivity, but was less specific, while a double diffusion test (DD5) based on sheep hydatid fluid was only 47% sensitive, although very specific (Verastegui *et al.*, 1992). Serology can also be used to assess the effectiveness of surgery or chemotherapy.

The diagnosis of echinococcosis can also be made by identification of hydatid sand in aspirates of cystic fluid, but because of the high risk of seeding new cysts or provoking anaphylaxis, aspiration is discouraged, except in PAIR procedures described below. In histological specimens, hooklets are acid-fast and the cyst capsule is periodic acid-Schiff positive.

8.5.1.6 Treatment

Treatment is generally deferred unless the patient is symptomatic or the cyst is a threat to an anatomical structure. Surgical removal of the cyst is considered the gold standard. Relative contraindications for surgery are the presence of multiple cysts or anatomical location. Radical surgery carries a higher risk of complications, but a lower rate of relapse. Use of protoscolicidal agents intraoperatively is controversial. Commonly used agents include 70–90% ethanol, 15–20% saline solution, and 0.5% cetrimide, and these should be left in the cystic cavity for at least 15 minutes. Patients should also receive preoperative and postoperative chemotherapy for several weeks with albendazole.

PAIR (Puncture, Aspiration, Injection, Reaspiration) is a less invasive treatment performed under ultrasound guidance. The physician inserts the needle into the cyst and aspirates the contents. A protoscolicidal agent, which kills the protoscolex larval form of the tapeworm, is then injected into the cavity, dwells for several minutes, and is reaspirated. All daughter cysts within the main cysts must be punctured as well. PAIR should only be performed with intensive care support because of the risk of anaphylaxis. Patients should receive chemotherapy before and after the procedure. PAIR reduces complication rates from 28 to 5–10% compared to surgery (Aygun *et al.*, 2001, Men *et al.*, 1999, Pelaez *et al.*, 2000). PAIR is contraindicated if the cyst is superficial, contains thick internal septae, or communicates with the biliary tree.

Medical treatment alone is rarely curative, but is an option for patients who are not surgical candidates. Albendazole 10mg/kg divided into two daily doses is the best, and when used alone for 12 weeks to 6 months cures 30% of patients and improves in 50% of the remaining patients. Some physicians recommend two-week rest periods off the drug to decrease toxicity, but recent data does not suggest that this decreases side effects (McManus, 2003). Praziquantel 25 mg/kg daily may also provide some improvement. Studies in sheep suggest that weekly use of praziquantel is also effective and may cause fewer side effects (Dueger *et al.*, 1999).

8.5.1.7 Control

Good animal husbandry practices and good hygiene and sanitation are currently the most practical and effective methods to control *E. granulosus* infection. Dogs should not be allowed to eat raw flesh from domestic animals, particularly sheep, and humans should use good hygiene to avoid contamination of food and water by dog feces. It is very important that dogs be kept away from places where sheep are slaughtered. Limiting the number of stray dogs living in the community may also decrease prevalence in humans. Infected dogs have been successfully treated with praziquantel. Programs based on periodic treatment of dogs and restriction of dogs from slaughterhouses have successfully eradicated *E. granulosus* in Iceland and have greatly decreased the prevalence of the disease in New Zealand, Tasmania, Cyprus, Chile, and Argentina. Vaccine development holds great promise for future control of the disease. In particular, the recombinant vaccine EG95 has been effective in trials in sheep in Australia, New Zealand, and Argentina and it seems to provide some protection in goats and cattle as well (Lightowers, 1999, 2001, 2003). Studies on dog vaccines are in their infancy, but show promise for decreasing egg excretion. Dog vaccination with an effective vaccine, in combination with periodic pharmacological treatment, would be the most efficacious method to control hydatid disease.

8.5.2 *Echinococcus multilocularis*

8.5.2.1 Biology

The definitive hosts are foxes, especially red and arctic foxes, wild dogs, and domestic dogs. Rodents are the main intermediate hosts, although the cestode can also infect pigs, monkeys, horses, and European beavers. Canines become infected with *E. multilocularis* when they eat the larval form of the worm in an infected rodent. The larva attaches to the small intestine and matures over about 4 weeks. Adult tapeworms are 1.2–4.5 mm long and have 2–6 proglottids, with an average of 5 proglottids. The round or oval eggs, measuring 30–36 μm in diameter, are released from the gravid proglottid and expelled in feces. Eggs can survive for extended periods of time in the environment. The life cycle is very similar to that of *E. granulosus* (Fig. 8.7).

However, the cysts of *E. multilocularis* are different from the cysts of *E. granulosus*. *E. multilocularis* larvae remain in a proliferative state, forming multilocular cysts with vesicles that grow out from the germinal layer to invade adjacent tissues. The cysts vary in size between 1 and 20 mm. Protoscolices and hydatid sand are rarely seen in humans. Cysts are very aggressive, destroying the liver and even metastasizing hematogenously to other organs. The cyst of *E. multilocularis* contains a laminated and a germinal layer, but the laminated layer does not provide an effective host-parasite barrier. Cellular immunity appears to be more important than humoral immunity, and the parasite may be capable of modifying the host's immune response to enhance its own odds of survival (Vuitton, 2003).

8.5.2.2 Epidemiology

Echinococcus multilocularis is found only in the northern hemisphere, in arctic, subarctic, or mountainous areas of Canada, the United States, central and northern Europe, and Asia. It predominantly participates in a sylvatic cycle in foxes and wild rodents, though domestic dogs and cats can also harbor the parasite. In recent years, there is evidence that the infection has the potential to emerge in urban populations; as foxes move into more urban habitats and come into greater contact with domestic animals. Alveolar echinococcosis is a rare disease, but is medically important because it is frequently fatal when untreated. Prevalence in humans varies with location and is not always correlated with the presence of foxes or other hosts. Men and women have similar prevalence rates. Fur trappers and their families have the highest risk of infection due to contact with contaminated fur.

8.5.2.3 Transmission
Humans are accidental intermediate hosts of *E. multilocularis*, and they acquire the infection by ingesting eggs from contaminated food or by contact with contaminated furs.

8.5.2.4 Clinical Relevance and Diagnosis
Alveolar echinococcosis has a long asymptomatic period lasting 5–15 years. The cysts grow at an average of 15 ml per year and can ultimately involve areas 15–20 cm in diameter. About 10% of cysts metastasize hematogenously. Liver cysts, usually in the right lobe, are found in 99% of patients, and 13% of patients have multiorgan disease. Patients usually become symptomatic between 50 and 70 years of age and can present with a variety of symptoms. One third present with right upper quadrant or epigastric pain and one third with cholestatic jaundice. Other signs and symptoms include hepatomegaly, weight loss, and fatigue. Patients can form amyloids in the liver and develop biliary cirrhosis. Complications include abscess formation, cholangitis, sepsis, portal hypertension, and Budd-Chiari syndrome. Metastatic disease can cause a myriad of other organ-specific signs, symptoms, and complications. The cysts provoke an inflammatory reaction, but liver enzymes are rarely significantly elevated. The levels of all types of immunoglobins, particularly IgE, increase, but they are usually not marked eosinophilia.

Imaging studies, especially ultrasonography, are the most useful diagnostic tools. Ultrasonography is 88–98% sensitive and more than 95% specific for alveolar echinococcosis (Raether and Hänel, 2003). On plain film chest X-rays, lung metastases appear as multiple small solid masses at the peripheries of the lobes. On CT, liver masses are irregular, heterogeneous, and hypodense with central necrosis and calcifications, and they do not enhance with contrast. MRI is less sensitive because it does not detect calcifications, but it is useful to distinguish cystic and alveolar lesions in the brain. Serological testing is also useful for diagnosis. Specific IgE levels are 73.6% sensitive for diagnosis and can be used to follow treatment success as well (Wellinghausen and Kern, 2001). Em2 is a species-specific native antigen ELISA that is useful for screening, diagnosis, and monitoring treatment. This test is even better when used with II/3–10, a recombinant antigen; demonstrating 97% sensitivity and 99% specificity (Gottstein *et al.*, 1993).

8.5.2.5 Treatment
Surgery is the cornerstone of treatment for alveolar echinococcosis. When possible, the lesion is usually treated like hepatic cancer with radical resection of tissue containing the cysts. Early diagnosis is essential because early treatment carries a better prognosis, though even after surgery, 10–20% of patients have a recurrence. If the cyst cannot be removed in entirety, improving biliary drainage may increase quality of life. Patients should receive preoperative and postoperative chemotherapy, usually with albendazole. Because albendazole is only parasitostatic, patients need up to 2 years of therapy after surgery. Use of up to 20 mg/kg per day postoperatively has been associated with 80% survival at 10 years. Liver transplant has also been used

to treat alveolar echinococcosis. While it has been successful in many cases, there is considerable risk of proliferation of undiagnosed metastatic lesions or remaining cystic material due to post-transplant immunosuppressive therapy.

Albendazole, mebendazole, and praziquantel have all been used as chemotherapeutic agents against *E. multilocularis* infection. Albendazole is superior to mebendazole. Praziquantel is the most scolicidal, but does not prevent cyst expansion; albendazole is best to control cyst growth. Although cure is difficult with medical therapy alone, medication may be able to convert patients who are not suitable for surgery into operative candidates.

8.5.2.6 Control

Echinococcus multilocularis is very difficult to control because it is predominantly sylvatic. Limiting the number of stray dogs in the community and treating infected dogs with praziquantel can reduce human incidence, and placing praziquantel baits for foxes can decrease infection in this population. Elimination of *E. multilocularis* has only been successful on the Japanese island of Rebun, but this was only accomplished by elimination of all dogs and foxes from the island.

8.6 HYMENOLEPIS

8.6.1 *Hymenolepis nana*

8.6.1.1 Biology

Hymenolepis nana, known as the dwarf tapeworm, is a member of the order Cyclophyllidea and is the only cestode that does not require two distinct hosts, as the life cycle can occur entirely in humans. With a length of only 2–5 cm, it is the smallest cestode that infects humans. Its scolex has four suckers and one row of hooks. When mature, the strobila consists of 150–200 proglottids that are wider than they are long. The round or oval eggs, 40–50 μm in diameter, are transparent and white with a double membrane. The life span of the tapeworm is 4–10 weeks, but because of autoinfection, infected persons may harbor worms for much longer periods of time.

Hymenolepis nana has a direct life cycle (Fig 8.8), meaning that humans can serve as the only hosts for the parasite—it does not require a separate intermediate host although rats and insects often participate in transmission. *H. nana* is acquired by ingestion of eggs or metacestodes in food or on fomites; autoinfection can also occur when gravid proglottids release eggs that hatch inside the gut. Eggs hatch in the small intestine, liberating the oncosphere embryo, which then penetrates the lamina propria of the intestinal villi. The oncosphere encysts inside the villi for 4–5 days and then reenters the lumen of the small intestine. The worm then attaches to the small intestine wall with its scolex and matures over 3–4 weeks. If the metacestode (cysticercoid) form is eaten, it will attach directly to the small intestine wall without penetrating the lamina propria and will mature into an adult tapeworm in about 2 weeks. Gravid proglottids are released from the distal end of the worm and disintegrate in the small intestine so that only fertilized eggs are

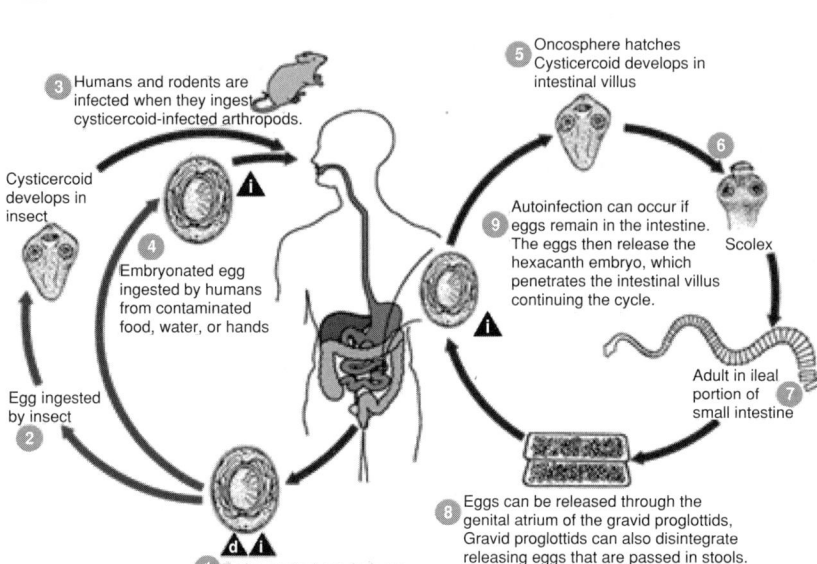

Figure 8.8. *Hymenolepis nana* life cycle. http://www.dpd.cdc.gov/dpdx

released in feces. They are infectious from the moment they are expelled in human feces.

8.6.1.2 Epidemiology

Hymenolepis nana is a ubiquitous human parasite, found worldwide, but it is especially common in Asia, Africa, Central and South America, and southern and eastern Europe. Children are the most commonly infected, possibly due to the immaturity of their immune systems. It is especially common in institutionalized children, where up to 8% may be infected (Yoeli *et al.*, 1972), and in immunocompromised or malnourished people. In tropical nations, where infection is most common, prevalence is about 1% (Botero *et al.*, 1998), but in certain endemic areas it can exceed 25% (King, 2000).

8.6.1.3 Transmission

Transmission occurs mainly through a fecal-oral route or by accidental ingestion of infected insects. Both eggs and metacestode larvae are infective for humans. The eggs may be eaten by rodents or beetle larvae, or may contaminate food when people do not follow good hygienic practices in the kitchen. Food, especially cereals and grains, becomes contaminated with human or rat feces containing eggs, or with mealworm larvae containing metacestodes, which is then eaten by humans.

8.6.1.4 Clinical Aspects and Diagnosis

Infection with *H. nana* is usually asymptomatic. Both cellular and humoral immunity appear important for clearance of the infection. In adults, the infection is cleared quickly because most adults rapidly mount a protective immune response, but children can have more prolonged infections. Very heavily infected people may have symptoms such as loss of appetite and abdominal pain. Diarrhea can result from damage to the intestinal mucosa.

Definitive diagnosis of *H. nana* infection is by identification of the eggs in stool microscopy. If the entire worm is recovered, diagnosis can be made by inspection of the scolex or by expressing eggs from gravid proglottids and identifying them as *H. nana*. There is an enzyme-linked immunosorbent (ELISA) assay that is approximately 80% sensitive, but it is not specific and frequently cross-reacts with other parasites (Castillo *et al.*, 1991).

8.6.1.5 Treatment

The infection is treated with either niclosamide or praziquantel. The cysticercoid form in the lamina propria is somewhat resistant to medical therapy, so patients may require higher doses or longer courses of the medications than are used for other helminthic infections. Praziquantal is advantageous because it kills both adult and larval forms and can be administered in a single dose of 25 mg/kg. The recommended dose of niclosamide is 2 g for adults or 1.5 g for children for the first dose, followed by 1g for adults or 0.5 g for children daily for 1 week in order to insure that any emerging cysts are killed. Because of the tapeworm's relative resistance to chemotherapy and the potential for autoinfection, the patient's stool should be checked for eggs 1 and 3 months after completion of treatment. If the stool contains eggs, the patient should receive another course of treatment.

8.6.1.6 Control

Good personal hygiene and sanitation to prevent contamination of food with human feces are the most effective ways to control *H. nana* infection. Controlling rodent populations may also decrease transmission by eliminating reservoir hosts. In epidemics, mass chemotherapy has been effective to prevent spread of the infection.

REFERENCES

Allan, J. C., Avila, G., Garcia-Noval, J., Flisser, A., and Craig, P. S., 1990, Immunodiagnosis of taeniasis by coproantigen detection, *Parasitology* **101**:473–477.

Allan, J. C., Velasquez-Tohom, M., Garcia-Noval, J., Torres-Alvarez, R., Yurrita, P., Fletes, C., de Mata, F., Soto de Alfara, H., and Craig, P. S., 1996a, Epidemiology of intestinal taeniasis in four rural Guatemalan communities, *Ann. Trop. Med. Parasitol.* **90**:157–165.

Allan, J. C., Velasquez-Tohom, M., Torres-Alvarez, R., Yurrita, P., and Garcia-Noval, J., 1996b, Field trial of the coproantigen-based diagnosis of *Taenia solium* taeniasis by enzyme-linked immunosorbent assay, *Am. J. Trop. Med. Hyg.* **54**:352–356.

Aygun, E., Sahin, M., Odev, K., *et al.*, 2001, The management of liver hydatid cysts by percutaneous drainage, *Can. J. Surg.* **44**:203–209.

Bai, Y., Chang, N., Jiang, C., *et al.*, 2002, Survey on cystic echinococcosis in Tibetans, West China, *Acta Tropica*. **82**:381–385.
Bern, C., Garcia, H. H., Evans, C., Gonzalez, A. E., Verastegui, M., Tsang, V. C., and Gilman, R. H., 1999, Magnitude of the disease burden from neurocysticercosis in a developing country, *Clin. Infect. Dis*. **29**:1203–1209.
Botero, R. D., Restrepo, M. A., Bedoya, E. V. I., Restrepo, I. M., Leiderman, W. E., Betancur, M. J. A., Gómez, C. I., and Vélez, G. L. A., 1998, *Parasitosis Humana*, 3rd edn. Corporación para Investigaciones Biologicas, Medillín, Colombia.
Cabrera, C., Tantalean, V., Manuel y Rojas, M., and Ricardo, 2001, *Diphyllobothrium pacificum* (Nybelin 1931) Margolis, 1956 en Canis familiaris de la ciudad de Chincha, Peru, *Bol. Chil. Parasitol*. **56**:26–28.
Carangelo, B., Erra, S., Del Basso Del Caro, M. L., Bucciero, A., Vizioli, L., Panagiotopoulos, K., and Cerillo, A., 2001, Neurocysticercosis: Case report, *J. Neurosurg. Sci*. **45**:43–46.
Castillo, R. M., Grado, P., Carcamo, C., Miranda, E., Montenegro, T., Guevara, A., and Gilman, R. H., 1991, Effect of treatment on serum antibody to *Hymenolepis nana* detected by enzyme-linked immunosorbent assay, *Clin. Microbiol*. **29**:413–414.
Castrodale, L., Beller, M., Wilson, J., Schantz, P. M., McManus, D. P., Zhang, L. H., Fallico, F. G., and Sacco, F. D., 1999, Two atypical cases of cystic chinococcosis (*Echinococcus granulosus*) in Alaska, *Am. J. Trop. Med. Hyg*. **66**:325–337.
Chopra, J. S., Kaur, U., and Mahajan, R. C., 1981, Cysticercosis and epilepsy: A clinical and serological study, *Trans. R. Soc. Trop. Med. Hyg*. **75**:518–520.
Commission on Tropical Diseases of the International League against Epilepsy, 1994, Relationship between epilepsy and tropical diseases, *Epilepsia* **35**:89–93.
Cysticercosis Working Group in Peru, 1993, The marketing of cysticercotic pigs in the sierra of Peru, *Bull. World Health Org*. **71**:223–228.
Cruz, M., Davis, A., Dixon, H., Pawlowski, Z. S., and Proano, J., 1989, Operational studies on the control of *Taenia solium* taeniasis/cysticercosis in Ecuador, *Bull. World Health Org*. **67**:401–407.
Cruz, M. E., Schantz, P. M., Cruz, I., Espinosa, P., Preux, P. M., Cruz, A., Benitez, W., Tsang, V. C., Fermoso, J., and Dumas, M., 1999, Epilepsy and neurocysticercosis in an Andean community, *Int. J. Epidemiol*. **28**:799–803.
Del Brutto, O. H., Santibanez, R., Noboa, C. A., Aguirre, R., Diaz, E., and Alarcon, T. A., 1992, Epilepsy due to neurocysticercosis: Analysis of 203 patients, *Neurology* **42**:389–392.
Del Brutto, O. H., Rajshekhar, B., White, A. C. Jr., Tsang, V. C., Nash, T. E., Takayanagui, O. M., Schantz, P. M., Evans, C. A. W., Flisser, A., Correa, D., Botero, D., Allan, J. C., Sarti, E., Gonzalez, A. E., Gilman, R. H., and García, H. H., 2001, Proposed diagnostic criteria for neurocysticercosis, *Neurology* **57**:177–183.
Dixon, H. B., and Lipscomb, F. M., 1961, Cysticercosis: An analysis and follow-up of 450 cases, Medical Research Council, London.
Dueger, E. L., Moro, P. L., and Gilman, R. H., 1999, Oxfendazole treatment of sheep with naturally acquired hydatid disease, *Antimicrob. Agents Chemother*. **43**:2263–2267.
Fernandez-Aranda, F., Solano, R., Badia, A., and Jimenez-Murcia, S., 2001, Binge eating disorder onset by unusual parasitic intestinal disease: A case report, *Int. J. Eat. Disord*. **30**:107–109.
Flisser, A., Viniegra, A. E., Aguilar-Vega, L., Garza-Rodriguez, A., Maravilla, P., and Avila, G., 2004, Portrait of human tapeworms, *J. Parasitol*. **90**:914–916.
Galan-Puchades, M. T., and Fuentes, M. V., 2000, The Asian *taenia* and the possibility of cysticercosis, *Korean J. Parasitol*. **38**:1–7.
García, H. H., Talley, A., Gilman, R. H., Zorrilla, L., and Pretell, J., 1999, Epilepsy and neurocystecercosis in a village in Huaraz, Perú. *Clin. Neurol. Neurosurg*. **101**:225–228.

García, H. H., Gonzalez, A. E., Evans, C. A. W., and Gilman, R. H., for the Cysticercosis Working Group in Peru, 2003a, *Taenia solium* cysticercosis, *Lancet* **361**:547–556.

García, H. H., Gonzalez, A. E., and Gilman, R. H., for the Cysticercosis Working Group in Peru, 2003b, Diagnosis, treatment, and control of *Taenia solium* cysticercosis, *Curr. Opin. Infect. Dis.* **16**:411–419.

García, H. H., Pretell, E. J., Gilman, R. H., Maratinez, S. M., Moulton, L. H., Del Brutto, O. H., Herrera, G., Evans, C. A. W., and Gonzalez, A. E. for the Cysticercosis Working Group in Peru, 2004, A trial of antiparasitic treatment to reduce the rate of seizures due to cerebral cysticercosis, *N. Engl. J. Med.* **350**:249–258.

Garcia-Noval, J., Moreno, E., de Mata, F., Soto de Alfaro, H., Fletes, C., Craig, P. S., and Allan, J. C., 2001, An epidemiological study of epilepsy and epileptic seizures in two rural Guatemalan communities, *Am. Trop. Med. Parasitol.* **95**:167–175.

Gekeler, F., Eichenlaub, S., Mendoza, E. G., Sotelo, J. Hoelscher, M., and Loscher, T., 2002, Sensitivity and specificity of ELISA and immunoblot for diagnosing neurocysticercosis, *Eur. J. Clin. Microbiol., Infect. Dis.* **21**:227–229.

Gilman, R. H., Del Brutto, O. H., García, H. H., and Martinex, M., 2000, Prevalence of taeniasis among patients with neurocysticercosis is related to severity of infection, *Neurology* **55**:1062.

Gonzales, A. E., Falcon, N., Gavidia, C., García, H. H., Tsang, V. C., Bernal, T., Romero, M., and Gilman, R. H., 1997, Treatment of porcine cysticercosis with oxfendazole: A dose-reponse trial, *Vet. Rec.* **141**:420–422.

Gonzales, A. E., García, H. H., Gilman, R. H., Tsang, V. C. W., and Cysticercosis Working Group in Peru, 2003, Control of *Taenia solium, Acta Tropica* **87**:103–109.

Gottstein, B., Jacquier, P., Bresson-Hadni, S., and Eckert, J., 1993, Improved primary diagnosis of alveolar echinococcosis in humans by an enzyme-linked immunosorbent assay using the Em2^{plus} antigen, *J. Clin. Microbiol.* **31**:373–376.

Gottstein, B. and Reichen, J., 1996, Echinococcosi/Hydatidosis, In Cook, G. C. (ed), *Manson's Tropical Diseases*, 20th edn. WB Saunders, London, pp. 1486–1508.

Hoberg, E. P., 2002, *Taenia* tapeworms: Their biology, evolution, and socioeconomic significance, *Microbe Infect.* **4**:859–866.

Huang, B. Li, G. Jia, F., Lui, F., Ge, L. Li. W., and Cheng, Y., 2002, Determination of specific IgG4 for diagnosis and therapeutic evaluation of cerebral cysticercosis, *Chin. Med. J.* **115**:580–583.

King, C. H., 2000, Cestodes (Tapeworms), In Mandell, G. L., Bennett, J. E., and Dolin, R. (eds), *Principles and Practice of Infectious Diseases*, 5th edn., Churchill Livingstone, New York, pp. 2956–2965.

Larreiu, E. J., Casta, M. T., Del Carpio, M., *et al.*, 2002, A case-control study of the risk factors for cystic echinococcosis among the children of Rio Negro province in Argentina, *Ann. Trop. Med. Parasitol.* **96**:43–52.

Lightowlers, M. W., and Gauci, C. G., 2001, Vaccines against cysticercosis and hydatidosis, *Vet. Parasitol.* **101**:337–352.

Lightowlers, M., Jensen, O., Fernandez, E., Iriarte, J. A., Woollard, D. J., Gauci, C. G., Jenkins, D. J., and Heath, D. D., 1999, Vaccination trials in Australia and Argentina confirm the effectiveness of the EG95 hydatid vaccine in sheep, *Int. J. Parasitol.* **29**:531–534.

Lightowlers, M. W., Colebrook, A. L., Gauci, C. G., Kyngdon, C. T., Monkhouse, J. L., Vallejo Rodriquez, C., Read, A. J., Rolfe, R. A., and Sato, C., 2003, Vaccination against cestode parasites: Anti-helminth vaccines that work and why, *Vet. Parasitol.* **115**: 83–123.

Masci, J. R., 2004, Tapeworm infestation, In Ferri, F. F. (ed), *Ferri: Ferri's Clinical Advisor: Instant Diagnosis and Treatment*, 2004 edn. Mosby, Providence, RI, pp. 826.

Mayta, H., Talley, A., Gilman, R. H., Jimenez, J., Verastegui, M., Ruiz, M., Garcia, H. H., and Gonzalez, A. E., 2000, Differentiating *Taenia solium* and *Taenia saginata* infections by simple hematoxylin-eosin staining and PCR-restriction enzyme analysis, *J. Clin. Microbiol.* **38**:133–137.
Medina Flores, J., Tantaleán Vidaurre, V., León Rivera, M., and Cano Rosales, M., 2002, *Diphyllobothrium pacificum* en niños del Perú, *Diagnostico* **41**.
Men, S., Hekimoglu, B., Yucesoy, C., Arda, I., and Baan, I., 1999, Percutaneous treatment of hepatic hydatid cysts: An alternative to surgery, *Am. J. Roentgenol.* **172**:83–89.
Nicoletti, A., Bartoloni, A., Reggio, A., Bartelesi, F., Roselli, M., Sofia, V., Rosado Chavez, J., Gamboa Barahona, H., Paradisi, F., Cancrini, G., Tsang V. C., and Hall, A. J., 2002, Epilepsy, cysticercosis, and toxocariasis: A population-based case-control study in rural Bolivia, *Neurology* **58**:1256–1261.
Odashima, N. S., Takayanagui, O. M., and Figueiredo, J. F., 2002, Enzyme-linked immunosorbent assay (ELISA) for the detection of IgG, IgM, IgE, and IgA against *Cysticercosis cellulosae* in cerebrospinal fluid of patients with neurocysticercosis, *Arq. Neuropsiquiatr.* **60**:400–405.
Ong, S., Talan, D. A., Moran, G. J., Mower, W., Newdow, M., Tsang, V. C., Pinner, R. W., and EMERGEncy ID NET Study Group, 2002, Neurocysticercosis in radiographically imaged seizure patients in U.S. emergency departments, *Emerg. Infect. Dis.* **8**:608–613.
Osuntokun, B. O., and Schoenberg, B. S., 1982, Research protocol for measuring the prevalence of neurological disorders in developing countries: Results of a pilot study in Nigeria, *Neuroepidemiology* **1**:143–153.
Padma, M. V., Behari, M., Misra, N. K., and Ahuja, G. K., 1995, Albendazole in neurocysticercosis, *Natl. Med. J. India* **8**:255–258.
Pawlowski, I., Eckert, J., Vuitton, D., *et al.*, 2001, Echinococcosis in humans: Clinical aspects, diagnosis, and treatment, In Eckert, J., Gemmell, M., Meslin, F.-X., and Pawlowski, Z. (eds), *WHO Manual on Echinococcosis in Humans and Aanimals: A Public Health Problem of GlobalConcern*, World Organisation or Animal Health, Paris, pp. 20–71.
Pelaez, B., Kugler, C., Correa, D., *et al.*, 2000, PAIR as percutaneous treatment of hydatid liver cysts, *Acta Trop.* **75**:197–202.
Proano-Narvaez, J. V., Meza-Lucas, A., Mata-Ruiz, O., Garcia-Jeronimo, R. C., and Correa, D., 2002, Laboratory diagnosis of human neurocysticercosis: Double-blind comparison of enzyme-linked immunosorbent assay and electroimmunotransfer blot asay, *J. Clin. Microbiol.* **40**:2115–2188.
Raether, W., and Hänel, H., 2003, Epidemiology, clinical manifestations and diagnosis of zoonotic cestode infections: An update, *Parasitol. Res.* **91**:412–438.
Richards Jr., F.O., 2003, Human parasites and vectors: Cestodes, In Long, S. S., Pickering, L. K., and Prober, C. G. (ed), *Long: Principles and Practice of Pediatric Infectious Diseases*, 2nd edn. Elsevier, New York, pp. 1351–1363.
Sagua, H., Niera, I., Araya, J., and Gonzalez, J., 2001, New cases of *Diphyllobothrium pacificum* (Nybelin, 1931) Margolis, 1956 human infection in North of Chile, probably related with El Nino phenomenon, 1975–2000. *Bol. Chil. Parasitol.* **56**:22–25.
Sánchez, A. L., Lindbäck, J., Schantz, P. M., Sone, M., Sakai, H., Medina, M. T., and Ljungström, I., 1999, A population-based, case-control study of *Taenia solium* taeniasis and cysticercosis, *Ann. Trop. Med. Parasitol.* **93**:247–258.
Sarti, E., Schantz, P. M., Avila, G., Ambrosio, J., Medina-Santillán, R., and Flisser, A., 2000, Mass treatment against human taeniasis for the control of cysticercosis: A population-based intervention study, *Trans. Royal Soc. Trop. Med. Hyg.* **94**:85–89.
Schantz, P. M., More, A. C., and Munoz, J. L., 1992, Neurocysticercosis in an orthodox Jewish community in New York City. *N. Eng. J. Med.* **327**:692–295.

Schantz, P. M., Sarti, E., Plancarte, A., Wilson, M., Criales, J., Roberts, J., and Flisser, A., 1994, Community-based epidemiological investigations of cysticercosis due to *Taenia solium*: Comparison of serological screening tests and clinical findings in two populations in Mexico, *Clin. Infect. Dis.* **18**:879–885.

Siles-Lucas, M., and Gottstein, B., 2001, Molecular tools for the diagnosis of cystic and alveolar echinococcosis, *Trop. Med. Int. Health* **6**:463–475.

Sotiraki, S., Himonas, C., and Korkoliakou, P., 2003, Hydatidosis-echinococcosis in Greece, *Acta Tropica.* **85**:197–201.

Thompson, R. C. A., and McManus, D. P., 2001, Aetiology: Parasites and life cycles, In Eckert, J., Gemmell, M., Meslin, F.-X., and Pawlowski, Z. (eds), *WHO/OIE Manual on Echinococcosis in Humans and Animals: A PublicHealth Problem of Global Concern*, World Organisation for Animal Health, Paris, pp. 1–19.

Tsang, V. C., Brand, J. A., and Boyer, A. E., 1989, An enzyme-linked immunoelectrotransfer blot assay and glycoprotein antigens for diagnosing human cysticercosis (*Taenia solium*), *J. Infect. Dis.* **159**:50–59.

Verastegui, M., Gilman, R. H., García, H. H., Gonzalez, A. E., Arana, Y., Jeri, C., Tuero, I., Gavidia, C. M., Levine, M., Tsang, V. C., and the Cysticercosis Working Group in Peru, 2003, Prevalence of antibodies to unique *Taenia solium* oncosphere antigens in taeniasis and human and porcine cysticercosis, *Am. J. Trop. Med. Hyg.* **69**:438–444.

Verastegui, M., Moro, P., Guevara, A., Rodriguez, T., Miranda, E., and Gilman, R. H., 1992, Enzyme-linked immunoelectrotransfer blot test for diagnosis of human hydatid disease, *J. Clin. Microbiol.* **30**:1557–1561.

Vuitton, D. A., 2003, The ambiguous role of immunity in echinococcosis: Protection of the host or of the parasite? *Acta Tropica.* **85**:119–132.

Wellinghausen, N., and Kern, P., 2001, A new ImmunoCAP assay for detection of *Echinococcus multilocularis*-specific IgE, *Acta Tropica.* **79**:123–127.

White, A. C. Jr., and Weller, P. F., 2004, Cestodes, In Kasper, D. L., Braunwald, E., Fauci, A. S., Hauser, S. L., Longo, D. L., Jameson, J. L., Isselbacher, K. L. (eds), *Harrison's Principles of Internal Medicine*, 16th edn.

World Health Organization, 1992, Report of the WHO Working Group on clinical medicine and chemotherapy of alveolar and cystic echinococcus, WHO/CDS/VPU/93.118.

Yoeli, M., Most, H., Hammond, J., and Scheinesson, G. P., 1972, Parasitic infections in a closed community. Results of a 10-year survey in Willowbrook State School, *Trans. R Soc. Trop. Med. Hyg.* **66**:764–766.

CHAPTER 9

Waterborne Parasites and Diagnostic Tools

Gregory D. Sturbaum and George D. Di Giovanni

9.1 PREFACE

The purpose of parasite detection is not limited to curing disease in an infected individual but is crucial in the prevention and spread of disease. For these reasons, detection methods must be held at the highest standard of both sensitivity and specificity so that false-negative and false-positive results are avoided. Achieving these two requirements has been challenging, despite the development of various detection methodologies over the last several years. For parasitic organisms, those spread by food and water alike, microscopic based detection and diagnostic techniques are still revered as the gold standard. However, the advent and employment of molecular methodologies has proven to surpass microscopy in three major aspects: sensitivity, specificity, and the ability to speciate.

The choice and application of the appropriate detection or diagnostic technique begins with asking questions surrounding the target organism itself and the analytical objective. This task can be reduced to three essential questions. First, what is the matrix type to be assayed for the presence of the parasite; is it food (berries, lettuce, meat/tissue); is it from an infected individual (blood, fecal, or tissue); or is it environmental (water, wastewater, soil, etc.)? Second, what is the form of the parasite? Is it the trophozoite or cyst, as in the case of *Giardia lamblia*, the oocyst as for *Toxoplasma gondii*, or the nurse-cell for *Trichinella spiralis*? Finally, what is the purpose for detecting the organism's presence? Is it for patient diagnosis so that disease may be treated; is it source tracking for watershed management; or possibly epidemiological fingerprinting? Answering these questions is essential so that appropriate recovery and diagnostic techniques are used to confidently and consistently ascertain the answer. Once these questions have been answered, then the molecular technique or combination of techniques can be selected. Depending on the technique(s) selected, different analytes are targeted for recovery and purification from the sample matrix. These analytes can be divided into three groupings, the whole organism, nucleic acids (DNA and RNA), and proteins (including enzymes).

The parasites *Cryptosporidium parvum* and *G. lamblia* (syn. *G. intestinalis*, *G. lamblia*) are proven to (1) be endemic throughout the world, causing infections in both developed and developing countries; (2) be capable of causing outbreaks with significant illness and economic effects; (3) have a low infectious dose; and (4) utilize multiple transmission routes including ingestion of contaminated water and food, as well as direct fecal-oral transmission. In addition, *C. parvum* also poses special challenges to the water industry. It is small in size (~ 5 μm) making it difficult to remove via filtration; resistant to common disinfection processes, including chlorination; and environmentally hardy, remaining infectious for months. For these reasons, this chapter briefly discusses transmission routes and reported incidences of these two parasites while largely focusing on the multiple diagnostic techniques that have been developed for their detection. Other parasites will also be alluded to, as many life cycle particulars are shared between the different genera. For example, parasites that use contaminated water as a direct conduit for transmission or are foodborne include *Ascaris lumbricoides, A. suum, Balantidium coli, Blastocystis hominis, Cyclospora cayetanensis, Fasciola hepatica, Fasciolopsis buski*, the Microspora group, and *T. gondii*. In addition, it is noteworthy that many of the methods used for detecting parasites (helminth and protozoa alike) in environmental samples have derived from protocols developed in the clinical setting. Therefore, the molecular techniques discussed in this chapter, with modification to isolation procedures, are used in the detection of the foodborne parasites as well. Examples of these parasites include *C. parvum, G. lamblia, Diphyllobothrium latum, Paragonimus westermani, Taenia saginata* and *T. soilium, Trichinella spiralis*, and *T. gondii*.

Two important notes should be made at this time. First, it should be stated that molecular techniques are powerful tools and have different levels of complexities within the answers they provide. When used properly, the appropriate technique, even those that are less sophisticated, will give the information needed to answer the proposed question. However, sometimes a novel technique is applied to a question where the need for more complex information is not necessary (i.e., more information than needed is collected). And second, since the advent of the polymerase chain reaction (PCR) in 1985, and the multitude of subsequent derivative techniques, parasite detection methodologies have evolved more rapidly than the necessary and required quality assurance (QA) and quality control (QC) procedures and strategies. Proper QA/QC evaluation of these new techniques is discussed in the last section of this chapter.

9.2 PARASITES

The biology of many of the parasites are discussed elsewhere in other chapters. The purpose of these brief paragraphs is to outline important biological factors that need to be considered when developing a detection technique.

Cryptosporidium spp.: Taxonomy places the obligate intracellular parasite *Cryptosporidium* in the phylum Apicomplexa and class Sporozoa, which includes many medically important human and veterinary parasites such as *Cyclospora cayetanensis, Eimeria* spp., *Isospora* spp., *Plasmodium* spp., *Sarcocystis* spp., and *T. gondii* (Levine, 1984; Morgan *et al.*, 1999). The species *C. parvum* was originally described in 1912 (Tyzzer, 1912), however it wasn't until 1971 that the parasite was perceived important due to the economic impact as a result of bovine diarrhea (Panciere *et al.*, 1971). Later, cryptosporidiosis gained worldwide interest in 1982 with the emergence of the human immunodeficiency virus (HIV) and AIDS (Anonymous, 1982). Currently, 14 *Cryptosporidium* species are recognized based on vertebrate host infectivity, morphology and DNA classification. Of these, two species, *C. hominis* and *C. parvum*, are the major cause of human infection and outbreaks including ingestion of contaminated food and water (Table 9.1). These two species are morphologically similar and therefore proper classification relies upon

Table 9.1. Cryptosporidiosis outbreaks due to contaminated drinking water.[a]

Year	Location	Number of cases	Suspected cause	Reference
1984	Braun Station, TX	2006	Contaminated well water	(D'Antonio et al., 1985)
1987	Carrollton, GA	12,960	Treatment deficiencies	(Hayes et al., 1989)
1989	Ayrshire, UK	27	Treatment deficiencies	(Smith et al., 1989)
1990–1991	Isle of Thanet, UK	47	Treatment deficiencies	(Joseph et al., 1991)
1991	Berks County, PA	551	Treatment deficiencies	(Moore et al., 1993)
1992	York Shire, UK	125	Contaminated tap water	(Furtado et al., 1998)
1992	Jackson County, OR	15,000	Treatment deficiencies	(Frost et al., 1998; Moore et al., 1993)
1993	Milwaukee, WI	403,000	Treatment deficiencies	(MacKenzie et al., 1994)
1993	Waterloo, Canada	>1000	Contaminated tap water	(Rose et al., 1997)
1993	Las Vegas, NV	103	Unknown	(Goldstein et al., 1996)
1993	Wessex, UK	27	Contaminated tap water	(Furtado et al., 1998)
1994	Kanagawa, Japan	461	Contaminated drinking water	(Kuroki et al., 1996)
1994	Walla Walla, WA	104	Sewage contaminated well	(Dworkin et al., 1996)
1995	Northern Italy	294	Contaminated water tanks	(Frost et al., 2000; Pozio et al., 1997)
1995	Gainesville, FL	77	Contaminated tap water	(Anonymous, 1996)
1996	Ogose, Japan	>9000	Unfiltered ground water	(Yamazaki et al., 1997)
1996	New York, NY	>30	Contaminated apple cider	(Anonymous, 1997b)
1997	Shoal Lake, Ontario	100	Unfiltered lake water	(Anonymous, 1997a)
1998	Brushy Creek, TX	32	Sewage contaminated well	(Anonymous, 1998)
1999	Northwest England, UK	360	Unfiltered surface water	(Anonymous, 1999)
2000	Belfast, Northern Ireland	129	Contaminated drinking water	(Glaberman et al., 2002)
2000	Lancashire, England	58	Contaminated drinking water	(Howe et al., 2002)
2001	Belfast, Northern Ireland	230	Contaminated drinking water	(Glaberman et al., 2002)

[a] Table adapted from Fayer et al., 2000.

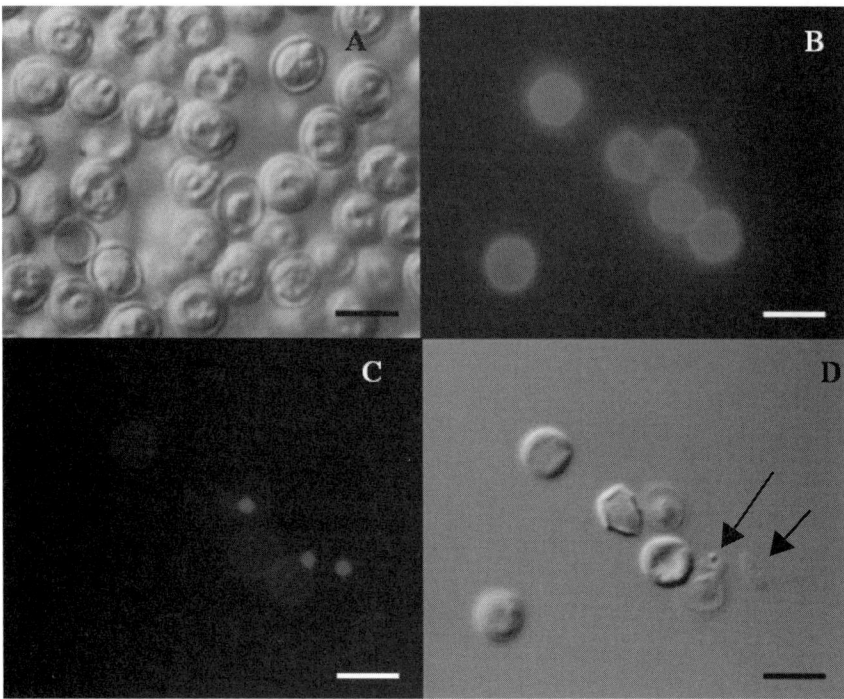

Figure 9.1. *Cryptosporidium parvum* photomicrographs at 1000X magnification. Panel A: Purified oocysts, differential interference contrast (DIC) microscopy; Panel B: Purified oocysts, immunofluorescent microscopy; Panel C: Purified oocysts stained with DAPI highlighting excysted sporozoite nuclei; Panel D: Purified oocysts, differential interference contrast (DIC) microscopy highlighting excysted sporozoite (arrowhead). Bar represents 5 μm.

molecular methods. Confounding matters, *C. hominis* has evolved to preferentially infect humans (Xiao *et al.*, 2002), making laboratory infectivity and disinfection studies with *C. hominis* nearly impossible to perform due to a lack of a consistent and standardized supply of oocysts. The characteristics of *C. parvum* (and by association *C. hominis*) complicates matters with two major aspects having a critical role for proper identification. First, the small size of *C. hominis* and *C. parvum* oocysts (4–6 μm) confound microscopic identification in fecal and environmental matrices since algae and yeast spores appear as the same size and shape and may cross-react with FITC labeled anti-*Cryptosporidium* monoclonal antibodies. In addition, as the oocysts for both species are similar in size and shape, speciation relies upon molecular methods. Second, oocyst concentrations in environmental samples are low thus making collection, concentration, and recovery techniques critical first steps for the application of molecular methods. In separate studies, a geometric mean of 2.7 oocysts/L (range: <0.007 to 484 oocysts/L, n = 66) was documented in source waters (LeChevallier *et al.*, 1991) and <1 oocysts/L (range: 0.001 to 0.48 oocysts/L, n = 158) is reported in drinking waters (Rose *et al.*, 1997). These combined aspects make proper detection of *Cryptosporidium* spp. vital for the protection of public health and are technically very challenging.

Giardia lamblia: Of the six recognized *Giardia* species, *G. lamblia* (syn *G. intestinalis*, *G. duodenalis*) is the only species infectious to humans (Thompson, 2000). This primitive flagellate protozoan also infects a wide range of vertebrate hosts including pets, livestock, and wild animals, thus increasing the potential for zoonotic transmission (Thompson, 2000). The cysts of *G. lamblia* isolates are morphologically similar, however molecular characterization has determined seven different genotypes or assemblages, A through G (Monis *et al.*, 1999). Of these, humans are only infected with Assemblages A and B (Homan *et al.*, 1992; Maryhofer *et al.*, 1995; Nash and Mowatt, 1992). Assemblages C and D are dog specific (Hopkins *et al.*, 1997; Monis *et al.*, 1998). Assemblage E is found in hoofed livestock (cattle, goats, pigs, and sheep) (Ey *et al.*, 1997). Finally, Assemblages F and G are infectious for cats and rats, respectively (Monis *et al.*, 1999). As with *Cryptosporidium* spp., objects of similar size and shape (e.g., algal cells) confound proper identification of *Giardia* cysts in fecal and environmental samples due to the use of non-specific dyes and cross-reacting FITC labeled monoclonal antibodies. Again, it is observed that molecular techniques are required for epidemiological and source tracking studies.

9.3 CONCENTRATION AND ISOLATION TECHNIQUES

Whether a microscopic or molecular technique is employed, a general first step is to execute an isolation protocol that both concentrates and purifies the parasite from the matrix (berries, meat, water, etc.), while reducing potential cross reactors or inhibiting substances. In general, the majority of the isolation techniques target the environmentally resistant stage of a parasite's life cycle (e.g., cyst, oocyst, and ova) based on the inherent robustness of this form rather than the active trophozoite and larval stages (cercariae and other infective larvae). The robustness of this resistant stage allows for exposure to chemicals such as surfactants or alcohols, as well as the separation steps such as centrifugal flotation or immunomagnetic separation (IMS) while having limited physical effect on the organism and subsequent microscopic or molecular detection. However, for some strictly foodborne parasites such as *Trichinella spiralis*, the direct carnivorous nature of the transmission route leaves detection protocols relying upon tissue sectioning for histology or tissue mincing for molecular protocols (such as ELISA or PCR).

The majority of collection/concentration techniques employed today are adopted and modified from isolation protocols developed within the clinical setting. For fecal samples, different protocols include an initial treatment step dilution with phosphate buffered saline (PBS) or a preservative followed by sieving through metal mesh sieves or cheesecloth to remove large debris. This semi-purified sample would then be subjected to an organic solvent such as phenol or diethyl ether in order to do a primary removal of lipids. Purification and concentration steps would include selective centrifugation such as zinc sulphate (Ryley *et al.*, 1976), diethyl-ether/Percoll (Garcia *et al.*, 1979; Garcia and Shimizu, 1981; Horen, 1983), Percoll density gradients (Waldman *et al.*, 1986), sequential discontinuous Percoll®/Sucrose (Arrowood and Sterling, 1987), Sheather's sugar flotation (Current et al., 1983; Sheather, 1923), or for highly purified organisms, sucrose/cesium-chloride gradient (Arrowood and Donaldson, 1996). While none of these flotation techniques provide debris free parasite recovery, all assist in concentrating target organisms.

For inherently low parasite concentrations found in environmental samples, the goal is to concentrate and purify at the same time. Current detection protocols for *Cryptosporidium* spp., and *T. gondii*, as well as other parasites, have all evolved from techniques developed for *Giardia*. An outbreak of giardiasis in Aspen, CO, in 1965 (Craun, 1986) initiated the development of a microscopic based method that combined water filtration to concentrate *Giardia* cysts, followed by zinc sulfate floatation for separation and iodine staining for microscopic identification (Quinones et al., 1988; Reference-Method, 1992). This basic detection strategy of concentration, separation/isolation, and microscopic detection is generally followed currently (oo)cyst detection in water samples, but with modern improvements. Filtration of water samples is used to concentrate water particulates, including (oo)cysts. Types of samples include treated drinking water, ground water, surface waters (lakes/reservoirs, irrigation ditches, rivers, etc.), storm water, and sewage treatment effluent. Various approaches to large volume water sample concentration have been used. Adapted from the *Giardia* recovery methods, a 10-inch polypropylene cartridge filter with a 1 μm nominal porosity was the filter of choice of the US Environmental Protection Agency (USEPA)-mandated Information Collection Rule (ICR) (Musial et al., 1987; USEPA, 1996). However, this filter was later demonstrated to have low recovery efficiencies (LeChevallier et al., 1995; Shepherd and Wyn-Jones, 1996) and therefore, other filtration/concentration techniques were developed, including membrane filtration (Ongerth and Stibbs, 1987), calcium carbonate flocculation (for small volumes) (Vesey et al., 1993), continuous flow centrifugation (Borchardt and Spencer, 2002; Renoth et al., 1996; Zuckerman et al., 1999), flow cytometry (Vesey et al., 1994a, 1994b), wound fiberglass cartridge filters (Kaucner and Stinear, 1998), foam filters (McCuin and Clancy, 2003), and capsule filters (Matheson et al., 1998). The current USEPA Method 1623 for detection of *Cryptosporidium* and *Giardia* in water has approved various filter systems for (oo)cyst recovery, typically from 10 to 50 L surface water samples (USEPA, 2003). For example, the Envirochek™ High Volume (HV) Sampling Capsule (Pall Gelman Laboratory, Ann Arbor, MI), a self-contained 1 μm absolute porosity pleated membrane capsule filter, has a reported recovery rate of $>70\%$ (Matheson et al., 1998). The other approved filter system is based on compression of foam disks to achieve the appropriate porosity to collect (oo)cysts on the surface of the filter. For example, the Filta-Max® (IDEXX Laboratories, Inc., Westbrook, Maine) recovery efficiency in tap water was $48.4\% \pm 11.8\%$, and in raw source waters, the range was 19.5 to 54.5% (McCuin and Clancy, 2003).

Once the water sample has been filtered and the water particulates, including (oo)cysts, have been captured on the surface of the filter, a surfactant is applied to release the particulates back into solution as individual particles for downstream purification protocols. Different surfactants have been used, including sodium dodecyl sulfate (SDS), Laureth 12, Tween 20, and Tween 80. Generally, these surfactants are mixed solely or in combinations with a buffering solution such as EDTA,

PBS, and/or Tris. For use while sampling finished drinking waters, during which scaling (residue accumulation) can occur on the surface of a membrane filter and potentially trap target organisms (e.g., *C. parvum* oocysts), the addition of sodium hexametaphosphate to the elution solution has been proposed to breakdown/dissolve scaling to release any trapped organisms (Fireman and Reitemeier, 1944; Kasper, 1993). Once the (oo)cysts have been resuspended, a concentration step is performed solely or in conjunction with a separation step to further remove sample debris as well as concentrate the (oo)cysts. Two techniques that are performed without a separation step are calcium carbonate flocculation and continuous flow centrifugation. The recovery efficiencies for calcium carbonate flocculation and continuous flow centrifugation have been reported as > 70% (Shepherd and Wyn-Jones, 1996) and >90% (Borchardt and Spencer, 2002), respectively. However time constraints (>4 and >2 h required, respectively) restrict the use of these techniques for day-to-day applications. The selective centrifugation techniques developed for fecal material combine concentration and separation in one step and were directly transferred to use with environmental samples. However, these selective centrifugation techniques have been shown to have moderate recovery efficiencies and while appropriate for use with fecal samples in which (oo)cyst concentrations are generally high, the low (oo)cyst concentrations in environmental samples prove to be an issue. The most common techniques used for (oo)cyst separation from environmental samples include Percoll/sucrose flotation (Arrowood and Sterling, 1987; LeChevallier *et al.*, 1995), Percoll/Percoll step gradient (Nieminski *et al.*, 1995), flow cytometry (Vesey *et al.*, 1994a), and immunomagnetic separation (IMS) (Johnson *et al.*, 1995). Of these, IMS has gained the most notoriety and is currently the separation method of choice for the USEPA and United Kingdom Drinking Water Inspectorate (UKDWI).

9.4 DETECTION METHODOLOGIES

9.4.1 Microscopic Techniques

9.4.1.1 Strengths and Weaknesses
Microscopic identification has several positive attributes that preserve its status as the "Gold" standard even in light of flourishing molecular techniques. Under the qualified eye of a technician, the physical observance of a life cycle stage is convincing proof that the patient or environmental sample has that particular parasite present. To assist in detection and identification, several methods can be used including various stains, fluorescence, and/or Nomarski differential interference contrast microscopy (DIC). These different methods highlight and allow the microscopist to visualize defining characteristics such as internal structures (i.e., hooks, median bodies, nuclei, etc.) or abnormal characteristics such as algal chloroplasts. However, two qualities make microscopy a less than optimal detection technique. First, as stated above, demands on the technician are not limited to knowing all the different parasites and the respective different life cycle stages but also all the potential cross-reactors that might be present in any given sample. Months, if not

years, are required for a technician to become proficient in properly identifying all the different parasites. Second, microscopy lacks the ability to speciate, even with the best-qualified technicians. As in the case of *Taenia* spp., the eggs of *T. saginata* and *T. soilium* are identical via microscopy, however, the corresponding diseases are drastically different and type of treatment depends on proper identification of the species.

9.4.1.2 Dye Stains

A multitude of dye stains for brightfield, negative, and fluorochrome microscopy have been used to detect various parasite ova and (oo)cysts in stool specimens (Fig. 9.1). These techniques have also been used for detection in environmental samples with limited success. Examples for bright field staining for *C. parvum* detection include safranin methylene blue stain (Baxby *et al.*, 1984), Ziehl-Neelsen modified acid-fast (Henricksen and Pohlenz, 1981), and DMSO-carbolfuchsin (Pohjola *et al.*, 1985). Parasite staining with these techniques is not 100% uniform and the discernment from fecal debris is difficult and require's experienced microscopists. The advantage of using bright field stains, for example the employment of the Ziehl-Neelsen modified acid-fast to detect *C. parvum* oocysts, is that other organisms such as *Cyclospora cayetanensis* and *Isospora* spp. are also stained and thus have a greater potential of being detected. Other techniques include negative staining with nigrosin (Pohjola, 1984) and malachite green (Elliot *et al.*, 1999) where the background is stained.

As observed with the non-specific bright field stains, fluorescent dye stains such as acridine orange (Ma and Soave, 1983), propidium iodide (PI) and 4,6 diamidino-2′-phenylindole (DAPI) (Kawamoto *et al.*, 1987) have a higher propensity for false-positive results and have variable staining characteristics depending on the organism's viability status. However, the use of these dyes, especially DAPI, can greatly assist in identification when used concurrently with other microscopic techniques such as fluorescence and DIC (see below).

9.4.1.3 Fluorescent Labeling

The detection of *Cryptosporidium* spp. in stool specimens was greatly enhanced with the introduction of immunofluorescent assays in 1985 and 1986 (Fig. 9.1) (Casemore *et al.*, 1985; Garcia *et al.*, 1987; McLauchlin *et al.*, 1987; Sterling and Arrowood, 1986; Stibbs and Ongerth, 1986). Both sensitivity and specificity were shown to outperform conventional staining techniques (mentioned above) (Arrowood and Sterling, 1989; Garcia *et al.*, 1992; Grigoriew *et al.*, 1994). Immunofluorescent assays for the detection of *Giardia* spp. were introduced at approximately the same time and again were shown to have high sensitivity and specificity (Alles *et al.*, 1995; Grigoriew *et al.*, 1994; Nieminski *et al.*, 1995; Riggs *et al.*, 1983; Rose *et al.*, 1989; Sauch, 1985; Winiecka-Krusnell and Linder, 1995). In addition, one study found that *G. muris* and *G. lamblia* cysts subjected to freezing and thawing remained detectable via immunofluorescent microscopy, whereas the cyst walls often became distorted and therefore were not detected by bright field microscopy (Erlandsen *et al.*, 1990).

9.4.1.4 USEPA Approved Microscopic Methods

Among the diverse examples of concentration, separation, and detection methodologies, one combination has been adopted by the USEPA for the detection of *Cryptosporidium* spp. and *Giardia* spp. in surface water (USEPA, 2003). Method 1623 combines filtration (membrane or foam) followed by elution, concentration by centrifugation, and oocyst separation via IMS. Detection and identification uses the triple combination of immunofluorescence, DAPI staining, and DIC microscopy. Reported recovery efficiencies range from 36 to 75% (DiGiorgio *et al.*, 2002; LeChevallier *et al.*, 2003; McCuin and Clancy, 2003). (Oo)cysts in environmental samples are identified based on three general (oo)cyst morphological characteristics. First, (oo)cysts are detected at 200X magnification via immunofluorescence labeling of the (oo)cyst wall. Fluorescence intensity, (oo)cyst wall thickness, and characteristic folds in the wall are observed and assessed. Second, 4,6 diamidino-2'-phenylindole (DAPI) staining assists in identification, location, and number of nuclei, if present. Finally, DIC microscopy facilitates (oo)cyst internal morphology including identification of sporozoites and nuclei for *Cryptosporidium*, and nuclei, median bodies and axonemes for *Giardia*. The triple combination of fluorescence, DAPI staining and internal structure assessment assists the microscopist to decide if the object is an (oo)cyst or not. Again, it is important to note that this methodology is not 100% specific and cross-reacting algae and other debris can interfere, thus making the identification process very subjective. Table 9.2 lists a summary of advantages and disadvantages of the various detection methods.

9.4.2 Nucleic Acid Techniques

9.4.2.1 Strengths and Weaknesses

The greatest strengths of nucleic acid techniques over microscopic techniques are specificity and the ability to determine the species and potentially subspecies/genotype of the parasite. This feature is notably important when considering the different species of *Cryptosporidium*, *Entamoeba*, *Giardia*, and *Taenia*. With each of these parasites, the transmissible form is microscopically indistinguishable between the different species within its genera. As stated previously, correct identification of species has implications for pathogenesis, physician treatment, epidemiological studies, and source tracking, and therefore this advantage of molecular techniques is highly desirable.

Another strength of many molecular-based protocols is that many have improved over the years to be less labor intensive and at the same time shown to be as sensitive, if not more, as microscopic protocols. Examples of specific recovery efficiencies and lower limits of detection (LLD) are reviewed later in this chapter. In addition, the subjective nature of microscopy is eliminated with molecular based protocols. Not to be confused with specificity (discussed in the following paragraph), molecular based detection techniques give a yes or no answer. With microscopy, detection is reliant on the technician's experience as well as the ability to stay focused while analyzing the sample.

However, while certain molecular techniques are showing promise, each is not without its limitations. Limitations of most include concerns over specificity, risk

Table 9.2. Advantages and disadvantages of detection methods for parasites.

Method	Target	Advantage(s)	Disadvantage(s)	Required time	Speciation	Viability	Setup cost ($)[a]	Cost per sample ($)
Microscopic detection	Whole organism	Specific; Quantitative	Microscopist dependent; Subjective	1–4 days	No	No	M	35–100
IFA Microscopy	Whole organism	Specific; Quantitative	Microscopist dependent; Subjective	1–4 days	No	No	M	35–300
Culture and cell culture	Whole organism	Viability assessment; Semi-quantitative	Subject to inhibitors; Many organisms are not cultivable	1–7 days	No	Yes	H	25–50
PCR	DNA	Simple; Specific	Single target/organism reactions; Subject to inhibitors; Non-quantitative	1–2 days	Yes, via DNA seq. and/or RFLP	No	L	<10
Nested PCR	DNA	Increased sensitivity; Specific	Higher chance for contamination; Time consuming; Single target/organism reactions; Subject to inhibitors; Non-quantitative	2–3 days	Yes, via DNA seq. and/or RFLP	No	L	<20
Multiplex PCR	DNA	Multiple targets/organisms	Difficult to optimize; Subject to inhibitors; Non-quantitative	1–2 days	Possible but difficult	No	L	<20

Method	Target	Advantages	Disadvantages	Time	Multiplex	Quantification	Cost	Setup (US$ ×1000)[a]
Real-Time PCR	DNA/RNA	Multiple targets/organisms; Quantification assessment	Difficulty working with RNA; Difficult to optimize; Subject to inhibitors	<1 day	Yes, via multiple probes	No	M	10–25
RT-PCR[b]	RNA	Viability assessment	Single target/organism reactions; Difficulty working with RNA; Difficult to optimize; Subject to inhibitors	1–2 days	Yes, via DNA seq. and/or RFLP	Yes	M	10–25
CC-PCR	DNA	Viability assessment	Single target/organism reactions; Difficulty working with environmental samples; Subject to inhibitors	3–7 days	Yes, via DNA seq. and/or RFLP	Yes	H	25–50
DNA Array	DNA/RNA	Multiple targets/organisms per reaction	Difficulty working with environmental samples; Subject to inhibitors	2–3 days	Yes	Yes	H	25–50

[a] Setup cost for laboratory equipment (US$): (L) Low, <10,000; (M) Moderate, 10,000–35,000; (H) High, >35,000
[b] RT-PCR, Reverse Transcription PCR

of laboratory contamination, and sensitivity. First, specificity is of concern because of the "yes or no" answer and the response for any test is based on that simple answer. Therefore, techniques that have not been properly evaluated for specificity are of great concern and lead to confounded results. The general trend for specificity testing is to challenge the new technique with organisms that are readily available. These would include a suite of fecal bacteria including *E. coli* and other organisms that are closely related to the target organisms. This list should also be extended to distantly related organisms that might also be present in any given fecal or environmental sample. It is also recommended that the developed method be amenable to confirmatory tests, such as oligonucleotide probe hybridization and sequence analysis of PCR amplicons. Second, contamination and false-positive results are of concern with any detection method but with molecular techniques, especially those that exponentially amplify a specific nucleic acid target, there is an increased risk due to potential laboratory contamination with amplified product. Proper quality assurance (QA) and quality control (QC) measures must be established and practiced with vigilance in order to reduce the risk of contamination. Third, while nucleic acid based detection methods continue to gain popularity in the research and commercial laboratory settings, methods for liberation and recovery of high quality nucleic acid are still in need of improvement.

9.4.2.2 Nucleic Acid Recovery Techniques
As observed with the microscopic based protocols, a limitation in parasite detection, notably within environmental samples, is the low concentration of ova and (oo)cysts. Therefore, molecular detection techniques are also limited by current concentration and isolation protocols that target the environmentally stable form of the parasite. The same concentration and isolation techniques outlined earlier are also used as precursor steps followed by a molecular detection technique. Following Method 1623 as an example, the combination of filtration for concentration and IMS for isolation has become the norm for detecting *Cryptosporidium* spp. and *Giardia* spp. in water samples with a molecular detection technique. For example, in combination with capsule filtration followed by IMS, nested PCR was able to detect eight *C. parvum* oocysts seeded into treated waters (Monis and Saint, 2001); real-time PCR had a detection limit of five *C. parvum* oocysts seeded into purified water and eight oocysts seeded into raw water sample (Fontaine and Guillot, 2003a); and reverse transcription-PCR was able to detect ten oocysts seeded into tap water (Hallier-Soulier and Guillot, 2003). In 2003, an IMS-PCR system was introduced to detect *Enterocytozoon bieneusi* (Sorel *et al.*, 2003). Development of a filtration/IMS recovery system for *T. gondii* has also been outlined (Dumetre and Darde, 2003).

The second challenge is liberating, capturing, and purifying the nucleic acid from the organisms and the matrix in which the parasite is suspended. The ova and (oo)cysts serve as perfect package delivery systems that contain and protect the nucleic acid from the surrounding matrix. However, the problem arises when trying to break open these environmentally resistant stages and gain access to their nucleic acids for detection. A spectrum of different approaches has been used, from simple boiling to complex automated systems. The most common nucleic acid liberating schemes include: (1) simple boiling; (2) freeze/thawing (Johnson *et al.*, 1995; Sluter *et al.*, 1997) in the presence of an inhibitor removing substance in conjunction with an inhibitor remover such as InstaGene™Chelex resin (BioRad) (Higgins *et al.*, 2001b; Sturbaum *et al.*, 1998), Maximator® (Connex), and GeneReleaser® (Bioventures) (Kramer *et al.*, 2002); (3) exposure to guanidinium thiocyanate in the presence of silica (Boom *et al.*, 1990); (4) exposure to digestive enzymes such as proteinase-K (Abbaszadegan *et al.*, 1991; Gross *et al.*, 1992); microwave heating (Goodwin and Lee, 1993); a variety of commercially available spin columns are available such as the QIAamp® DNA Mini Kit (QIAGEN Inc., Valencia, CA); (6) novel approaches such as the FTA filter paper (McOrist *et al.*, 2002; Orlandi and Lampel, 2000; Subrungruang *et al.*, 2004); and (7) automated silica-based membrane capture systems such as the MagNA Pure LC System (Roche Diagnostics Corp.) (Knepp *et al.*, 2003; Wolk *et al.*, 2002).

Yet, even though different combinations of these listed techniques are promising, limitations remain due to the fact that molecular based techniques are subject to inhibition by substances that are introduced during the preparation process, or naturally occur and are coextracted with the nucleic acid (Johnson *et al.*, 1995; Sluter *et al.*, 1997). Naturally occurring substances such as humic acids (Ijzerman *et al.*, 1997; Kreader, 1996), chemicals such as SDS commonly used in elution and lysis buffers, and residual alcohol used during the nucleic acid purification have been shown to inhibit and reduce the activity of the enzymes used in various detection methods. One approach to guard against false-negative results due to inhibitory substances is the introduction of an internal positive control (IPC) (Kaucner and Stinear, 1998; Stinear *et al.*, 1996).

Currently, a consensus method has not been agreed upon for the best combination of parasite recovery and nucleic acid liberation/purification method for any one given parasite, let alone a consensus method that potentially can be used for multiple parasites. The latter is a lofty goal that will be difficult to achieve due to the multiple characteristics of each parasite and the different matrixes to which the detection protocol will need to be applied.

9.4.2.3 DNA Based Protocols

9.4.2.3.1 Polymerase Chain Reaction (PCR)

PCR is the selective exponential amplification of a nucleic acid locus and countless numbers of studies employing PCR as a detection tool for parasites in clinical and environmental matrixes have been published since the technique's introduction in 1985 (Mullis *et al.*, 1986; Saiki *et al.*, 1985). The nucleic acid locus is targeted by oligonucleotide primers that flank the region to be amplified. Amplification is achieved via a heat stable DNA polymerase that synthesizes a new strand of DNA determined by the primer locations during thermal cycling, typically between three temperatures for denaturation, primer annealing, and elongation. The choice of the targeted locus is of critical concern and issues such as single versus multicopy targets for sensitivity, level of genetic relatedness for specificity, and amount of observed genetic variability need to be considered.

PCR-based protocols for *Cryptosporidium* spp. include targeting the 18S rRNA (Johnson *et al.*, 1995; Leng *et al.*, 1996; Lowery *et al.*, 2000; Sturbaum *et al.*, 2002; Xiao *et al.*, 1999), *hsp*70 gene (Gobet and Toze, 2001; Kaucner and Stinear, 1998;

Monis and Saint, 2001), beta tubulin (Perz and Le Blancq, 2001), an unknown locus (Laxer *et al.*, 1991) as well as multiple other protocols (Awad-el-Kariem *et al.*, 1994; Gibbons and Awad-El-Kariem, 1999; Gile *et al.*, 2002; Hallier-Soulier and Guillot, 2000; Morgan *et al.*, 1998; Scorza *et al.*, 2003; Xiao *et al.*, 2001). Various studies have determined that the lower limit of detection (LLOD) of *C. parvum* with PCR in different matrixes is as low as a single oocyst (Table 9.3). It is necessary to keep in mind that the sensitivity of these different protocols is only as good as the method that is used to determine the actual number of oocysts seeded into the matrix. This is discussed in the final section of this chapter. A choice of *G. lamblia* PCR detection protocols is also noted in the literature. Examples include PCR protocols targeting the rRNA genes (Hopkins *et al.*, 1999; Hopkins *et al.*, 1997; van Keulen *et al.*, 1995), two genes encoding cysteine-rich trophozoite surface proteins (Ey *et al.*, 1993), and the triose phosphate isomerase gene (*tim*) (Lu *et al.*, 1998). The LLOD with PCR detection is also as low as a single cyst in purified water and five cysts in environmental water (Caccio, 2003; Rochelle *et al.*, 1997a). While many PCR protocols are available to detect *T. gondii* in clinical specimens (Dupon *et al.*, 1995; Dupouy-Camet *et al.*, 1993; Lamoril *et al.*, 1996; Savva *et al.*, 1990; Schoondermark-van de Ven *et al.*, 1993; Weiss *et al.*, 1991), only recently has interest been shown for *T. gondii* detection in environmental samples (Dumetre and Darde, 2003; Ellis, 1998; Kourenti and Karanis, 2004). A few PCR protocols have been described detecting *Encephalitozoon intestinalis* and *Enterocytozoon bieneusi* spores in formalinized fecal samples and environmental water samples (Dowd *et al.*, 1998; Sorel *et al.*, 2003; Thurston-Enriquez *et al.*, 2002). One cestode PCR protocol allows for diagnostic differential detection of *Taenia saginata* and *Taenia solium* (Gonzalez *et al.*, 2000).

9.4.2.3.2 Nested PCR
A second round PCR amplification of a DNA segment internal to the priming sites of a first round PCR product (amplicon) is referred to as "nested PCR." The nested PCR amplicon is generated by performing a second round of PCR using the first round PCR amplicons as the template DNA and primers designed to "nest" within the DNA template extremities. The use of one of the first round PCR primers and one new second round primer is referred to as "semi-nested PCR." The major strength of nested PCR is increased sensitivity, while the major weakness is the greatly increased risk of contamination and false-positives. Several nested PCR protocols have been published for use with environmental samples. For example, in combination with IMS, nested PCR targeting the 18S rRNA gene was able to detect eight *C. parvum* oocysts seeded into treated waters (Monis and Saint, 2001). A sensitivity study carried out with micromanipulated *C. parvum* oocysts determined that a single oocyst was detected using nested PCR 38% of the time (19 out of 50 replicates) and 10 micromanipulated oocysts were detected in all 50 replicates (Sturbaum *et al.*, 2001). The same nested PCR protocol, in conjunction with IMS, was then evaluated at the five, ten, and fifteen oocysts levels seeded into two different environmental source waters (Sturbaum *et al.*, 2002). The seeded oocysts were detected in both source waters at all three seeding levels (Sturbaum *et al.*, 2002). However, false positive results were noted as naturally occurring *C. muris*, a nontarget species, was

Table 9.3. *Cryptosporidium parvum* detection methodologies and associated lower limits of detection (LLOD).[a]

Detection	Oocyst concentrations	Determined by	LLOD[a]	Reference
Oocysts/PCR/18S rRNA[b]	900, 90, 9, 1	Serial Dilution	90 oocysts	(Johnson et al., 1995)
Oocysts/RT-PCR/*hsp70*[c]	1000, 100,50,25,12,1	Serial Dilution	1 oocyst	(Stinear et al., 1996)
Oocysts/PCR/18S rRNA	5000, 1000,500, 50, 5	Serial Dilution	50 oocysts	(Rochelle et al., 1997a)
PCR/Unknown region	5000, 1000,500, 50, 5	Serial Dilution	50 oocysts	(Rochelle et al., 1997a)
PCR/Unknown region	5000, 1000,500, 50, 5	Serial Dilution	5 oocysts	(Rochelle et al., 1997a)
Oocysts/IMS[d]/IFA[e]	100	Flow Cytometry	100 oocysts	(Reynolds et al., 1999)
PCR/Unknown region	100, 50, 10, 1	Serial Dilution	1 oocyst[f]	(Wu et al., 2000)
Oocysts/membrane dissolution/IFA	4×10^4, 100	Serial Dilution	<100 oocysts	(McCuin et al., 2000)
Oocysts/PCR-18s rRNA	10, 7, 5, 4, 3, 2, 1	Micromanipulation	1 oocyst	(Sturbaum et al., 2001)
Oocysts/IMS/PCR-18s rRNA	1000, 500, 250, 100, 10, 1	Serial Dilution[1]	10 oocysts	(Lowery et al., 2000)
Oocysts/IMS/IFA	10, 5	Serial Dilution	5 oocysts	(McCuin et al., 2001)
Oocysts/IMS/RT-PCR/*hsp70*	10000, 1000, 100, 8	Serial Dilution	8 oocysts	(Monis and Saint, 2001)
Oocysts/IMS/PCR-18s rRNA	15, 10, 5	Flow Cytometry	5 oocysts	(Sturbaum et al., 2002)
Hollow fiber ultrafilter/IFA	10000, 1000, 600	Serial Dilution	ND[g]	(Kuhn and Oshima, 2001)
Oocysts/Flocculation/IFA	1×10^f, 1000, 10, 5, 1	Serial Dilution	10 oocysts	(Karanis and Kimura, 2002)
Oocysts/IMS/Real-Time PCR	10000, 1000, 100, 10, 5	Serial Dilution	10 oocysts	(Hallier-Soulier and Guillot, 2003)
DNA/Real-Time PCR	1000, 100, 10, 1	Serial Dilution	1 oocyst[f]	(Guy et al., 2003)
Oocysts/Real-Time PCR	775, 75	Serial Dilution	75 oocysts	(Fontaine and Guillot, 2003a)

[a] lower limits of detection (LLOD)
[b] 18S rRNA, Small Subunit ribosomal RNA
[c] *hsp70*, 70 KDa heat shock protein
[d] IMS, Immunomagnetic Separation
[e] IFA, Immunofluorescent microscopy
[f] as determined by serial dilution of *C. parvum* previously isolated DNA
[g] ND, not determined

detected in one source water, and DNA from an algal species, *Gymnodinium fuscum*, was amplified from the other source water (Sturbaum *et al.*, 2002). These results only emphasize the importance of primer design and specificity testing.

Multiple other protocols using nested or hemi-nested PCR have been outlined for detection of *G. lamblia*, *C. parvum*, and *T. gondii* in both clinical and environmental samples (Amar *et al.*, 2002; Balatbat *et al.*, 1996; Deng *et al.*, 1997; Fischer *et al.*, 1998; Gibbons and Awad-El-Kariem, 1999; Jones *et al.*, 2000; Kato *et al.*, 2003; Kostrzynska *et al.*, 1999; Mayer and Palmer, 1996; Nichols *et al.*, 2003; Ostergaard *et al.*, 1993; Ward *et al.*, 2002; Zhu *et al.*, 1998).

9.4.2.3.3 Real-time PCR
When compared with PCR protocols that require 6–8 h to conduct and nested PCR protocols that require 8–12 h to complete, real-time PCR is superior in that results can be produced in less than 2 h. This time frame is achieved with real-time PCR via: (1) the rapid heating and cooling processes unique to each platform; (2) the relatively small amplicon size (<130 base pairs); and (3) measurable signal (or lack thereof) is generated via fluorescence during the actual temperature cycling. For an excellent review of the different real-time PCR platforms and fluorescent probe variations see Wolk, 2001 (Wolk *et al.*, 2001). In addition, multiple probes, each with its own fluorescent signal, can be used in the same reaction vial, which allows for the detection of multiple organisms or genotypes at the same time. Finally, due to the nature of signal amplification and measurement during temperature cycling, real-time PCR is being used for estimating starting DNA template concentration, which can then be used to estimate the number of organisms present in any given sample.

TaqMan probe chemistry is commonly used for real-time quantitative PCR. TaqMan quantitative PCR requires the use of primers similar to those used in conventional PCR; however, unlike conventional PCR, TaqMan quantitative PCR also requires an oligonucleotide probe labeled with 5′ reporter and 3′ quencher fluorescent dyes and a thermal cycler equipped with a fluorometer (Heid *et al.*, 1996). During each cycle of the PCR, if the target of interest is present, the probe specifically anneals to the target amplicon between the forward and reverse primer sites. Due to 5′ nuclease activity of *Taq* DNA polymerase, the probe is cleaved during the polymerization step (PCR product formation) of the PCR resulting in an increase in reporter fluorescence detected by the instrument. Target signal increases in direct proportion to the concentration of the PCR product being formed. The threshold cycle (C_T) is the fractional PCR cycle number at which a significant increase in target signal fluorescence above baseline is first detected for a sample. By using amplification standards consisting of known quantities of target nucleic acid or organisms to generate a standard curve, the starting copy number of nucleic acid targets or target organisms for each sample can be estimated.

Real-time PCR protocols developed for *Cryptosporidium* spp. and *G. lamblia* are demonstrating sensitivity and are able to distinguish different species and genotypes (Amar *et al.*, 2003, 2004; Fontaine and Guillot, 2003a, b; Guy *et al.*, 2003; Higgins *et al.*, 2001a; Limor *et al.*, 2002; MacDonald *et al.*, 2002; Tanriverdi *et al.*, 2003; Tanriverdi *et al.*, 2002). Protocols for *T. gondii* (Buchbinder *et al.*, 2003; Jauregui *et al.*, 2001; Reischl *et al.*, 2003) *Encephalitozoon intestinalis* (Wolk *et al.*,

2002) and *Entamoeba histolytica* and *Entamoeba dispar* (Blessmann *et al.*, 2002) detection have also been described. One study outlines the simultaneous detection of *Entamoeba histolytica*, *G. lamblia*, and *C. parvum* in stool samples at the same time (Verweij *et al.*, 2004).

9.4.2.3.4 Multiplex PCR

The incorporation of multiple oligonucleotide primer sets into the same PCR reaction tube permits amplification of multiple loci at the same time. As with real-time PCR, different organisms, species, and/or genotypes can be detected within a single sample. Limitations to be considered are the nucleotide ratios (guanine, cytosine, adenine, and thymine; GC:AT) of the different parasite genomes and the biased amplification nature of PCR. First, the GC:AT ratio, which is unique to each organism, has a direct affect on annealing temperatures of PCR primers. Second, even if the multiple targets are present, there is no guarantee that PCR will amplify all the targets consistently, equally or at all (Reed *et al.*, 2002), especially when there are large differences in concentrations of different targets.

Considering these limitations, multiplex PCR has been applied with success for detecting *C. parvum*, *Eimeria spp.*, *E. histolytica*, *E. dispar*, *G. lamblia*, *T. gondii*, and *Taenia* spp. in clinical as well as environmental samples (Abe *et al.*, 2002; Evangelopoulos *et al.*, 2000; Fernandez *et al.*, 2003; Gomez-Couso *et al.*, 2004; Guay *et al.*, 1993; Lindergard *et al.*, 2003; Patel *et al.*, 1999; Rochelle *et al.*, 1997a; Yamasaki *et al.*, 2004).

9.4.2.3.5 DNA Arrays

The unique feature that DNA microarray technology provides is that literally thousand of probes representing entire genomes or a mixture of specific target genes from a variety of organisms can be immobilized on a chip. Therefore multiple target loci from multiple target organisms can be screened for their presence at the same time from the same sample. Two formats of DNA microarray technology are currently available. The first format, called DNA microarray, immobilizes cDNA probes (generally 500 to 5000 nucleotide base pairs in length) that are placed or "spotted" directly onto a solid surface, generally glass and sometimes nylon substrates, via robotics (Ekins and Chu, 1999). The second format, DNA chips (Affymetrix, Inc.) differs in that probes of 20 to 80 base pairs in length are synthesized *in vitro* directly onto the chip or spotted onto the chip via robotics. With both formats, the chip is then probed with previously labeled target RNA, cDNA, or DNA either separately or in a mixture and signal from hybridization between the immobilized probe and the target is measured and recorded.

While, this technology is promising, however, it is rather sophisticated to perform and can be costly with designed chips costing as much as $500 to $5000. As with other detection methodologies, DNA array detection is only as good as the upstream sample processing protocol for the concentration and purification of the target organism from the matrix, whether clinical or environmental.

DNA array technology has been adapted to detect and discriminate between different *Cryptosporidium* species and genotypes (Straub *et al.*, 2002) as well as detecting genotypes of *E. histolytica*, *E. dispar*, *G. lamblia*, and *C. parvum* on the same chip (Wang *et al.*, 2004).

9.4.2.4 RNA

9.4.2.4.1 Reverse Transcription PCR (RT-PCR)
Employing messenger RNA (mRNA) as the target template, reverse transcription PCR (RT-PCR) has the advantages of lower inhibitory substance contamination issues as well as the ability of addressing viability questions. Reverse transcriptase is the enzyme that is able to make a copy of complementary DNA (cDNA) from mRNA. This newly synthesized strand of cDNA is then used in a PCR protocol as template. Due to continual degradation of mRNA, this molecule is suitable for viability assessment, as only viable organisms are able to synthesize mRNA.

This technique has been used solely or in combination with water filtration and IMS recovery for the detection of *C. parvum* and other *Cryptosporidium* species in clinical and environmental samples (Hallier-Soulier and Guillot, 2003; Rochelle *et al.*, 1999; Rochelle *et al.*, 1997b).

9.4.2.4.2 NASBA
A second RNA based technique is nucleic acid sequence-based amplification (NASBA) (Cook, 2003; Romano *et al.*, 1997). NASBA is an isothermal amplification method, which, like reverse transcriptase PCR, uses RNA as the target molecule (DNA can also be used with modifications to the protocol) (Wolk *et al.*, 2001). RNA is reverse transcribed into cDNA that is then used as a template to produce more RNA transcripts via RNA polymerase. NASBA, in conjunction with other novel detection protocols, is currently being used to detect *C. parvum* in water samples (Baeumner *et al.*, 2001; Esch *et al.*, 2001a; Esch *et al.*, 2001b) and is predicted to gain importance for detection of other food-borne and waterborne parasites (Caccio, 2003; Cook, 2003).

9.4.2.5 Innovative Techniques
Innovative techniques such as fiber optics and biosensors (Snowden and Anslyn, 1999; Wang, 2000), transcription mediated amplification (TMA) (Walker *et al.*, 1992), Invader® technology (Third Wave Technologies, Inc.) (Kwiatkowski *et al.*, 1999), Q-beta replicase (Gene-Trak Systems) (Cahill *et al.*, 1991; Pritchard and Stefano, 1990), the Ligase Chain Reaction (Barany, 1991), and Laboratory Multi-Analyte Profiling with suspension arrays (Luminex, Inc.) are demonstrating insightful thought and promise as new detection methodologies. A review for several of these different approaches is given by Wolk, 2001 (Wolk *et al.*, 2001).

9.4.3 Immunological-based Techniques

9.4.3.1 Strengths and Weaknesses
Non-microscopic based detection methods include immunological and molecular based techniques. Examples of immunological-based techniques include enzyme-linked immunosorbent assays (ELISA) (Anusz *et al.*, 1990; Chapman *et al.*, 1990; Knowles and Gorham, 1993; Ungar, 1990), reverse passive haemagglutination (Farrington *et al.*, 1994), and solid-phase qualitative immunochromatographic assays (Garcia and Shimizu, 2000). Used primarily for clinical diagnosis, numerous antigen detection assays have been developed for parasites in stools including

C. parvum (Garcia and Shimizu, 1997; Newman *et al.*, 1993; Rosenblatt and Sloan, 1993), *G. lamblia* (Boone *et al.*, 1999; Ungar *et al.*, 1984; Vinayak *et al.*, 1991), and *E. histolytica/E. dispar* (Gonzalez-Ruiz *et al.*, 1994; Haque *et al.*, 1993; Jelinek *et al.*, 1996; Schunk *et al.*, 2001). A single enzyme immunoassay (EIA), the Triage Parasite Panel (Biosite Diagnostics, Inc., San Diego, CA), has been developed to simultaneously detect *C. parvum*, *G. lamblia*, and *E. histolytica/E. dispar* (Garcia *et al.*, 2000; Sharp *et al.*, 2001). The test immobilizes organism specific proteins that are present in a stool specimen onto a membrane via antibodies. Incubation with an antibody-enzyme conjugate followed by the addition of the substrate results in a presence/absence marker that is detected visually. These various immunological-based techniques have proven effective with reported sensitivity and specificity results at >90% when compared to O & P examination (Garcia and Shimizu, 1997, 2000; Garcia *et al.*, 2000; Jelinek *et al.*, 1996), however specific reports do cite false-positive and false-negative results (Doing *et al.*, 1999; Sharp *et al.*, 2001). Taken together, questionable results from an immunological-based technique should be confirmed via a second assay.

9.4.3.2 Speciation and Genotyping Techniques

9.4.3.2.1 Strengths and Weaknesses
While some of the above listed molecular techniques have speciation and genotyping capabilities inherent in the methodology (such as real-time PCR and the use of multiple probes in the same reaction tube), the majority of the techniques require an additional step for further molecular characterization. The goal of these various techniques is to detect variations in the genetic code as low as single nucleotide polymorphisms (SNPs). However, it should be noted that detection of a SNP does not necessarily depict a new genotype or species. Classification schemes for new parasite genotypes and species need to be based on morphology, host specificity, and multiple nucleic acid differences identified within coding regions of the genome.

9.4.3.2.2 Techniques
Examples of speciation and genotyping techniques include restriction fragment length polymorphism (RFLP), random amplified DNA polymorphism (RAPD), single strand confirmation polymorphism (SSCP), direct DNA sequencing, amplified fragment length polymorphism (AFLP), denaturing gradient gel electrophoresis (DGG), real-time PCR, and microsatellite analysis, as well as many others. While each of these techniques is able to successfully detect variation and SNPs, two are used with high fidelity, RFLP, and direct DNA sequencing. This claim is based on the number of variables that can be introduced for any of these given techniques. For RFLP and direct DNA sequencing, with proper QA and QC checks such as the use of proof-reading Taq polymerase, only one variable is introduced which is the PCR amplicon itself. If one suspects that multiple alleles are present, the PCR amplicon can be subject to cloning followed by direct DNA sequencing. With the financial cost being <$30 and the potential for valuable information to be gained, direct DNA sequencing is considered the gold standard when detecting variation in nucleotide sequences.

It should be noted that techniques that analyze the entire genome, such as RAPD or AFLP analysis, are designed for fingerprinting pure cultures and not fecal or environmental samples. Current separation and purification schemes do not isolate the target organism with 100% fidelity and extraneous microbes (e.g., bacteria) will compromise the analysis.

9.4.4 Viability Techniques

9.4.4.1 Strengths and Weaknesses

Several *in vitro* surrogate methods have been proposed as convenient, user-friendly alternatives to animal infectivity assays for determining the viability and/or infectivity for a number of parasites. Of the listed detection techniques, only one has the ability to assess viability and none are able to assess infectivity. To be clear, viability differs from infectivity. Viability requires that the organism be intact and capable of metabolic activity. Infectivity includes the definition of viability and extends this capacity to the ability to cause disease in animal models, or invade and multiply for *in vitro* cell culture models. The need for establishing viability and/or infectivity is two-fold. First, methodologies such as drug therapy for treatment and UV exposure or radiation for disinfection need to be accurately verified. To install confidence in a treatment or disinfection methodology, the viability and/or infectivity validation techniques need to be accurate and absolute. Second, organisms, such as *C. parvum* oocysts, detected during routine monitoring or in an outbreak situation should be evaluated for viability/infectivity to assess the public health risk. The ideal detection method would not only have high sensitivity and specificity, but would also be able to evaluate the organism's viability/infectivity status.

9.4.4.2 Dye Permeability Assays

Dye permeability assays (inclusion or exclusion) have been used for years to assess viability in a number of parasites including *C. parvum* (Arrowood *et al.*, 1991; Campbell *et al.*, 1992; McCuin *et al.*, 2000), *Entamoeba* cysts (Kawamoto *et al.*, 1987) and trophozoites (Cano-Mancera and Lopez-Revilla, 1988), *Giardia* spp. (deRegnier *et al.*, 1989; Labatiuk *et al.*, 1991; Sauch *et al.*, 1991; Schupp and Erlandsen, 1987a, b; Smith and Smith, 1989), *Schistosoma mansoni* schistosomula (Van der Linden and Deelder, 1984), and *Taenia saginata* (Owen, 1985). The most commonly used dyes include fluorescein diacetate (FDA), propidium iodide (PI), 4'-6-diamidino-2-phenylindole (DAPI), Hoechst 33258, acridine orange, 2-(*p*-iodophenyl)-3-(*p*-nitrophenyl)-5-phenyl tetrazolium chloride (INT), and trypan blue (Korich *et al.*, 1993). While the fluorogenic dyes, FDA, DAPI, and PI, were shown to accurately assess *Giardia* cyst (Sauch *et al.*, 1991; Schupp and Erlandsen, 1987a, b) and *C. parvum* oocyst (Campbell *et al.*, 1992; Jenkins *et al.*, 1997) viability when compared with *in vitro* excystation, DAPI and PI were shown to overestimate *C. parvum* oocyst viability when compared with the neonatal mouse infectivity model (Black *et al.*, 1996; Korich *et al.*, 1993). However, separate studies demonstrated that viable *C. parvum* oocysts, which excluded the fluorogenic dyes SYTO-9 and SYTO-59 (while non-viable oocysts were permeable and fluoresced), were infectious in neonatal CD-1 mice (Belosevic *et al.*, 1997; Neumann *et al.*, 2000b).

9.4.4.3 In vitro excystation

In vitro excystation for *Cryptosporidium* spp. (Fayer and Leek, 1984; Robertson *et al.*, 1993; Sundermann *et al.*, 1987; Woodmansee, 1987) and *Giardia* spp.(Bingham *et al.*, 1979; Bingham and Meyer, 1979; Isaac-Renton *et al.*, 1992; Rice and Schaefer, 1981; Schaefer *et al.*, 1984; Smith and Smith, 1989), and sporulation for *C. cayetanensis* (Ortega *et al.*, 1993; Siripanth *et al.*, 2002) and *T. gondii* (Lindsay *et al.*, 2002; Lindsay *et al.*, 2003), are principle methods to assess viability. However, while excystation has been shown to be an effective measure of viability for *Giardia* spp. (Bingham *et al.*, 1979; Hoff *et al.*, 1985), it is not a reliable measure for *C. parvum* oocyst viability (Black *et al.*, 1996), and for both organisms it should not be used as a surrogate for infectivity (Black *et al.*, 1996; Bukhari *et al.*, 2000; Labatiuk *et al.*, 1991; Neumann *et al.*, 2000a). Finally, environmental triggers that influence *C. cayetanensis* and *T. gondii* sporulation have yet to be characterized (Lindsay *et al.*, 2002; Ortega *et al.*, 1993; Siripanth *et al.*, 2002), and therefore, sporulation as a measurement of oocyst viability should remain questionable.

9.4.4.4 FISH

Utilizing fluorescently labeled oligonucleotide or peptide nucleic acid probes, fluorescence *in situ* hybridization (FISH) (Nath and Johnson, 1998) has been used in detection assays as a measure of viability for both *C. parvum* and *G. lamblia* (Deere *et al.*, 1998a; Deere *et al.*, 1998b; Dorsch and Veal, 2001; Jenkins *et al.*, 2003; Smith *et al.*, 2004; Vesey *et al.*, 1998). The ability of FISH to determine viability is based on the presence of the target nucleic acid, which in certain reports is rRNA (Smith *et al.*, 2004; Vesey *et al.*, 1998), and is present only in viable organisms. For example, after loss of viability, rRNA synthesis will cease and be subject to degradation and therefore the FISH probe will not have the rRNA target to hybridize with. In a separate study, *C. parvum* oocyst viability was assessed by comparing FISH to cell culture and mouse infection (Jenkins *et al.*, 2003). The results showed that oocyst viability decreased proportionately over a 9-month period according to all methods, however, cell culture and mouse infection had the best agreement (Jenkins *et al.*, 2003). A unique advantage of FISH is that it may be combined with immunofluorescent microscopy or flow cytometry.

9.4.4.5 Reverse Transcription PCR

A final molecular technique for viability assessment is reverse transcription PCR (RT-PCR). Viability can be measured with RT-PCR because the target, mRNA, is degraded and thus constantly requires de novo synthesis. This results in nonviable organisms being undetected. Confirmation that residual RNA has been degraded should be included with all detection and/or viability protocols. Several studies report the reproducible RT-PCR detection of *C. parvum* and *G. lamblia* in reagent grade as well as environmental waters (Fontaine and Guillot, 2003b; Hallier-Soulier and Guillot, 2003; Jenkins *et al.*, 2003; Jenkins *et al.*, 2000; Kaucner and Stinear, 1998; Stinear *et al.*, 1996; Widmer *et al.*, 1999). However, in a single report directly comparing mouse infectivity, cell culture and RT-PCR targeting *C. parvum* amyloglucosidase mRNA, the RT-PCR assay underestimated oocyst viability (Jenkins *et al.*, 2003). Therefore, while in theory, RT-PCR is able to assess viability, additional research is needed to support this claim.

9.4.4.6 Cell Culture and Animal Infectivity Models

Animal infectivity models remain the current gold standard to assess viability and infectivity and all surrogate methods should be compared to animal infectivity. However, animal infectivity models are expensive, time consuming, and require experienced personnel and facilities (Rochelle *et al.*, 2002). In addition, not all parasites have been adapted to an animal model. For example, multiple animal models have been tested without success for *C. cayetanensis* (Eberhard *et al.*, 1999, 2000) infection and only gnotobiotic piglets have been shown to be susceptible to *C. hominis* infection (Widmer *et al.*, 2000a). While many of the surrogate techniques are being evaluated with *C. parvum* and *G. lamblia*, animal infectivity remains the technique of choice when evaluating viability/infectivity of *T. gondii* oocysts (Jauregui *et al.*, 2001; Lindsay *et al.*, 2002; Lindsay *et al.*, 2003).

The best surrogate for animal testing is *in vitro* cell culture. Cell culture (often abbreviated as "CC" when included in method acronyms) is well established with multiple food-borne and waterborne parasites. Complete life cycle development for *Encephalitozoon intestinalis* and other microsporidia spp.(Huffman *et al.*, 2002; Li *et al.*, 2003; Visvesvara, 2002; Wolk *et al.*, 2000) as well as *T. gondii* (Hoff *et al.*, 1977; Kniel *et al.*, 2002; Lindsay *et al.*, 1991) is documented and specifically for *E intestinalis*, cell culture is used to generate large numbers of spores (Wolk *et al.*, 2000). For *Cryptosporidium* spp., several cell culture methods have been described and even though final development of oocysts, the end product of the complete *Cryptosporidium* life cycle, is not robust; cell culture in combination with PCR or other molecular techniques has proven to be very reliable and a good measure of viability (Di Giovanni *et al.*, 1999; Gennaccaro *et al.*, 2003; Hijjawi *et al.*, 2001; Hijjawi *et al.*, 2002; Joachim *et al.*, 2003; Keegan *et al.*, 2003; LeChevallier *et al.*, 2003; Rochelle *et al.*, 1996, 1997b; Slifko *et al.*, 1997; Upton *et al.*, 1994, 1995; Widmer *et al.*, 2000b). For a sample to be considered positive, all that is needed is initial life cycle stages to begin development, and these stages can then be targeted by molecular methods. *C. parvum* oocysts recovered by IMS have been detected and determined viable by both PCR (Di Giovanni *et al.*, 1999; Jenkins *et al.*, 2003; LeChevallier *et al.*, 2003; Rochelle *et al.*, 1996) and RT-PCR (Rochelle *et al.*, 1999; Rochelle *et al.*, 1997b; Rochelle *et al.*, 2002). Cell culture has been shown to be equivalent to the "gold standard" mouse infectivity for disinfection studies (Rochelle *et al.*, 2002). Recently, CC-PCR has been used to determine the risk posed by infectious *C. parvum* and *C. hominis* in finished drinking water (Aboytes *et al.*, 2004). Another *Cryptosporidium* cell culture technique relies upon the immunofluorescent assay detection of foci of infection, referred to as the focus detection method (FDM) (Slifko *et al.*, 1997). The FDM method has been used for disinfection trials (Slifko *et al.*, 2002; Slifko *et al.*, 2000) and the detection of naturally occurring infectious *Cryptosporidium* in wastewater (Gennaccaro *et al.*, 2003; Quintero-Betancourt *et al.*, 2003). Despite the significant progress made in the area of *Cryptosporidium* cell culture and detection methods, optimization of culture conditions, pretreatment of oocysts for cell culture and further evaluation of detection methods is needed (Di Giovanni and Aboytes, 2003) and are the focus of current research efforts.

9.5 PROPER EVALUATION (QA/QC)

In addition to Good Laboratory Practices (GLP) and QA and QC checks, certain concerns need to be addressed with any detection method. These include specificity issues and proper determination of the lower limit of detection (LLOD).

Specificity evaluation of new methods is critical, as methods will be challenged with multiple unknown organisms in clinical and especially environmental samples. As discussed above, the majority of the concentration and isolation techniques are nonspecific and even those that incorporate specificity measures, such as IMS, may allow nonspecific organisms to be carried over to subsequent detection methods. Moreover, detection techniques that circumvent isolation protocols altogether, such as direct PCR on fecal material, are subject to more potential cross-contaminating organisms.

Understanding that the evaluation of a detection method cannot include every possible organism, certain criteria should be considered when selecting the organisms that are used for specificity testing. These include physical shape (for microscopic based techniques), genetic relatedness (for microscopic and molecular based techniques) and potential co-occurrence or presence in the sample matrix. However, in many cases, detection methods are only evaluated with organisms that are easily accessible without taking these criteria into consideration.

For example, in consideration of the potential co-occurrence or presence in the sample matrix, those techniques that are used with fecal samples should be evaluated with a higher proportion of coliform bacteria; while those techniques used with source waters should include microbiota such as algae. If a technique is to be used with sewage effluent or run-off water, then both groups of organisms should be included in the specificity evaluation analysis.

For physical shape, two criteria should be considered. First, genetically unrelated organisms, which are similar in same size and shape of the target organism, should be evaluated. Examples of such organisms include the algal species *Oocystis* spp. for *Giardia* and *Chlorella* spp. for *C. parvum*. Second, genetically related species, especially those within the same genus, need to be evaluated such as the 14 now recognized *Cryptosporidium* spp. For example, in the case of a new *Cryptosporidium* microscopic technique, all 14 different species should be evaluated for cross-reaction and the ability to distinguish the oocysts.

For genetic relatedness, criteria that should be considered include multiple species within the targeted genus, related organisms outside the genus, and homologous genetic loci. Methods for those genera that have multiple confounding species (*Cryptosporidium* spp., *Giardia* spp., the microsporidia group, *Taenia* spp.) should be thoroughly evaluated, especially if they will be applied to environmental samples. For example, in one study, storm waters contained four different *Cryptosporidium* spp. and 12 different genotypes based on detection of rRNA gene and subsequent DNA sequencing (Xiao *et al.*, 2000). Further, distantly related organisms to the target organism have largely not been included in any specificity trials. Again, using *C. parvum* as an example, the genus *Cryptosporidium* is genetically placed in the classification group, the Alveolata. Based on rRNA, the Alveolata is a robust, monophyletic taxon consisting of three phyla: Apicomplexa (e.g., *Cryptosporidium, Toxoplasma*), Ciliophora (ciliates) and Dinozoa (dinoflagellates) (Cavalier-Smith, 1993). Using a nested PCR primer set targeting the 18S rRNA and previously tested for *Cryptosporidium* spp. specificity, DNA from a common fresh water dinoflagellate, *Gymnodinium* spp., was nonspecifically amplified and detected via this *C. parvum* detection technique (Sturbaum *et al.*, 2002). This example further exemplifies the final point of DNA target selection. When choosing a locus for molecular detection, consideration must be given to the conserved nature of the locus, as well as the copy number per genome. In the case of the cross-reacting dinoflagellate, the 18S rRNA gene is highly conserved, and therefore it was not surprising that a related, nontarget organism was amplified, especially using a sensitive nested PCR primer set with environmental samples. As a second example, the Apicomplexa members *Cryptosporidium, Eimeria,* and *Toxoplasma* all produce oocysts that are present in the environment ubiquitously and concentration techniques will nonspecifically collect all three genera. Of the possible molecular loci as targets for detecting *C. parvum* in the environment, the thrombospondin-related anonymous protein (TRAP) gene has emerged as a potential candidate (Spano *et al.*, 1998a). However, the TRAP genes are conserved not only within the *Cryptosporidium* genus but also between the different members of the Apicomplexa (Kappe *et al.*, 1999; Spano *et al.*, 1998a; Spano *et al.*, 1998b; Sulaiman *et al.*, 1998). Therefore, when using a TRAP gene as the target loci in an environmental detection protocol, the protocol needs to incorporate a subsequent confirmation step to verify the amplicon came from the target organism.

Proper determination of the LLOD is the second major aspect of method development that is generally not evaluated thoroughly. First, to determine the LLOD of any detection system, accurate numbers of organisms need to be used. Historically for many parasites, the combination of hemocytometer counting followed by serial dilution was used to generate low numbers of target organisms. This technique, which is based on an estimated value for the initial count, introduces a standard error with every dilution introduced. Therefore the final estimated number of target organisms is always ± an accumulated standard error. While this procedure is acceptable for estimating high levels of organisms (>100), it should not be used to generate suspensions with low numbers of organisms.

To generate low numbers of accurately counted organisms, micromanipulation and flow cytometric enumeration and sorting (flow cytometry) have emerged (Reynolds *et al.*, 1999; Sturbaum *et al.*, 2001; Tanriverdi *et al.*, 2002). Micromanipulation, while determined to be accurate, is time consuming and labor intensive and therefore is recommended for use with small "in-house" research projects in which large numbers of replicates are not needed (Sturbaum *et al.*, 2002). Flow cytometry triumphs in cases where high numbers of replicates with a low standard deviation (±1 organism) are needed (Chesnot *et al.*, 2002; Reynolds *et al.*, 1999). Flow cytometry has been used to enumerate and sort oocysts of *C. parvum, G. lamblia* cysts (Ferrari and Veal, 2003; Vesey *et al.*, 1994a), *E. intestinalis* spores (Hoffman *et al.*, 2003), and *T. gondii* oocysts (Everson *et al.*, 2002) from both reagent grade water and environmental sources. In addition to the small standard deviation, flow cytometry can be used to directly sort (oo)cysts into the test vial of interest, and thus eliminating any additional standard errors that occur during a dispensing step. For instance, (oo)cysts can be directly sorted into a thin-walled PCR tube or the Leighton tube that is used in EPA Method 1623.

Currently, the EPA has approved flow cytometry for establishing low numbers of *C. parvum* oocysts and *G. lamblia* cysts that are used in QC recovery trials for Method 1623. The method requires that spiking suspensions contain between 100 and 500 (oo)cysts. Currently, commercial sources of flow cytometer-counted (oo)cysts are the Wisconsin State Laboratory of Hygiene Flow Cytometry Unit (WSLH) and BioTechnology Frontiers, (BTF). WSLH prepares and distributes (oo)cysts that are live and infectious while BTF prepares and distributes (oo)cysts that have been inactivated via irradiation. BTF also prepares and distributes oocysts that have been permanently stained with a fluorochrome similar to Texas Red. The utility of this product is an immunofluorescent assay internal positive control that can be used to measure step-by-step losses that occur throughout processing as well as cross-contamination.

The final aspect of proper LLOD determination is directed at molecular based methods. The majority of manuscripts reporting recovery efficiencies and the LLOD use serial dilutions of whole organisms or previously extracted DNA. Again, serial dilutions of whole organisms incorporate a standard error from which the actual number of organisms being seeded and tested for detection/recovery is not known. Second, working with previously extracted DNA does not incorporate the difficulty of liberating DNA from the organism or unintentional co-extraction of PCR inhibitors. In general, nucleic acid is mass extracted from large quantities of organisms ($>10^5$) and subsequently diluted to achieve a low number equivalent ($<10^2$) or even a single organism. This DNA "equivalent" is then used in a recovery assay to determine the LLOD. Following this protocol completely avoids the difficulty of liberating DNA from the low numbers of organisms typically present in purified environmental sample matrices. The best way to determine the LLOD of a given molecular detection method is to use whole organisms (ideally flow cytometry-enumerated) and evaluate the entire method, including concentration, purification, and nucleic acid liberation and recovery with a variety of different sample matrices.

REFERENCES

Abbaszadegan, M., Gerba, C. P., and Rose, J. B., 1991, Detection of *Giardia* cysts with a cDNA probe and applications to water samples, *Appl. Environ. Microbiol.* **57**:927–931.

Abe, N., Kimata, I., and Iseki, M., 2002, Usefulness of multiplex-PCR for identification of *Entamoeba histolytica* and *Entamoeba dispar*, *Kansenshogaku Zasshi* **76**:921–927.

Aboytes, R., Di Giovanni, G. D., Abrams, F. A., Rheinecker, C., McElroy, W., Shaw, N., and LeChevallier, M. W., 2004, Detection of infectious *Cryptosporidium* in filtered drinking water, *JAWWA* **96**:88–98.

Alles, A. J., Waldron, M. A., Sierra, L. S., and Mattia, A. R., 1995, Prospective comparison of direct immunofluorescence and conventional staining methods for detection of *Giardia* and *Cryptosporidium* spp. in human fecal specimens, *J. Clin. Microbiol.* **33**:1632–1634.

Amar, C. F., Dear, P. H., Pedraza-Diaz, S., Looker, N., Linnane, E., and McLauchlin, J., 2002, Sensitive PCR-restriction fragment length polymorphism assay for detection and genotyping of *Giardia duodenalis* in human feces, *J. Clin. Microbiol.* **40**:446–452.

Amar, C. F., Dear, P. H., and McLauchlin, J., 2003, Detection and genotyping by real-time PCR/RFLP analyses of *Giardia duodenalis* from human faeces, *J. Med. Microbiol.* **52**:681–683.

Amar, C. F., Dear, P. H., and McLauchlin, J., 2004, Detection and identification by real time PCR/RFLP analyses of *Cryptosporidium* species from human faeces, *Lett. Appl. Microbiol.* **38**:217–222.

Anonymous, 1982, Cryptosporidiosis: Assessment of chemotherapy of males with acquired immune deficiency syndrome (AIDS), *Morb. Mortal. Wkly. Rep.* **31**:589–592.

Anonymous, 1996, Outbreak of cryptosporidiosis at a day camp—Florida, July-August 1995, *Morb. Mortal. Wkly. Rep.* **45**:442–444.

Anonymous, 1997a, Outbreak at a first nation community on Shoal Lake, Ontario, Canada, *Cryptosporidium Caps Newslett.* **2**:2–3.

Anonymous, 1997b, Outbreaks of Escherichia coli O157:H7 infection and cryptosporidiosis associated with drinking unpasteurized apple cider—Connecticut and New York, October 1996, *Morb. Mortal. Wkly. Rep.* **46**:4–8.

Anonymous, 1998, Outbreak in Texas—sewage leak suspected as source of groundwater contamination, *Cryptosporidium Caps Newslett.* **3**:1–2.

Anonymous, 1999, Outbreak in NW England—water samples found positive, *Cryptosporidium Caps Newslett.* **4**:1.

Anusz, K. Z., Mason, P. H., Riggs, M. W., and Perryman, L. E., 1990, Detection of *Cryptosporidium parvum* oocysts in bovine feces by monoclonal antibody capture enzyme-linked immunosorbent assay, *J. Clin. Microbiol.* **28**:2770–2774.

Arrowood, M. J., and Sterling, C. R., 1987, Isolation of *Cryptosporidium* oocysts and sporozoites using discontinuous sucrose and isopycnic Percoll gradients, *J. Parasitol.* **73**:314–319.

Arrowood, M. J., and Sterling, C. R., 1989, Comparison of conventional staining methods and monoclonal antibody-based methods for *Cryptosporidium* oocyst detection, *J. Clin. Microbiol.* **27**:1490–1495.

Arrowood, M. J., and Donaldson, K., 1996, Improved purification methods for calf-derived *Cryptosporidium parvum* oocysts using discontinuous sucrose and cesium chloride gradients, *J. Eukaryot. Microbiol.* **43**:89S.

Arrowood, M. J., Jaynes, J. M., and Healey, M. C., 1991, Hemolytic properties of lytic peptides active against the sporozoites of *Cryptosporidium parvum*, *J. Protozool.* **38**:161S–163S.

Awad-el-Kariem, F. M., Warhurst, D. C., and McDonald, V., 1994, Detection and species identification of *Cryptosporidium* oocysts using a system based on PCR and endonuclease restriction, *Parasitology* **109**(Pt 1):19–22.

Baeumner, A. J., Humiston, M. C., Montagna, R. A., and Durst, R. A., 2001, Detection of viable oocysts of *Cryptosporidium parvum* following nucleic acid sequence based amplification, *Anal. Chem.* **73**:1176–1180.

Balatbat, A. B., Jordan, G. W., Tang, Y. J., and Silva, J., Jr., 1996, Detection of *Cryptosporidium parvum* DNA in human feces by nested PCR, *J. Clin. Microbiol.* **34**:1769–1772.

Barany, F., 1991, The ligase chain reaction in a PCR world, *PCR Methods Appl.* **1**:5–16.

Baxby, D., Blundell, N., and Hart, C. A., 1984, The development and performance of a simple, sensitive method for the detection of *Cryptosporidium* oocysts in faeces, *J. Hyg. (London).* **93**:317–323.

Belosevic, M., Guy, R. A., Taghi-Kilani, R., Neumann, N. F., Gyurek, L. L., Liyanage, L. R., Millard, P. J., and Finch, G. R., 1997, Nucleic acid stains as indicators of *Cryptosporidium parvum* oocyst viability, *Int. J. Parasitol.* **27**:787–798.

Bingham, A. K., and Meyer, E. A., 1979, *Giardia* excystation can be induced *in vitro* in acidic solutions, *Nature* **277**:301–302.

Bingham, A. K., Jarroll, E. L., Jr., Meyer, E. A., and Radulescu, S., 1979, *Giardia* sp.: Physical factors of excystation *in vitro*, and excystation vs eosin exclusion as determinants of viability, *Exp. Parasitol.* **47**:284–291.

Black, E. K., Finch, G. R., Taghi-Kilani, R., and Belosevic, M., 1996, Comparison of assays for *Cryptosporidium parvum* oocysts viability after chemical disinfection, *FEMS Microbiol. Lett.* **135**:187–189.

Blessmann, J., Buss, H., Nu, P. A., Dinh, B. T., Ngo, Q. T., Van, A. L., Alla, M. D., Jackson, T. F., Ravdin, J. I., and Tannich, E., 2002, Real-time PCR for detection and differentiation of *Entamoeba histolytica* and *Entamoeba dispar* in fecal samples, *J. Clin. Microbiol.* **40**:4413–4417.

Boom, R., Sol, C. J., Salimans, M. M., Jansen, C. L., Wertheim-van Dillen, P. M., and van der Noordaa, J., 1990, Rapid and simple method for purification of nucleic acids, *J. Clin. Microbiol.* **28**:495–503.

Boone, J. H., Wilkins, T. D., Nash, T. E., Brandon, J. E., Macias, E. A., Jerris, R. C., and Lyerly, D. M., 1999, TechLab and alexon *Giardia* enzyme-linked immunosorbent assay kits detect cyst wall protein 1, *J. Clin. Microbiol.* **37**:611–614.

Borchardt, M. A., and Spencer, S. K., 2002, Concentration of *Cryptosporidium*, microsporidia and other water-borne pathogens by continuous separation channel centrifugation, *J. Appl. Microbiol.* **92**:649–656.

Buchbinder, S., Blatz, R., and Rodloff, A. C., 2003, Comparison of real-time PCR detection methods for B1 and P30 genes of *Toxoplasma gondii*, *Diagn. Microbiol Infect. Dis.* **45**:269–271.

Bukhari, Z., Marshall, M. M., Korich, D. G., Fricker, C. R., Smith, H. V., Rosen, J., and Clancy, J. L., 2000, Comparison of *Cryptosporidium parvum* viability and infectivity assays following ozone treatment of oocysts, *Appl. Environ. Microbiol.* **66**:2972–2980.

Caccio, S. M., 2003, Molecular techniques to detect and identify protozoan parasites in the environment, *Acta Microbiol. Pol.* **52**(Suppl):23–34.

Cahill, P., Foster, K., and Mahan, D. E., 1991, Polymerase chain reaction and Q beta replicase amplification, *Clin. Chem.* **37**:1482–1485.

Campbell, A. T., Robertson, L. J., and Smith, H. V., 1992, Viability of *Cryptosporidium parvum* oocysts: Correlation of *in vitro* excystation with inclusion or exclusion of fluorogenic vital dyes, *Appl. Environ. Microbiol.* **58**:3488–3493.

Cano-Mancera, R., and Lopez-Revilla, R., 1988, Maintenance of integrity, viability, and adhesion of *Entamoeba histolytica* trophozoites in different incubation media, *J. Protozool.* **35**:470–475.

Casemore, D. P., Armstrong, M., and Sands, R. L., 1985, Laboratory diagnosis of cryptosporidiosis, *J. Clin. Pathol.* **38**:1337–1341.

Cavalier-Smith, T., 1993, Kingdom Protozoa and its 18 phyla, *Microbiol. Rev.* **57**:593–994.

Chapman, P. A., Rush, B. A., and McLauchlin, J., 1990, An enzyme immunoassay for detecting *Cryptosporidium* in faecal and environmental samples, *J. Med. Microbiol.* **32**:233–237.

Chesnot, T., Marly, X., Chevalier, S., Estevenon, O., Bues, M., and Schwartzbrod, J., 2002, Optimised immunofluorescence procedure for enumeration of *Cryptosporidium parvum* oocyst suspensions, *Water Res.* **36**:3283–3288.

Cook, N., 2003, The use of NASBA for the detection of microbial pathogens in food and environmental samples, *J. Microbiol. Methods* **53**:165–174.

Craun, G. F., 1986, Waterborne giardiasis in the United States 1965—84, *Lancet* **2**:513–514.

Current, W. L., Reese, N. C., Ernst, J. V., Bailey, W. S., Heyman, M. B., and Weinstein, W. M., 1983, Human cryptosporidiosis in immunocompetent and immunodeficient persons. Studies of an outbreak and experimental transmission, *N. Engl. J. Med.* **308**:1252–1257.

D'Antonio, R. G., Winn, R. E., Taylor, J. P., Gustafson, T. L., Current, W. L., Rhodes, M. M., Gary, G. W., Jr., and Zajac, R. A., 1985, A waterborne outbreak of cryptosporidiosis in normal hosts, *Ann. Intern. Med.* **103**:886–888.

Deere, D., Vesey, G., Ashbolt, N., Davies, K. A., Williams, K. L., and Veal, D., 1998a, Evaluation of fluorochromes for flow cytometric detection of *Cryptosporidium parvum* oocysts labelled by fluorescent in situ hybridization, *Lett. Appl. Microbiol.* **27**:352–356.

Deere, D., Vesey, G., Milner, M., Williams, K., Ashbolt, N., and Veal, D., 1998b, Rapid method for fluorescent *in situ* ribosomal RNA labelling of *Cryptosporidium parvum*, *J. Appl. Microbiol.* **85**:807–818.

Deng, M. Q., Cliver, D. O., and Mariam, T. W., 1997, Immunomagnetic capture PCR to detect viable *Cryptosporidium parvum* oocysts from environmental samples, *Appl. Environ. Microbiol.* **63**:3134–3138.

deRegnier, D. P., Cole, L., Schupp, D. G., and Erlandsen, S. L., 1989, Viability of *Giardia* cysts suspended in lake, river, and tap water, *Appl. Environ. Microbiol.* **55**:1223–1229.

Di Giovanni, G. D., and Aboytes, R., 2003, Detection of infectious *Cryptosporidium parvum* oocysts in environmental water samples, In Thompson, R. C. A., Armson, A., and Ryan, U. M. (eds), *Cryptosporidium: From Molecules to Disease*, Elsevier Science, New York, pp. 213–224.

Di Giovanni, G. D., Hashemi, F. H., Shaw, N. J., Abrams, F. A., LeChevallier, M. W., and Abbaszadegan, M., 1999, Detection of infectious *Cryptosporidium parvum* oocysts in surface and filter backwash water samples by immunomagnetic separation and integrated cell culture-PCR, *Appl. Environ. Microbiol.* **65**:3427–3432.

DiGiorgio, C. L., Gonzalez, D. A., and Huitt, C. C., 2002, *Cryptosporidium* and *Giardia* recoveries in natural waters by using environmental protection agency method 1623, *Appl. Environ. Microbiol.* **68**:5952–5955.

Doing, K. M., Hamm, J. L., Jellison, J. A., Marquis, J. A., and Kingsbury, C., 1999, False-positive results obtained with the Alexon ProSpecT *Cryptosporidium* enzyme immunoassay, *J. Clin. Microbiol.* **37**:1582–1583.

Dorsch, M. R., and Veal, D. A., 2001, Oligonucleotide probes for specific detection of *Giardia lamblia* cysts by fluorescent in situ hybridization, *J. Appl. Microbiol.* **90**:836–842.

Dowd, S. E., Gerba, C. P., Enriquez, F. J., and Pepper, I. L., 1998, PCR amplification and species determination of microsporidia in formalin-fixed feces after immunomagnetic separation, *Appl. Environ. Microbiol.* **64**:333–336.

Dumetre, A., and Darde, M. L., 2003, How to detect *Toxoplasma gondii* oocysts in environmental samples? *FEMS Microbiol. Rev.* **27**:651–661.

Dupon, M., Cazenave, J., Pellegrin, J. L., Ragnaud, J. M., Cheyrou, A., Fischer, I., Leng, B., and Lacut, J. Y., 1995, Detection of *Toxoplasma gondii* by PCR and tissue culture in cerebrospinal fluid and blood of human immunodeficiency virus-seropositive patients, *J. Clin. Microbiol.* **33**:2421–2426.

Dupouy-Camet, J., de Souza, S. L., Maslo, C., Paugam, A., Saimot, A. G., Benarous, R., Tourte-Schaefer, C., and Derouin, F., 1993, Detection of *Toxoplasma gondii* in venous blood from AIDS patients by polymerase chain reaction, *J. Clin. Microbiol.* **31**:1866–1869.

Dworkin, M. S., Goldman, D. P., Wells, T. G., Kobayashi, J. M., and Herwaldt, B. L., 1996, Cryptosporidiosis in Washington State: An outbreak associated with well water, *J. Infect. Dis.* **174**:1372–1376.

Eberhard, M. L., Nace, E. K., and Freeman, A. R., 1999, Survey for *Cyclospora cayetanensis* in domestic animals in an endemic area in Haiti, *J. Parasitol.* **85**:562–563.

Eberhard, M. L., Ortega, Y. R., Hanes, D. E., Nace, E. K., Do, R. Q., Robl, M. G., Won, K. Y., Gavidia, C., Sass, N. L., Mansfield, K., Gozalo, A., Griffiths, J., Gilman, R., Sterling, C. R., and Arrowood, M. J., 2000, Attempts to establish experimental *Cyclospora cayetanensis* infection in laboratory animals, *J. Parasitol.* **86**:577–582.

Ekins, R., and Chu, F. W., 1999, Microarrays: Their origins and applications, *Trends Biotechnol.* **17**:217–218.

Elliot, A., Morgan, U. M., and Thompson, R. C., 1999, Improved staining method for detecting *Cryptosporidium* oocysts in stools using malachite green, *J. Gen. Appl. Microbiol.* **45**:139–142.

Ellis, J. T., 1998, Polymerase chain reaction approaches for the detection of *Neospora caninum* and *Toxoplasma gondii*, *Int. J. Parasitol.* **28**:1053–1060.

Erlandsen, S. L., Sherlock, L. A., and Bemrick, W. J., 1990, The detection of *Giardia muris* and *Giardia lamblia* cysts by immunofluorescence in animal tissues and fecal samples subjected to cycles of freezing and thawing, *J. Parasitol.* **76**:267–271.

Esch, M. B., Baeumner, A. J., and Durst, R. A., 2001a, Detection of *Cryptosporidium parvum* using oligonucleotide-tagged liposomes in a competitive assay format, *Anal. Chem.* **73**:3162–3167.

Esch, M. B., Locascio, L. E., Tarlov, M. J., and Durst, R. A., 2001b, Detection of viable *Cryptosporidium parvum* using DNA-modified liposomes in a microfluidic chip, *Anal. Chem.* **73**:2952–2958.

Evangelopoulos, A., Spanakos, G., Patsoula, E., Vakalis, N., and Legakis, N., 2000, A nested, multiplex, PCR assay for the simultaneous detection and differentiation of *Entamoeba histolytica* and *Entamoeba dispar* in faeces, *Ann. Trop. Med. Parasitol.* **94**:233–240.

Everson, W. V., Ware, M. W., Dubey, J. P., and Lindquist, H. D., 2002, Isolation of purified oocyst walls and sporocysts from *Toxoplasma gondii*, *J. Eukaryot. Microbiol.* **49**:344–349.

Ey, P. L., Darby, J. M., Andrews, R. H., and Maryhofer, G., 1993, *Giardia intestinalis*: Detection of major genotypes by restriction analysis of gene amplification products, *Int. J. Parasitol.* **23**:591–600.

Ey, P. L., Mansouri, M., Kulda, J., Nohynkova, E., Monis, P. T., Andrews, R. H., and Maryhofer, G., 1997, Genetic analysis of *Giardia* from hoofed farm animals reveals artiodactyl-specific and potentially zoonotic genotypes, *J. Eukaryot. Microbiol.* **44**:626–635.

Farrington, M., Winters, S., Walker, C., Miller, R., and Rubenstein, D., 1994, *Cryptosporidium* antigen detection in human feces by reverse passive hemagglutination assay, *J. Clin. Microbiol.* **32**:2755–2759.

Fayer, R., and Leek, R. G., 1984, The effects of reducing conditions, medium, pH, temperature, and time on *in vitro* excystation of *Cryptosporidium*, *J. Protozool.* **31**:567–569.

Fernandez, S., Pagotto, A. H., Furtado, M. M., Katsuyama, A. M., Madeira, A. M., and Gruber, A., 2003, A multiplex PCR assay for the simultaneous detection and discrimination of the seven *Eimeria* species that infect domestic fowl, *Parasitol.* **127**:317–325.

Ferrari, B. C., and Veal, D., 2003, Analysis-only detection of *Giardia* by combining immunomagnetic separation and two-color flow cytometry, *Cytometry* **51A**:79–86.

Fireman, M., and Reitemeier, R. F., 1944, Prevention of calcium carbonate precipitation in soil solutions and waters by sodium hexametaphosphate, *Soil Sci.* **2**:35–41.

Fischer, P., Taraschewski, H., Ringelmann, R., and Eing, B., 1998, Detection of *Cryptosporidium parvum* in human feces by PCR, *Tokai J. Exp. Clin. Med.* **23**:309–311.

Fontaine, M., and Guillot, E., 2003a, An immunomagnetic separation-real-time PCR method for quantification of *Cryptosporidium parvum* in water samples, *J. Microbiol. Methods.* **54**:29–36.

Fontaine, M., and Guillot, E., 2003b, Study of 18S rRNA and rDNA stability by real-time RT-PCR in heat-inactivated *Cryptosporidium parvum* oocysts, *FEMS Microbiol. Lett.* **226**:237–243.

Frost, F. J., Calderon, R. L., Muller, T. B., Curry, M., Rodman, J. S., Moss, D. M., and de la Cruz, A. A., 1998, A two-year follow-up survey of antibody to *Cryptosporidium* in Jackson County, Oregon following an outbreak of waterborne disease, *Epidemiol. Infect.* **121**:213–217.

Frost, F. J., Fea, E., Gilli, G., Biorci, F., Muller, T. M., Craun, G. F., and Calderon, R. L., 2000, Serological evidence of *Cryptosporidium* infections in southern Europe, *Eur. J. Epidemiol.* **16**:385–390.

Furtado, C., Adak, G. K., Stuart, J. M., Wall, P. G., Evans, H. S., and Casemore, D. P., 1998, Outbreaks of waterborne infectious intestinal disease in England and Wales, 1992–5, *Epidemiol. Infect.* **121**:109–119.

Garcia, L. S., and Shimizu, R., 1981, Comparison of clinical results for the use of ethyl acetate and diethyl ether in the formalin-ether sedimentation technique performed on polyvinyl alcohol-preserved specimens, *J. Clin. Microbiol.* **13**:709–713.

Garcia, L. S., and Shimizu, R. Y., 1997, Evaluation of nine immunoassay kits (enzyme immunoassay and direct fluorescence) for detection of *Giardia lamblia* and *Cryptosporidium parvum* in human fecal specimens, *J. Clin. Microbiol.* **35**:1526–1529.

Garcia, L. S., and Shimizu, R. Y., 2000, Detection of *Giardia lamblia* and *Cryptosporidium parvum* antigens in human fecal specimens using the ColorPAC combination rapid solid-phase qualitative immunochromatographic assay, *J. Clin. Microbiol.* **38**:1267–1268.

Garcia, L. S., Brewer, T. C., and Bruckner, D. A., 1979, A comparison of the formalin-ether concentration and trichrome-stained smear methods for the recovery and identification of intestinal protozoa, *Am. J. Med. Technol.* **45**:932–935.

Garcia, L. S., Brewer, T. C., and Bruckner, D. A., 1987, Fluorescence detection of *Cryptosporidium* oocysts in human fecal specimens by using monoclonal antibodies, *J. Clin. Microbiol.* **25**:119–121.

Garcia, L. S., Shum, A. C., and Bruckner, D. A., 1992, Evaluation of a new monoclonal antibody combination reagent for direct fluorescence detection of *Giardia* cysts and *Cryptosporidium* oocysts in human fecal specimens, *J. Clin. Microbiol.* **30**:3255–3257.

Garcia, L. S., Shimizu, R. Y., and Bernard, C. N., 2000, Detection of *Giardia lamblia*, *Entamoeba histolytica/Entamoeba dispar*, and *Cryptosporidium parvum* antigens in human fecal specimens using the triage parasite panel enzyme immunoassay, *J. Clin. Microbiol.* **38**:3337–3340.

Gennaccaro, A. L., McLaughlin, M. R., Quintero-Betancourt, W., Huffman, D. E., and Rose, J. B., 2003, Infectious *Cryptosporidium parvum* oocysts in final reclaimed effluent, *Appl. Environ. Microbiol.* **69**:4983–4984.

Gibbons, C. L., and Awad-El-Kariem, F. M., 1999, Nested PCR for the detection of *Cryptosporidium parvum*, *Parasitol. Today* **15**:345.

Gile, M., Warhurst, D. C., Webster, K. A., West, D. M., and Marshall, J. A., 2002, A multiplex allele specific polymerase chain reaction (MAS-PCR) on the dihydrofolate reductase gene for the detection of *Cryptosporidium parvum* genotypes 1 and 2, *Parasitology* **125**:35–44.

Glaberman, S., Moore, J. E., Lowery, C. J., Chalmers, R. M., Sulaiman, I., Elwin, K., Rooney, P. J., Millar, B. C., Dooley, J. S., Lal, A. A., and Xiao, L., 2002, Three drinking-water-associated cryptosporidiosis outbreaks, Northern Ireland, *Emerg. Infect. Dis.* **8**:631–633.

Gobet, P., and Toze, S., 2001, Sensitive genotyping of *Cryptosporidium parvum* by PCR-RFLP analysis of the 70-kilodalton heat shock protein (HSP70) gene, *FEMS Microbiol. Lett.* **200**:37–41.

Goldstein, S. T., Juranek, D. D., Ravenholt, O., Hightower, A. W., Martin, D. G., Mesnik, J. L., Griffiths, S. D., Bryant, A. J., Reich, R. R., and Herwaldt, B. L., 1996, Cryptosporidiosis: an outbreak associated with drinking water despite state-of-the-art water treatment, *Ann. Intern. Med.* **124**:459–468.

Gomez-Couso, H., Freire-Santos, F., Amar, C. F., Grant, K. A., Williamson, K., Ares-Mazas, M. E., and McLauchlin, J., 2004, Detection of *Cryptosporidium* and *Giardia* in molluscan shellfish by multiplexed nested-PCR, *Int. J. Food Microbiol.* **91**:279–288.

Gonzalez, L. M., Montero, E., Harrison, L. J., Parkhouse, R. M., and Garate, T., 2000, Differential diagnosis of *Taenia saginata* and *Taenia solium* infection by PCR, *J. Clin. Microbiol.* **38**:737–744.

Gonzalez-Ruiz, A., Haque, R., Rehman, T., Aguirre, A., Hall, A., Guhl, F., Warhurst, D. C., and Miles, M. A., 1994, Diagnosis of amebic dysentery by detection of *Entamoeba histolytica* fecal antigen by an invasive strain-specific, monoclonal antibody-based enzyme-linked immunosorbent assay, *J. Clin. Microbiol.* **32**:964–970.

Goodwin, D. C., and Lee, S. B., 1993, Microwave miniprep of total genomic DNA from fungi, plants, protists and animals for PCR, *Biotechniques* **15**:438, 441–432, 444.

Grigoriew, G. A., Walmsley, S., Law, L., Chee, S. L., Yang, J., Keystone, J., and Krajden, M., 1994, Evaluation of the Merifluor immunofluorescent assay for the detection of *Cryptosporidium* and *Giardia* in sodium acetate formalin-fixed stools, *Diagn. Microbiol. Infect. Dis.* **19**:89–91.

Gross, U., Roggenkamp, A., Janitschke, K., and Heesemann, J., 1992, Improved sensitivity of the polymerase chain reaction for detection of *Toxoplasma gondii* in biological and human clinical specimens, *Eur. J. Clin. Microbiol. Infect. Dis.* **11**:33–39.

Guay, J. M., Dubois, D., Morency, M. J., Gagnon, S., Mercier, J., and Levesque, R. C., 1993, Detection of the pathogenic parasite *Toxoplasma gondii* by specific amplification of ribosomal sequences using comultiplex polymerase chain reaction, *J. Clin. Microbiol.* **31**:203–207.

Guy, R. A., Payment, P., Krull, U. J., and Horgen, P. A., 2003, Real-time PCR for quantification of *Giardia* and *Cryptosporidium* in environmental water samples and sewage, *Appl. Environ. Microbiol.* **69**:5178–5185.

Hallier-Soulier, S., and Guillot, E., 2000, Detection of cryptosporidia and *Cryptosporidium parvum* oocysts in environmental water samples by immunomagnetic separation-polymerase chain reaction, *J. Appl. Microbiol.* **89**:5–10.

Hallier-Soulier, S., and Guillot, E., 2003, An immunomagnetic separation-reverse transcription polymerase chain reaction (IMS-RT-PCR) test for sensitive and rapid detection of viable waterborne *Cryptosporidium parvum*, *Environ. Microbiol.* **5**:592–598.

Haque, R., Kress, K., Wood, S., Jackson, T. F., Lyerly, D., Wilkins, T., and Petri, W. A., Jr., 1993, Diagnosis of pathogenic *Entamoeba histolytica* infection using a stool ELISA based on monoclonal antibodies to the galactose-specific adhesin, *J. Infect. Dis.* **167**:247–249.

Hayes, E. B., Matte, T. D., O'Brien, T. R., McKinley, T. W., Logsdon, G. S., Rose, J. B., Ungar, B. L., Word, D. M., Pinsky, P. F., Cummings, M. L., *et al.*, 1989, Large community outbreak of cryptosporidiosis due to contamination of a filtered public water supply, *N. Engl. J. Med.* **320**:1372–1376.

Heid, C. A., Stevens, J., Livak, K. J., and Williams, P. M., 1996, Real time quantitative PCR, *Genome Res.* **6**:986–994.

Henricksen, S. A., and Pohlenz, J. F., 1981, Staining of cryptosporidia by a modified Ziehl-Neelsen technique, *Acat Vet. Scand.* **22**:594–596.

Higgins, J. A., Fayer, R., Trout, J. M., Xiao, L., Lal, A. A., Kerby, S., and Jenkins, M. C., 2001a, Real-time PCR for the detection of *Cryptosporidium parvum*, *J. Microbiol. Methods* **47**:323–337.

Higgins, J. A., Jenkins, M. C., Shelton, D. R., Fayer, R., and Karns, J. S., 2001b, Rapid extraction of DNA From *Escherichia coli* and *Cryptosporidium parvum* for use in PCR, *Appl. Environ. Microbiol.* **67**:5321–5324.

Hijjawi, N. S., Meloni, B. P., Morgan, U. M., and Thompson, R. C., 2001, Complete development and long-term maintenance of *Cryptosporidium parvum* human and cattle genotypes in cell culture, *Int. J. Parasitol.* **31**:1048–1055.

Hijjawi, N. S., Meloni, B. P., Ryan, U. M., Olson, M. E., and Thompson, R. C., 2002, Successful *in vitro* cultivation of *Cryptosporidium andersoni*: Evidence for the existence of novel extracellular stages in the life cycle and implications for the classification of *Cryptosporidium*, *Int. J. Parasitol.* **32**:1719–1726.

Hoff, J. C., Rice, E. W., and Schaefer, F. W., 3rd., 1985, Comparison of animal infectivity and excystation as measures of *Giardia muris* cyst inactivation by chlorine, *Appl. Environ. Microbiol.* **50**:1115–1117.

Hoff, R. L., Dubey, J. P., Behbehani, A. M., and Frenkel, J. K., 1977, *Toxoplasma gondii* cysts in cell culture: New biologic evidence, *J. Parasitol.* **63**:1121–1124.

Hoffman, R. M., Marshall, M. M., Polchert, D. M., and Jost, B. H., 2003, Identification and characterization of two subpopulations of *Encephalitozoon intestinalis*, *Appl. Environ. Microbiol.* **69**:4966–4970.

Homan, W. L., van Enckevort, F. H., Limper, L., van Eys, G. J., Schoone, G. J., Kasprzak, W., Majewska, A. C., and van Knapen, F., 1992, Comparison of *Giardia* isolates from different laboratories by isoenzyme analysis and recombinant DNA probes, *Parasitol. Res.* **78**:316–323.

Hopkins, R. M., Meloni, B. P., Groth, D. M., Wetherall, J. D., Reynoldson, J. A., and Thompson, R. C., 1997, Ribosomal RNA sequencing reveals differences between the genotypes of *Giardia* isolates recovered from humans and dogs living in the same locality, *J. Parasitol.* **83**:44–51.

Hopkins, R. M., Constantine, C. C., Groth, D. A., Wetherall, J. D., Reynoldson, J. A., and Thompson, R. C., 1999, PCR-based DNA fingerprinting of *Giardia duodenalis* isolates using the intergenic rDNA spacer, *Parasitology* **118**(Pt 6):531–539.

Horen, W. P., 1983, Detection of *Cryptosporidium* in human fecal specimens, *J. Parasitol.* **69**:622–624.

Howe, A. D., Forster, S., Morton, S., Marshall, R., Osborn, K. S., Wright, P., and Hunter, P. R., 2002, *Cryptosporidium* oocysts in a water supply associated with a cryptosporidiosis outbreak, *Emerg. Infect. Dis.* **8**:619–624.

Huffman, D. E., Gennaccaro, A., Rose, J. B., and Dussert, B. W., 2002, Low- and medium-pressure UV inactivation of microsporidia *Encephalitozoon intestinalis*, *Water Res.* **36**:3161–3164.

Ijzerman, M. M., Dahling, D. R., and Fout, G. S., 1997, A method to remove environmental inhibitors prior to the detection of waterborne enteric viruses by reverse transcription-polymerase chain reaction, *J. Virol. Methods* **63**:145–153.

Isaac-Renton, J. L., Shahriari, H., and Bowie, W. R., 1992, Comparison of an *in vitro* method and an *in vivo* method of *Giardia* excystation, *Appl. Environ. Microbiol.* **58**:1530–1533.

Jauregui, L. H., Higgins, J., Zarlenga, D., Dubey, J. P., and Lunney, J. K., 2001, Development of a real-time PCR assay for detection of *Toxoplasma gondii* in pig and mouse tissues, *J. Clin. Microbiol.* **39**:2065–2071.

Jelinek, T., Peyerl, G., Loscher, T., and Nothdurft, H. D., 1996, Evaluation of an antigen-capture enzyme immunoassay for detection of *Entamoeba histolytica* in stool samples, *Eur. J. Clin. Microbiol. Infect. Dis.* **15**:752–755.

Jenkins, M. B., Anguish, L. J., Bowman, D. D., Walker, M. J., and Ghiorse, W. C., 1997, Assessment of a dye permeability assay for determination of inactivation rates of *Cryptosporidium parvum* oocysts, *Appl. Environ. Microbiol.* **63**:3844–3850.

Jenkins, M. C., Trout, J., Abrahamsen, M. S., Lancto, C. A., Higgins, J., and Fayer, R., 2000, Estimating viability of *Cryptosporidium parvum* oocysts using reverse transcriptase-polymerase chain reaction (RT-PCR) directed at mRNA encoding amyloglucosidase, *J. Microbiol. Methods* **43**:97–106.

Jenkins, M., Trout, J. M., Higgins, J., Dorsch, M., Veal, D., and Fayer, R., 2003, Comparison of tests for viable and infectious *Cryptosporidium parvum* oocysts, *Parasitol. Res.* **89**:1–5.

Joachim, A., Eckert, E., Petry, F., Bialek, R., and Daugschies, A., 2003, Comparison of viability assays for *Cryptosporidium parvum* oocysts after disinfection, *Vet. Parasitol.* **111**:47–57.

Johnson, D. W., Pieniazek, N. J., Griffin, D. W., Misener, L., and Rose, J. B., 1995, Development of a PCR protocol for sensitive detection of *Cryptosporidium* oocysts in water samples, *Appl. Environ. Microbiol.* **61**:3849–3855.

Jones, C. D., Okhravi, N., Adamson, P., Tasker, S., and Lightman, S., 2000, Comparison of PCR detection methods for B1, P30, and 18S rDNA genes of *T. gondii* in aqueous humor, *Invest. Ophthalmol. Vis. Sci.* **41**:634–644.

Joseph, C., Hamilton, G., O'Connor, M., Nicholas, S., Marshall, R., Stanwell-Smith, R., Sims, R., Ndawula, E., Casemore, D., Gallagher, P., and *et al.*, 1991, Cryptosporidiosis in the Isle of Thanet; An outbreak associated with local drinking water, *Epidemiol. Infect.* **107**:509–519.

Kappe, S., Bruderer, T., Gantt, S., Fujioka, H., Nussenzweig, V., and Menard, R., 1999, Conservation of a gliding motility and cell invasion machinery in Apicomplexan parasites, *J. Cell. Biol.* **147**:937–944.

Karanis, P., and Kimura, A., 2002, Evaluation of three flocculation methods for the purification of *Cryptosporidium parvum* oocysts from water samples, *Lett. Appl. Microbiol.* **34**:444–449.

Kasper, D. R., 1993, Pre- and post treatment processes for membrane water treatment systems, In *Proceedings AWWA Membrane Technology Conference*, Baltimore, MD, p. 105.

Kato, S., Lindergard, G., and Mohammed, H. O., 2003, Utility of the *Cryptosporidium* oocyst wall protein (COWP) gene in a nested PCR approach for detection infection in cattle, *Vet. Parasitol.* **111**:153–159.

Kaucner, C., and Stinear, T., 1998, Sensitive and rapid detection of viable *Giardia* cysts and *Cryptosporidium parvum* oocysts in large-volume water samples with wound fiberglass cartridge filters and reverse transcription-PCR, *Appl. Environ. Microbiol.* **64**:1743–1749.

Kawamoto, F., Mizuno, S., Fujioka, H., Kumada, N., Sugiyama, E., Takeuchi, T., Kobayashi, S., Iseki, M., Yamada, M., Matsumoto, Y., and *et al.*, 1987, Simple and rapid staining for detection of *Entamoeba* cysts and other protozoans with fluorochromes, *Jpn. J. Med. Sci. Biol.* **40**:35–46.

Keegan, A. R., Fanok, S., Monis, P. T., and Saint, C. P., 2003, Cell culture-Taqman PCR assay for evaluation of *Cryptosporidium parvum* disinfection, *Appl. Environ. Microbiol.* **69**:2505–2511.

Knepp, J. H., Geahr, M. A., Forman, M. S., and Valsamakis, A., 2003, Comparison of automated and manual nucleic acid extraction methods for detection of enterovirus RNA, *J. Clin. Microbiol.* **41**:3532–3536.

Kniel, K. E., Lindsay, D. S., Sumner, S. S., Hackney, C. R., Pierson, M. D., and Dubey, J. P., 2002, Examination of attachment and survival of *Toxoplasma gondii* oocysts on raspberries and blueberries, *J. Parasitol.* **88**:790–793.

Knowles, D. P., Jr., and Gorham, J. R., 1993, Advances in the diagnosis of some parasitic diseases by monoclonal antibody-based enzyme-linked immunosorbent assays, *Rev. Sci. Tech.* **12**:425–433.

Korich, D. G., Yozwiak, M. L., Marshall, M. M., Sinclair, N. A., and Sterling, C. R., 1993, Development of a test to assess *Cryptosporidium parvum* oocyst viability: Correlation with infectivity potential, *Am. Water Works Assoc. Res. Foundation.* Denver, CO.

Kostrzynska, M., Sankey, M., Haack, E., Power, C., Aldom, J. E., Chagla, A. H., Unger, S., Palmateer, G., Lee, H., Trevors, J. T., and De Grandis, S. A., 1999, Three sample preparation protocols for polymerase chain reaction based detection of *Cryptosporidium parvum* in environmental samples, *J. Microbiol. Methods* **35**:65–71.

Kourenti, C., and Karanis, P., 2004, Development of a sensitive polymerase chain reaction method for the detection of *Toxoplasma gondii* in water, *Water Sci. Technol.* **50**:287–291.

Kramer, F., Vollrath, T., Schnieder, T., and Epe, C., 2002, Improved detection of endoparasite DNA in soil sample PCR by the use of anti-inhibitory substances, *Vet. Parasitol.* **108**:217–226.

Kreader, C. A., 1996, Relief of amplification inhibition in PCR with bovine serum albumin or T4 gene 32 protein, *Appl. Environ. Microbiol.* **62**:1102–1106.

Kuhn, R. C., and Oshima, K. H., 2001, Evaluation and optimization of a reusable hollow fiber ultrafilter as a first step in concentrating *Cryptosporidium parvum* oocysts from water, *Water Res.* **35**:2779–2783.

Kuroki, T., Watanabe, Y., Asai, Y., Yamai, S., Endo, T., Uni, S., Kimata, I., and Iseki, M., 1996, An outbreak of waterborne Cryptosporidiosis in Kanagawa, Japan, *Kansenshogaku Zasshi* **70**:132–140.

Kwiatkowski, R. W., Lyamichev, V., de Arruda, M., and Neri, B., 1999, Clinical, genetic, and pharmacogenetic applications of the Invader assay, *Mol. Diagn.* **4**:353–364.

Labatiuk, C. W., Schaefer, F. W., 3rd, Finch, G. R., and Belosevic, M., 1991, Comparison of animal infectivity, excystation, and fluorogenic dye as measures of *Giardia muris* cyst inactivation by ozone, *Appl. Environ. Microbiol.* **57**:3187–3192.

Lamoril, J., Molina, J. M., de Gouvello, A., Garin, Y. J., Deybach, J. C., Modai, J., and Derouin, F., 1996, Detection by PCR of *Toxoplasma gondii* in blood in the diagnosis of cerebral toxoplasmosis in patients with AIDS, *J. Clin. Pathol.* **49**:89–92.

Laxer, M. A., Timblin, B. K., and Patel, R. J., 1991, DNA sequences for the specific detection of *Cryptosporidium parvum* by the polymerase chain reaction, *Am. J. Trop. Med. Hyg.* **45**:688–694.

LeChevallier, M., Norton, W., and Lee, R., 1991, Occurrence of *Giardia* and *Cryptosporidium* spp. in surface water supplies, *Appl. Environ. Microbiol.* **57**:2610–2616.

LeChevallier, M. W., Norton, W. D., Siegel, J. E., and Abbaszadegan, M., 1995, Evaluation of the immunofluorescence procedure for detection of *Giardia* cysts and *Cryptosporidium* oocysts in water, *Appl. Environ. Microbiol.* **61**:690–697.

LeChevallier, M. W., Di Giovanni, G. D., Clancy, J. L., Bukhari, Z., Bukhari, S., Rosen, J. S., Sobrinho, J., and Frey, M. M., 2003, Comparison of method 1623 and cell culture-PCR for detection of *Cryptosporidium* spp. in source waters, *Appl. Environ. Microbiol.* **69**:971–979.

Leng, X., Mosier, D. A., and Oberst, R. D., 1996, Differentiation of *Cryptosporidium parvum*, *C. muris*, and *C. baileyi* by PCR-RFLP analysis of the 18S rRNA gene, *Vet. Parasitol.* **62**:1–7.

Levine, N. D., 1984, Taxonomy and review of the coccidian genus *Cryptosporidium* (protozoa, apicomplexa), *J. Protozool.* **31**:94–98.

Li, X., Palmer, R., Trout, J. M., and Fayer, R., 2003, Infectivity of microsporidia spores stored in water at environmental temperatures, *J. Parasitol.* **89**:185–188.

Limor, J. R., Lal, A. A., and Xiao, L., 2002, Detection and differentiation of *Cryptosporidium* parasites that are pathogenic for humans by real-time PCR, *J. Clin. Microbiol.* **40**:2335–2338.

Lindergard, G., Nydam, D. V., Wade, S. E., Schaaf, S. L., and Mohammed, H. O., 2003, A novel multiplex polymerase chain reaction approach for detection of four human infective Cryptosporidium isolates: *Cryptosporidium parvum*, types H and C, *Cryptosporidium canis*, and *Cryptosporidium felis* in fecal and soil samples, *J. Vet. Diagn. Invest.* **15**:262–267.

Lindsay, D. S., Dubey, J. P., Blagburn, B. L., and Toivio-Kinnucan, M., 1991, Examination of tissue cyst formation by *Toxoplasma gondii* in cell cultures using bradyzoites, tachyzoites, and sporozoites, *J. Parasitol.* **77**:126–132.

Lindsay, D. S., Blagburn, B. L., and Dubey, J. P., 2002, Survival of nonsporulated *Toxoplasma gondii* oocysts under refrigerator conditions, *Vet. Parasitol.* **103**:309–313.

Lindsay, D. S., Collins, M. V., Mitchell, S. M., Cole, R. A., Flick, G. J., Wetch, C. N., Lindquist, A., and Dubey, J. P., 2003, Sporulation and survival of *Toxoplasma gondii* oocysts in seawater, *J. Eukaryot. Microbiol.* **50**(Suppl):687–688.

Lowery, C. J., Moore, J. E., Millar, B. C., Burke, D. P., McCorry, K. A., Crothers, E., and Dooley, J. S., 2000, Detection and speciation of *Cryptosporidium* spp. in environmental water samples by immunomagnetic separation, PCR and endonuclease restriction, *J. Med. Microbiol.* **49**:779–785.

Lu, S. Q., Baruch, A. C., and Adam, R. D., 1998, Molecular comparison of Giardia lamblia isolates, *Int. J. Parasitol.* **28**:1341–1345.

Ma, P., and Soave, R., 1983, Three-step stool examination for cryptosporidiosis in 10 homosexual men with protracted watery diarrhea, *J. Infect. Dis.* **147**:824–828.

MacKenzie, W. R., Hoxie, N. J., Proctor, M. E., Gradus, M. S., Blair, K. A., Peterson, D. E., Kazmierczak, J. J., Addiss, D. G., Fox, K. R., Rose, J. B. *et al.*, 1994, A massive outbreak in Milwaukee of *Cryptosporidium* infection transmitted through the public water supply, *N. Engl. J. Med.* **331**:161–167.

MacDonald, L. M., Sargent, K., Armson, A., Thompson, R. C., and Reynoldson, J. A., 2002, The development of a real-time quantitative-PCR method for characterisation of a *Cryptosporidium parvum in vitro* culturing system and assessment of drug efficacy, *Mol. Biochem. Parasitol.* **121**:279–282.

Matheson, Z., Hargy, T. M., McCuin, R. M., Clancy, J. L., and Fricker, C. R., 1998, An evaluation of the Gelman Envirochek capsule for the simultaneous concentration of *Cryptosporidium* and *Giardia* from water, *J. Appl. Microbiol.* **85**:755–761.

Mayer, C. L., and Palmer, C. J., 1996, Evaluation of PCR, nested PCR, and fluorescent antibodies for detection of Giardia and Cryptosporidium species in wastewater, *Appl. Environ. Microbiol.* **62**:2081–2085.

Maryhofer, G., Andrews, R. H., Ey, P. L., and Chilton, N. B., 1995, Division of *Giardia* isolates from humans into two genetically distinct assemblages by electrophoretic analysis of enzymes encoded at 27 loci and comparison with *Giardia muris*, *Parasitology* **111**(Pt 1):11–17.

McCuin, R. M., and Clancy, J. L., 2003, Modifications to United States Environmental Protection Agency methods 1622 and 1623 for detection of *Cryptosporidium* oocysts and *Giardia* cysts in water, *Appl. Environ. Microbiol.* **69**:267–274.

McCuin, R. M., Bukhari, Z., and Clancy, J. L., 2000, Recovery and viability of *Cryptosporidium parvum* oocysts and *Giardia intestinalis* cysts using the membrane dissolution procedure, *Can. J. Microbiol.* **46**:700–707.

McCuin, R. M., Bukhari, Z., Sobrinho, J., and Clancy, J. L., 2001, Recovery of *Cryptosporidium* oocysts and *Giardia* cysts from source water concentrates using immunomagnetic separation, *J. Microbiol. Methods* **45**:69–76.

McLauchlin, J., Casemore, D. P., Harrison, T. G., Gerson, P. J., Samuel, D., and Taylor, A. G., 1987, Identification of *Cryptosporidium* oocysts by monoclonal antibody, *Lancet* **1**: 51.

McOrist, A. L., Jackson, M., and Bird, A. R., 2002, A comparison of five methods for extraction of bacterial DNA from human faecal samples, *J. Microbiol. Methods* **50**:131–139.

Monis, P. T., and Saint, C. P., 2001, Development of a nested-PCR assay for the detection of *Cryptosporidium parvum* in finished water, *Water Res.* **35**:1641–1648.

Monis, P. T., Andrews, R. H., Maryhofer, G., Mackrill, J., Kulda, J., Isaac-Renton, J. L., and Ey, P. L., 1998, Novel lineages of *Giardia intestinalis* identified by genetic analysis of organisms isolated from dogs in Australia, *Parasitology* **116**(Pt 1):7–19.

Monis, P. T., Andrews, R. H., Maryhofer, G., and Ey, P. L., 1999, Molecular systematics of the parasitic protozoan *Giardia intestinalis*, *Mol. Biol. Evol.* **16**:1135–1144.

Moore, A. C., Herwaldt, B. L., Craun, G. F., Calderon, R. L., Highsmith, A. K., and Juranek, D. D., 1993, Surveillance for waterborne disease outbreaks–United States, 1991–1992, *MMWR CDC Surveill. Summ.* **42**:1–22.

Morgan, U. M., Pallant, L., Dwyer, B. W., Forbes, D. A., Rich, G., and Thompson, R. C., 1998, Comparison of PCR and microscopy for detection of Cryptosporidium parvum in human fecal specimens: Clinical trial, *J. Clin. Microbiol.* **36**:995–998.

Morgan, U. M., Xiao, L., Fayer, R., Lal, A. A., and Thompson, R. C., 1999, Variation in *Cryptosporidium*: Towards a taxonomic revision of the genus, *Int. J. Parasitol.* **29**:1733–1751.

Mullis, K., Faloona, F., Scharf, S., Saiki, R., Horn, G., and Erlich, H., 1986, Specific enzymatic amplification of DNA *in vitro*: The polymerase chain reaction, *Cold Spring Harb. Symp. Quant. Biol.* **51**(Pt 1):263–273.

Musial, C. E., Arrowood, M. J., Sterling, C. R., and Gerba, C. P., 1987, Detection of *Cryptosporidium* in water by using polypropylene cartridge filters, *Appl. Environ. Microbiol.* **53**:687–692.

Nash, T. E., and Mowatt, M. R., 1992, Identification and characterization of a *Giardia lamblia* group-specific gene, *Exp. Parasitol.* **75**:369–378.

Nath, J., and Johnson, K. L., 1998, Fluorescence *in situ* hybridization (FISH): DNA probe production and hybridization criteria, *Biotech. Histochem.* **73**:6–22.

Neumann, N. F., Gyurek, L. L., Finch, G. R., and Belosevic, M., 2000a, Intact *Cryptosporidium parvum* oocysts isolated after *in vitro* excystation are infectious to neonatal mice, *FEMS Microbiol. Lett.* **183**:331–336.

Neumann, N. F., Gyurek, L. L., Gammie, L., Finch, G. R., and Belosevic, M., 2000b, Comparison of animal infectivity and nucleic acid staining for assessment of *Cryptosporidium parvum* viability in water, *Appl. Environ. Microbiol.* **66**:406–412.

Newman, R. D., Jaeger, K. L., Wuhib, T., Lima, A. A., Guerrant, R. L., and Sears, C. L., 1993, Evaluation of an antigen capture enzyme-linked immunosorbent assay for detection of *Cryptosporidium* oocysts, *J. Clin. Microbiol.* **31**:2080–2084.

Nichols, R. A., Campbell, B. M., and Smith, H. V., 2003, Identification of *Cryptosporidium* spp. oocysts in United Kingdom noncarbonated natural mineral waters and drinking waters by using a modified nested PCR-restriction fragment length polymorphism assay, *Appl. Environ. Microbiol.* **69**:4183–4189.

Nieminski, E. C., Schaefer, F. W., 3rd, and Ongerth, J. E., 1995, Comparison of two methods for detection of *Giardia* cysts and *Cryptosporidium* oocysts in water, *Appl. Environ. Microbiol.* **61**:1714–1719.

Ongerth, J. E., and Stibbs, H. H., 1987, Identification of *Cryptosporidium* oocysts in river water, *Appl. Environ. Microbiol.* **53**:672–676.

Orlandi, P. A., and Lampel, K. A., 2000, Extraction-free, filter-based template preparation for rapid and sensitive PCR detection of pathogenic parasitic protozoa, *J. Clin. Microbiol.* **38**:2271–2277.

Ortega, Y. R., Sterling, C. R., Gilman, R. H., Cama, V. A., and Diaz, F., 1993, *Cyclospora* species—a new protozoan pathogen of humans, *N. Engl. J. Med.* **328**:1308–1312.

Ostergaard, L., Nielsen, A. K., and Black, F. T., 1993, DNA amplification on cerebrospinal fluid for diagnosis of cerebral toxoplasmosis among HIV-positive patients with signs or symptoms of neurological disease, *Scand. J. Infect. Dis.* **25**:227–237.

Owen, R. R., 1985, Improved *in vitro* determination of the viability of *Taenia saginata* embryos, *Ann. Trop. Med. Parasitol.* **79**:655–656.

Panciere, R. J., Thomassen, R. W., and Garner, F. M., 1971, Cryptosporidial infection in a calf, *Vet. Pathol.* **8**:479.

Patel, S., Pedraza-Diaz, S., and McLauchlin, J., 1999, The identification of *Cryptosporidium* species and *Cryptosporidium parvum* directly from whole faeces by analysis of a multiplex PCR of the 18S rRNA gene and by PCR/RFLP of the *Cryptosporidium* outer wall protein (COWP) gene, *Int. J. Parasitol.* **29**:1241–1247.

Perz, J. F., and Le Blancq, S. M., 2001, *Cryptosporidium parvum* infection involving novel genotypes in wildlife from lower New York State, *Appl. Environ. Microbiol.* **67**:1154–1162.

Pohjola, S., 1984, Negative staining method with nigrosin for the detection of cryptosporidial oocysts: A comparative study. *Res. Vet. Sci.* **36**:217–219.

Pohjola, S., Jokipii, L., and Jokipii, A. M., 1985, Dimethylsulphoxide-Ziehl-Neelsen staining technique for detection of cryptosporidial oocysts, *Vet. Rec.* **116**:442–443.

Pozio, E., Rezza, G., Boschini, A., Pezzotti, P., Tamburrini, A., Rossi, P., Di Fine, M., Smacchia, C., Schiesari, A., Gattei, E., Zucconi, R., and Ballarini, P., 1997, Clinical cryptosporidiosis and human immunodeficiency virus (HIV)-induced immunosuppression: Findings from a longitudinal study of HIV-positive and HIV-negative former injection drug users, *J. Infect. Dis.* **176**:969–975.

Pritchard, C. G., and Stefano, J. E., 1990, Amplified detection of viral nucleic acid at subattomole levels using Q beta replicase, *Ann. Biol. Clin. (Paris)* **48**:492–497.

Quinones, B. E., Hibler, C. P., and Hancock, C. M., 1988, Comparison of the Modified Reference Method and Indirect Fluorescent Antibody Technique for Detection of *Giardia* Cysts in Water, In Wallis, P. M., and Hammond, B. R. (eds), *Advances in Giardia Research*, University of Calgary Press, Calgary, Canada, pp. 215–217.

Quintero-Betancourt, W., Gennaccaro, A. L., Scott, T. M., and Rose, J. B., 2003, Assessment of methods for detection of infectious *Cryptosporidium* oocysts and *Giardia* cysts in reclaimed effluents, *Appl. Environ. Microbiol.* **69**:5380–5388.

Reed, C., Sturbaum, G. D., Hoover, P. J., and Sterling, C. R., 2002, *Cryptosporidium parvum* mixed genotypes detected by PCR-restriction fragment length polymorphism analysis, *Appl. Environ. Microbiol.* **68**:427–429.

Reference-Method, 1992, Pathogenic Protozoa: *Giardia lamblia*, In Greenberg, A., Clesceri, L.S., and Eaton A.D. (eds), *Standard Methods for the Examination of Water and*

Wastewater, 18th edn. American Public Health Association and American Water Works Association, Hanover, MD, pp. 124–128.

Reischl, U., Bretagne, S., Kruger, D., Ernault, P., and Costa, J. M., 2003, Comparison of two DNA targets for the diagnosis of Toxoplasmosis by real-time PCR using fluorescence resonance energy transfer hybridization probes, *BMC Infect. Dis.* **3**:7.

Renoth, S., Karanis, P., Schoenen, D., and Seitz, H. M., 1996, Recovery efficiency of *Cryptosporidium* from water with a crossflow system and continuous flow centrifugation: A comparison study, *Zentralbl. Bakteriol.* **283**:522–528.

Reynolds, D. T., Slade, R. B., Sykes, N. J., Jonas, A., and Fricker, C. R., 1999, Detection of *Cryptosporidium* oocysts in water: Techniques for generating precise recovery data, *J. Appl. Microbiol.* **87**:804–813.

Rice, E. W., and Schaefer, F. W., 3rd., 1981, Improved *in vitro* excystation procedure for *Giardia lamblia* cysts, *J. Clin. Microbiol.* **14**:709–710.

Riggs, J. L., Dupuis, K. W., Nakamura, K., and Spath, D. P., 1983, Detection of *Giardia lamblia* by immunofluorescence, *Appl. Environ. Microbiol.* **45**:698–700.

Robertson, L. J., Campbell, A. T., and Smith, H. V., 1993, *In vitro* excystation of *Cryptosporidium parvum*, *Parasitology* **106**(Pt 1):13–19.

Rochelle, P. A., Ferguson, D. M., Handojo, T. J., De Leon, R., Stewart, M. H., and Wolfe, R. L., 1996, Development of a rapid detection procedure for *Cryptosporidium*, using *in vitro* cell culture combined with PCR, *J. Eukaryot. Microbiol.* **43**:72S.

Rochelle, P. A., De Leon, R., Stewart, M. H., and Wolfe, R. L., 1997a, Comparison of primers and optimization of PCR conditions for detection of *Cryptosporidium parvum* and *Giardia lamblia* in water, *Appl. Environ. Microbiol.* **63**:106–114.

Rochelle, P. A., Ferguson, D. M., Handojo, T. J., De Leon, R., Stewart, M. H., and Wolfe, R. L., 1997b, An assay combining cell culture with reverse transcriptase PCR to detect and determine the infectivity of waterborne *Cryptosporidium parvum*, *Appl. Environ. Microbiol.* **63**:2029–2037.

Rochelle, P. A., De Leon, R., Johnson, A., Stewart, M. H., and Wolfe, R. L., 1999, Evaluation of immunomagnetic separation for recovery of infectious *Cryptosporidium parvum* oocysts from environmental samples, *Appl. Environ. Microbiol.* **65**:841–845.

Rochelle, P. A., Marshall, M. M., Mead, J. R., Johnson, A. M., Korich, D. G., Rosen, J. S., and De Leon, R., 2002, Comparison of *in vitro* cell culture and a mouse assay for measuring infectivity of *Cryptosporidium parvum*, *Appl. Environ. Microbiol.* **68**:3809–3817.

Romano, J. W., Williams, K. G., Shurtliff, R. N., Ginocchio, C., and Kaplan, M., 1997, NASBA technology: Isothermal RNA amplification in qualitative and quantitative diagnostics, *Immunol. Invest.* **26**:15–28.

Rose, J., Lisle, J., and LeChevallier, M., 1997, Ch. 4: Waterborne Cryptosporidiosis: Incidence, Outbreaks, and Treatment Strategies, In Fayer, R. (ed), *Cryptosporidium and Cryptosporidiosis*, CRC Press, Inc., New York, pp. 93–109.

Rose, J. B., Landeen, L. K., Riley, K. R., and Gerba, C. P., 1989, Evaluation of immunofluorescence techniques for detection of *Cryptosporidium* oocysts and *Giardia* cysts from environmental samples, *Appl. Environ. Microbiol.* **55**:3189–3196.

Rosenblatt, J. E., and Sloan, L. M., 1993, Evaluation of an enzyme-linked immunosorbent assay for detection of *Cryptosporidium* spp. in stool specimens, *J. Clin. Microbiol.* **31**:1468–1471.

Ryley, J. F., Meade, R., Hazelhurst, J., and Robinson, T. E., 1976, Methods in coccidiosis research: Separation of oocysts from faeces, *Parasitology* **73**:311–326.

Saiki, R. K., Scharf, S., Faloona, F., Mullis, K. B., Horn, G. T., Erlich, H. A., and Arnheim, N., 1985, Enzymatic amplification of beta-globin genomic sequences and restriction site analysis for diagnosis of sickle cell anemia, *Science* **230**:1350–1354.

Sauch, J. F., 1985, Use of immunofluorescence and phase-contrast microscopy for detection and identification of *Giardia* cysts in water samples, *Appl. Environ. Microbiol.* **50**:1434–1438.

Sauch, J. F., Flanigan, D., Galvin, M. L., Berman, D., and Jakubowski, W., 1991, Propidium iodide as an indicator of *Giardia* cyst viability, *Appl. Environ. Microbiol.* **57**:3243–3247.

Savva, D., Morris, J. C., Johnson, J. D., and Holliman, R. E., 1990, Polymerase chain reaction for detection of *Toxoplasma gondii*, *J. Med. Microbiol.* **32**:25–31.

Schaefer, F. W., 3rd, Rice, E. W., and Hoff, J. C., 1984, Factors promoting *in vitro* excystation of *Giardia muris* cysts, *Trans. R. Soc. Trop. Med. Hyg.* **78**:795–800.

Schoondermark-van de Ven, E., Galama, J., Kraaijeveld, C., van Druten, J., Meuwissen, J., and Melchers, W., 1993, Value of the polymerase chain reaction for the detection of *Toxoplasma gondii* in cerebrospinal fluid from patients with AIDS, *Clin. Infect. Dis.* **16**:661–666.

Schunk, M., Jelinek, T., Wetzel, K., and Nothdurft, H. D., 2001, Detection of *Giardia lamblia* and *Entamoeba histolytica* in stool samples by two enzyme immunoassays, *Eur. J. Clin. Microbiol. Infect. Dis.* **20**:389–391.

Schupp, D. G., and Erlandsen, S. L., 1987a, Determination of *Giardia muris* cyst viability by differential interference contrast, phase, or brightfield microscopy, *J. Parasitol.* **73**:723–729.

Schupp, D. G., and Erlandsen, S. L., 1987b, A new method to determine *Giardia* cyst viability: Correlation of fluorescein diacetate and propidium iodide staining with animal infectivity, *Appl. Environ. Microbiol.* **53**:704–707.

Scorza, A. V., Brewer, M. M., and Lappin, M. R., 2003, Polymerase chain reaction for the detection of *Cryptosporidium* spp. in cat feces, *J. Parasitol.* **89**:423–426.

Sharp, S. E., Suarez, C. A., Duran, Y., and Poppiti, R. J., 2001, Evaluation of the Triage Micro Parasite Panel for detection of *Giardia lamblia*, *Entamoeba histolytica/Entamoeba dispar*, and *Cryptosporidium parvum* in patient stool specimens, *J. Clin. Microbiol.* **39**:332–334.

Sheather, A. L., 1923, The detection of intestinal protozoa and mange parasites by floatation technique, *J. Comp. Pathol. Ther.* **36**:260–275.

Shepherd, K. M., and Wyn-Jones, A. P., 1996, An evaluation of methods for the simultaneous detection of *Cryptosporidium* oocysts and *Giardia* cysts from water, *Appl. Environ. Microbiol.* **62**:1317–1322.

Siripanth, C., Phraevanich, R., Suphadtanaphongs, W., Thima, N., and Radomyos, P., 2002, *Cyclospora cayetanensis*: Oocyst characteristics and excystation, *Southeast Asian J. Trop. Med. Public Health* **33**(Suppl 3):45–48.

Slifko, T. R., Friedman, D., Rose, J. B., and Jakubowski, W., 1997, An *in vitro* method for detecting infectious *Cryptosporidium* oocysts with cell culture, *Appl. Environ. Microbiol.* **63**:3669–3675.

Slifko, T. R., Raghubeer, E., and Rose, J. B., 2000, Effect of high hydrostatic pressure on *Cryptosporidium parvum* infectivity, *J. Food Prot.* **63**:1262–1267.

Slifko, T. R., Huffman, D. E., Dussert, B., Owens, J. H., Jakubowski, W., Haas, C. N., and Rose, J. B., 2002, Comparison of tissue culture and animal models for assessment of *Cryptosporidium parvum* infection, *Exp. Parasitol.* **101**:97–106.

Sluter, S. D., Tzipori, S., and Widmer, G., 1997, Parameters affecting polymerase chain reaction detection of waterborne *Cryptosporidium parvum* oocysts, *Appl. Microbiol. Biotechnol.* **48**:325–330.

Smith, A. L., and Smith, H. V., 1989, A comparison of fluorescein diacetate and propidium iodide staining and *in vitro* excystation for determining *Giardia intestinalis* cyst viability, *Parasitology* **99**(Pt 3):329–331.

Smith, H. V., Patterson, W. J., Hardie, R., Greene, L. A., Benton, C., Tulloch, W., Gilmour, R. A., Girdwood, R. W., Sharp, J. C., and Forbes, G. I., 1989, An outbreak of waterborne cryptosporidiosis caused by post-treatment contamination, *Epidemiol. Infect.* **103**:703–715.

Smith, J. J., Gunasekera, T. S., Barardi, C. R., Veal, D., and Vesey, G., 2004, Determination of *Cryptosporidium parvum* oocyst viability by fluorescence *in situ* hybridization using a ribosomal RNA-directed probe, *J. Appl. Microbiol.* **96**:409–417.

Snowden, T. S., and Anslyn, E. V., 1999, Anion recognition: Synthetic receptors for anions and their application in sensors, *Curr. Opin. Chem. Biol.* **3**:740–746.

Sorel, N., Guillot, E., Thellier, M., Accoceberry, I., Datry, A., Mesnard-Rouiller, L., and Miegeville, M., 2003, Development of an immunomagnetic separation-polymerase chain reaction (IMS-PCR) assay specific for *Enterocytozoon bieneusi* in water samples, *J. Appl. Microbiol.* **94**:273–279.

Spano, F., Putignani, L., Guida, S., and Crisanti, A., 1998a, *Cryptosporidium parvum*: PCR-RFLP analysis of the TRAP-C1 (thrombospondin-related adhesive protein of *Cryptosporidium*-1) gene discriminates between two alleles differentially associated with parasite isolates of animal and human origin, *Exp. Parasitol.* **90**:195–198.

Spano, F., Putignani, L., Naitza, S., Puri, C., Wright, S., and Crisanti, A., 1998b, Molecular cloning and expression analysis of a *Cryptosporidium parvum* gene encoding a new member of the thrombospondin family, *Mol. Biochem. Parasitol.* **92**:147–162.

Sterling, C. R., and Arrowood, M. J., 1986, Detection of *Cryptosporidium* sp. infections using a direct immunofluorescent assay, *Pediatr. Infect. Dis.* **5**:S139–S142.

Stibbs, H. H., and Ongerth, J. E., 1986, Immunofluorescence detection of *Cryptosporidium* oocysts in fecal smears, *J. Clin. Microbiol.* **24**:517–521.

Stinear, T., Matusan, A., Hines, K., and Sandery, M., 1996, Detection of a single viable *Cryptosporidium parvum* oocyst in environmental water concentrates by reverse transcription-PCR, *Appl. Environ. Microbiol.* **62**:3385–3390.

Straub, T. M., Daly, D. S., Wunshel, S., Rochelle, P. A., DeLeon, R., and Chandler, D. P., 2002, Genotyping *Cryptosporidium parvum* with an hsp70 single-nucleotide polymorphism microarray, *Appl. Environ. Microbiol.* **68**:1817–1826.

Sturbaum, G. D., Ortega, Y. R., Gilman, R. H., Sterling, C. R., Cabrera, L., and Klein, D. A., 1998, Detection of *Cyclospora cayetanensis* in wastewater, *Appl. Environ. Microbiol.* **64**:2284–2286.

Sturbaum, G. D., Reed, C., Hoover, P. J., Jost, B. H., Marshall, M. M., and Sterling, C. R., 2001, Species-specific, nested PCR-restriction fragment length polymorphism detection of single *Cryptosporidium parvum* oocysts, *Appl. Environ. Microbiol.* **67**:2665–2668.

Sturbaum, G. D., Klonicki, P. T., Marshall, M. M., Jost, B. H., Clay, B. L., and Sterling, C. R., 2002, Immunomagnetic separation (IMS)-fluorescent antibody detection and IMS-PCR detection of seeded *Cryptosporidium parvum* oocysts in natural waters and their limitations, *Appl. Environ. Microbiol.* **68**:2991–2996.

Subrungruang, I., Mungthin, M., Chavalitshewinkoon-Petmitr, P., Rangsin, R., Naaglor, T., and Leelayoova, S., 2004, Evaluation of DNA Extraction and PCR Methods for Detection of *Enterocytozoon bienuesi* in Stool Specimens, *J. Clin. Microbiol.* **42**:3490–3494.

Sulaiman, I. M., Xiao, L., Yang, C., Escalante, L., Moore, A., Beard, C. B., Arrowood, M. J., and Lal, A. A., 1998, Differentiating human from animal isolates of *Cryptosporidium parvum*, *Emerg. Infect. Dis.* **4**:681–685.

Sundermann, C. A., Lindsay, D. S., and Blagburn, B. L., 1987, *In vitro* excystation of *Cryptosporidium baileyi* from chickens, *J. Protozool.* **34**:28–30.

Tanriverdi, S., Tanyeli, A., Baslamisli, F., Koksal, F., Kilinc, Y., Feng, X., Batzer, G., Tzipori, S., and Widmer, G., 2002, Detection and genotyping of oocysts of *Cryptosporidium parvum* by real-time PCR and melting curve analysis, *J. Clin. Microbiol.* **40**:3237–3244.

Tanriverdi, S., Arslan, M. O., Akiyoshi, D. E., Tzipori, S., and Widmer, G., 2003, Identification of genotypically mixed *Cryptosporidium parvum* populations in humans and calves, *Mol. Biochem. Parasitol.* **130**:13–22.

Thompson, R. C., 2000, Giardiasis as a re-emerging infectious disease and its zoonotic potential, *Int. J. Parasitol.* **30**:1259–1267.

Thurston-Enriquez, J. A., Watt, P., Dowd, S. E., Enriquez, R., Pepper, I. L., and Gerba, C. P., 2002, Detection of protozoan parasites and microsporidia in irrigation waters used for crop production, *J. Food Prot.* **65**:378–382.

Tyzzer, E., 1912, *Cryptosporidium parvum* (sp. nov.), a coccidium found in the small intestine of the common mouse, *Arch. Protistenkd.* **26**:394.

Ungar, B. L., 1990, Enzyme-linked immunoassay for detection of *Cryptosporidium* antigens in fecal specimens, *J. Clin. Microbiol.* **28**:2491–2495.

Ungar, B. L., Yolken, R. H., Nash, T. E., and Quinn, T. C., 1984, Enzyme-linked immunosorbent assay for the detection of *Giardia lamblia* in fecal specimens, *J. Infect. Dis.* **149**:90–97.

Upton, S. J., Tilley, M., and Brillhart, D. B., 1994, Comparative development of *Cryptosporidium parvum* (Apicomplexa) in 11 continuous host cell lines, *FEMS Microbiol. Lett.* **118**:233–236.

Upton, S. J., Tilley, M., and Brillhart, D. B., 1995, Effects of select medium supplements on *in vitro* development of *Cryptosporidium parvum* in HCT-8 cells, *J. Clin. Microbiol.* **33**:371–375.

USEPA, 1996, National primary drinking regulation: Monitoring requirements for public drinking water suppliers: *Cryptosporidium*, *Giardia*, viruses, disinfection byproducts, water treatment plant data and other information requirements, *Fed. Regist.* **61**:24354–24388.

USEPA, 2003, Method 1623: *Cryptosporidium* and *Giardia* in Water by Filtration/IMS/FA. *U.S. Environmental Protection Agency, Office of Water*: EPA-821-R-801-025.

Van der Linden, P. W., and Deelder, A. M., 1984, *Schistosoma mansoni*: A diamidinophenylindole probe for *in vitro* death of schistosomula, *Exp. Parasitol.* **57**:125–131.

van Keulen, H., Homan, W. L., Erlandsen, S. L., and Jarroll, E. L., 1995, A three nucleotide signature sequence in small subunit rRNA divides human *Giardia* in two different genotypes, *J. Eukaryot. Microbiol.* **42**:392–394.

Verweij, J. J., Blange, R. A., Templeton, K., Schinkel, J., Brienen, E. A., van Rooyen, M. A., van Lieshout, L., and Polderman, A. M., 2004, Simultaneous detection of *Entamoeba histolytica*, *Giardia lamblia*, and *Cryptosporidium parvum* in fecal samples by using multiplex real-time PCR, *J. Clin. Microbiol.* **42**:1220–1223.

Vesey, G., Slade, J. S., Byrne, M., Shepherd, K., and Fricker, C. R., 1993, A new method for the concentration of *Cryptosporidium* oocysts from water, *J. Appl. Bacteriol.* **75**:82–86.

Vesey, G., Hutton, P., Champion, A., Ashbolt, N., Williams, K. L., Warton, A., and Veal, D., 1994a, Application of flow cytometric methods for the routine detection of *Cryptosporidium* and *Giardia* in water, *Cytometry* **16**:1–6.

Vesey, G., Narai, J., Ashbolt, N., Williams, K., and Veal, D., 1994b, Detection of specific microorganisms in environmental samples using flow cytometry, *Methods Cell. Biol.* **42**(Pt B):489–522.

Vesey, G., Ashbolt, N., Fricker, E. J., Deere, D., Williams, K. L., Veal, D. A., and Dorsch, M., 1998, The use of a ribosomal RNA targeted oligonucleotide probe for fluorescent labelling of viable *Cryptosporidium parvum* oocysts, *J. Appl. Microbiol.* **85**:429–440.

Vinayak, V. K., Dutt, P., and Puri, M., 1991, An immunoenzymatic dot-ELISA for the detection of *Giardia lamblia* antigen in stool eluates of clinical cases of giardiasis, *J. Immunol. Methods* **137**:245–251.

Visvesvara, G. S., 2002, *In vitro* cultivation of microsporidia of clinical importance, *Clin. Microbiol. Rev.* **15**:401–413.

Waldman, E., Tzipori, S., and Forsyth, J. R., 1986, Separation of *Cryptosporidium* species oocysts from feces by using a percoll discontinuous density gradient, *J. Clin. Microbiol.* **23**:199–200.

Walker, G. T., Fraiser, M. S., Schram, J. L., Little, M. C., Nadeau, J. G., and Malinowski, D. P., 1992, Strand displacement amplification–an isothermal, *in vitro* DNA amplification technique, *Nucleic Acids Res.* **20**:1691–1696.

Wang, J., 2000, From DNA biosensors to gene chips, *Nucleic Acids Res.* **28**:3011–3016.

Wang, Z., Vora, G. J., and Stenger, D. A., 2004, Detection and genotyping of *Entamoeba histolytica*, *Entamoeba dispar*, *Giardia lamblia*, and *Cryptosporidium parvum* by oligonucleotide microarray, *J. Clin. Microbiol.* **42**:3262–3271.

Ward, P. I., Deplazes, P., Regli, W., Rinder, H., and Mathis, A., 2002, Detection of eight *Cryptosporidium* genotypes in surface and waste waters in Europe, *Parasitology* **124**:359–368.

Weiss, L. M., Udem, S. A., Salgo, M., Tanowitz, H. B., and Wittner, M., 1991, Sensitive and specific detection of toxoplasma DNA in an experimental murine model: Use of *Toxoplasma gondii*-specific cDNA and the polymerase chain reaction, *J. Infect. Dis.* **163**:180–186.

Widmer, G., Orbacz, E. A., and Tzipori, S., 1999, beta-tubulin mRNA as a marker of *Cryptosporidium parvum* oocyst viability, *Appl. Environ. Microbiol.* **65**:1584–1588.

Widmer, G., Akiyoshi, D., Buckholt, M. A., Feng, X., Rich, S. M., Deary, K. M., Bowman, C. A., Xu, P., Wang, Y., Wang, X., Buck, G. A., and Tzipori, S., 2000a, Animal propagation and genomic survey of a genotype 1 isolate of *Cryptosporidium parvum*, *Mol. Biochem. Parasitol.* **108**:187–197.

Widmer, G., Corey, E. A., Stein, B., Griffiths, J. K., and Tzipori, S., 2000b, Host cell apoptosis impairs *Cryptosporidium parvum* development *in vitro*, *J. Parasitol.* **86**:922–928.

Winiecka-Krusnell, J., and Linder, E., 1995, Detection of *Giardia lamblia* cysts in stool samples by immunofluorescence using monoclonal antibody, *Eur. J. Clin. Microbiol. Infect. Dis.* **14**:218–222.

Wolk, D. M., Johnson, C. H., Rice, E. W., Marshall, M. M., Grahn, K. F., Plummer, C. B., and Sterling, C. R., 2000, A spore counting method and cell culture model for chlorine disinfection studies of *Encephalitozoon* syn. *Septata intestinalis*, *Appl. Environ. Microbiol.* **66**:1266–1273.

Wolk, D., Mitchell, S., and Patel, R., 2001, Principles of molecular microbiology testing methods, *Infect. Dis. Clin. North Am.* **15**:1157–1204.

Wolk, D. M., Schneider, S. K., Wengenack, N. L., Sloan, L. M., and Rosenblatt, J. E., 2002, Real-time PCR method for detection of *Encephalitozoon intestinalis* from stool specimens, *J. Clin. Microbiol.* **40**:3922–3928.

Woodmansee, D. B., 1987, Studies of *in vitro* excystation of *Cryptosporidium parvum* from calves, *J. Protozool.* **34**:398–402.

Wu, Z., Nagano, I., Matsuo, A., Uga, S., Kimata, I., Iseki, M., and Takahashi, Y., 2000, Specific PCR primers for *Cryptosporidium parvum* with extra high sensitivity, *Mol. Cell. Probes.* **14**:33–39.

Xiao, L., Alderisio, K., Limor, J., Royer, M., and Lal, A. A., 2000, Identification of species and sources of *Cryptosporidium oocysts* in storm waters with a small-subunit rRNA-based diagnostic and genotyping tool, *Appl. Environ. Microbiol.* **66**:5492–5498.

Xiao, L., Escalante, L., Yang, C., Sulaiman, I., Escalante, A. A., Montali, R. J., Fayer, R., and Lal, A. A., 1999, Phylogenetic analysis of *Cryptosporidium* parasites based on the small-subunit rRNA gene locus, *Appl. Environ. Microbiol.* **65**:1578–1583.

Xiao, L., Singh, A., Limor, J., Graczyk, T. K., Gradus, S., and Lal, A., 2001, Molecular characterization of *Cryptosporidium* oocysts in samples of raw surface water and wastewater, *Appl. Environ. Microbiol.* **67**:1097–1101.

Xiao, L., Sulaiman, I. M., Ryan, U. M., Zhou, L., Atwill, E. R., Tischler, M. L., Zhang, X., Fayer, R., and Lal, A. A., 2002, Host adaptation and host-parasite co-evolution in *Cryptosporidium*: Implications for taxonomy and public health, *Int. J. Parasitol.* **32**:1773–1785.

Yamasaki, H., Allan, J. C., Sato, M. O., Nakao, M., Sako, Y., Nakaya, K., Qiu, D., Mamuti, W., Craig, P. S., and Ito, A., 2004, DNA differential diagnosis of taeniasis and cysticercosis by multiplex PCR, *J. Clin. Microbiol.* **42**:548–553.

Yamazaki, T., Sasaki, N., Takahashi, S., Satomi, A., Hashikita, G., Oki, F., Itabashi, A., Hirayama, K., and Hori, E., 1997, Clinical features of Japanese children infected with *Cryptosporidium parvum* during a massive outbreak caused by contaminated water supply, *Kansenshogaku Zasshi* **71**:1031–1036.

Zhu, G., Marchewka, M. J., Ennis, J. G., and Keithly, J. S., 1998, Direct isolation of DNA from patient stools for polymerase chain reaction detection of *Cryptosporidium parvum*, *J. Infect. Dis.* **177**:1443–1446.

Zuckerman, U., Armon, R., Tzipori, S., and Gold, D., 1999, Evaluation of a portable differential continuous flow centrifuge for concentration of *Cryptosporidium* oocysts and *Giardia* cysts from water, *J. Appl. Microbiol.* **86**:955–961.

CHAPTER 10

Risk Assessment of Parasites in Food

Kristina D. Mena

10.1 PREFACE

Risk is the possibility or probability of an adverse event occurring due to a hazard or hazards. A hazard may be a physical, chemical, or microbial agent, such as a parasite. Determining whether an agent poses a threat—or the extent of that threat—to humans is not always straightforward. With the diversity of hazards present in the environment along with the multiple transmission routes possible, fully understanding the chances of exposure to that hazard as well as the subsequent human health significance are difficult at best.

Risk assessment has emerged as a methodology to address a wide variety of environmental hazards and their associated human health impacts. A risk assessment framework was first developed in the 1970s by the National Research Council (NRC) to systematically evaluate chemical hazards in the environment (NRC, 1983). In the 1980s and 1990s, this framework was applied to address microorganisms, particularly human health risks associated with waterborne pathogens (Haas, 1983; Regli et al., 1991; Rose and Sobsey, 1993; Rose et al., 1991). The application of this paradigm to food-borne microorganisms and food safety issues soon followed in the 1990s (Jaykus, 1996; Lammerding and Paoli, 1997; Rose et al., 1995). This approach is attractive to the food safety arena due to the Sanitary and Phyto-Sanitary (SPS) Agreement of the World Trade Organization (WTO) and the General Agreement on Tariffs and Trade (GATT) (Marks et al., 1998) and many organizations and governmental agencies—such as the US Department of Agriculture (USDA), Health Canada, and the Codex Alimentarius Commission (CAC), have applied this process (CAC, 1999). In addition, results of such an assessment can enhance the food industry's already prevalent HACCP (hazard analysis critical control point) programs, as they can aid in the identification of critical control points during food production and processing.

A goal of a microbial risk assessment is to provide an objective, science-based evaluation of a microbial hazard to risk managers for the subsequent development of strategies to minimize risk. Risk assessment, therefore, is the first component of the risk analysis process, followed by risk management and risk communication (NRC, 1994). The risk assessment component as developed by the NRC (1983) includes four steps: (1) hazard identification; (2) dose-response assessment; (3) exposure assessment; and (4) risk characterization. Although the framework is an iterative process, in some cases the food safety approach has applied a modified paradigm that addresses exposure prior to dose-response and has replaced the terms "dose-response assessment" with hazard characterization (Fig. 10.1).

The underreporting of food-borne illnesses in the United States makes it challenging to fully understand the public health significance of food-borne disease

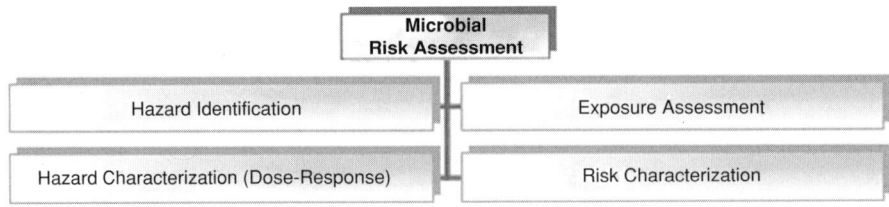

Figure 10.1. The Microbial Risk Assessment Framework.

agents, especially specific food-borne pathogens. Mead *et al.* (1999) estimates that 76 million food-borne illnesses occur in the United States annually leading to 325,000 hospitalizations and 5000 deaths. Protozoan parasites are microorganisms that can be transmitted to humans through either contaminated food or water, resulting in several clinical outcomes such as gastroenteritis. The following sections discuss how this framework can be applied to address parasites in food.

10.2 THE RISK ASSESSMENT FRAMEWORK

10.2.1 Defining the Hazard

This first step of the risk assessment framework provides qualitative information regarding the microorganism itself and its interaction with a host. Available and pertinent information from published epidemiological studies (such as outbreak investigations) as well as laboratory and field data are presented here to determine the extent—if any—of human health impact. Initially, a potential pathogen will be evaluated to determine the range of human illnesses (both acute and chronic) associated with exposure. Once the microorganism is deemed to be pathogenic, other characteristics are identified and reviewed including: (1) endemic and epidemic disease; (2) potential severity of human health consequences, including the determination of morbidity, mortality, and hospitalization ratios; and (3) population characterization, such as the identification of who is likely to be exposed and which subpopulations are more severely impacted.

Appropriately selecting pertinent information for the hazard identification step is critical to the overall risk assessment. Information learned through the completion of this step is incorporated throughout the risk assessment. It is important for the risk assessor to be able to critically review available information to determine its justification for inclusion in the assessment as well as be able to appropriately interpret study findings. Moreover, an understanding of the difficulties in determining a microorganism's ability to initiate infection and illness (*causation* as opposed to *association*) is necessary to adequately evaluate an agent as a (potential) hazard, particularly when addressing food-borne agents. Information available in the peer-reviewed literature regarding food-borne microorganisms is often descriptions of food-borne outbreak investigations that lack crucial information including the identification of the specific pathogen, food vehicle, and number of illness cases.

Table 10.1. Food-borne Protozoan Parasites of Human Health Concern (Haas *et al.*, 1999).

Parasite	Health Outcome
Cryptosporidium parvum	Gastroenteritis
Cyclospora	Gastroenteritis
Entamoeba histolytica	Gastroenteritis
	Intestinal tissue abscesses
Giardia lamblia	Gastroenteritis
	Chronic joint pain
	Lactose intolerance
Toxoplasma gondii	Congenital malformations
	Mental retardation
	Seizures

10.2.1.1 Important Food-borne Parasites

Table 10.1 provides a list of protozoan parasites that may be transmitted through food (and perhaps water and/or person-to-person). Clinical symptoms usually include gastroenteritis but more severe consequences may result, particularly in the immunocompromised. Although most risk assessments of food-borne pathogens focus on bacterial agents, risk assessment approaches have been undertaken for *Cryptosporidium parvum*, *Giardia lamblia*, and *Toxoplasma gondii*. The focus on the former two was on the waterborne route of transmission and the production of microbially safe drinking water (Haas *et al.*, 1996; Lammerding and Paoli, 1997; Rose *et al.*, 1991; Teunis and Havelaar, 2002). Unfortunately, the protozoa more commonly associated with food, such as *T. gondii* and *Cyclospora*, are lacking critical (dose-response) data for risk assessments to be conducted. Components of a risk assessment process have been explored for *T. gondii* and specific data needs were realized and are discussed in the following sections of this chapter.

10.2.2 Exposure Assessment

In the exposure assessment step, the risk assessor determines the intensity and frequency of human exposure to the hazard (parasite). Information is collected or determined regarding the source of exposure, the number of pathogens in the source, the extent (duration) of exposure, the population exposed, and perhaps events leading to exposure. This information may be obtained during the hazard identification step for perhaps a general model development or some of the information may be obtained from an epidemiological food-borne outbreak investigation leading to the development of a model that is situation-specific. Scenario trees have been built and applied to describe specific potential situations or sequences of events that need to occur for a particular outcome to take place (Jaykus, 1996; Marks *et al.*, 1998). Jaykus (1996) describes two types of "trees"—"fault trees" and "event trees"— where one is describing the series of events (probabilities) that would occur to lead to the preidentified "fault" and where the other attempts to predict the events that would need to occur for a food-borne contamination incident to lead to human disease. These "scenario" approaches can provide useful information that can be incorporated in HACCP programs. In addition, predictive microbiology—which has been applied to food-borne bacteria—develops mathematical models to predict the number of microorganisms present throughout a food production process (McMeekin et al., 1993). The mathematical models incorporate variables related to food composition and food processing that may potentially impact a microbe's growth or die-off.

A probabilistic model has been developed for *T. gondii*, although not specifically addressing food-borne transmission (Cassin et al., 1996). This model explores other factors related to exposure such as maternal exposure during pregnancy, effect of drug therapies, population age and immunity profiles, and cat ownership. The goal is to evaluate the role of various risk factors in the incidence of toxoplasmosis to then develop appropriate reduction strategies.

Other exposure-related factors—such as food composition, parasite survivability in the environment, and parasite resistance to treatment—also need to be addressed as they can impact the magnitude and duration of exposure. Food production, processing, and consumption also influence exposure and involve a variety of players including growers, manufacturers, distributors, and consumers. Moreover, consumers are the final point of contact with the potentially contaminated food and associated characteristics such as demographics and sociocultural factors influence food preparation practices and consumption patterns, which will also impact an exposure assessment. In addition, the (in many cases) inevitable role of secondary transmission needs to be considered during this step of the risk assessment.

10.2.2.1 Data Gaps and Challenges
Having quantitative data on the occurrence of specific pathogens (parasites) in food is of utmost importance and perhaps presents the biggest challenge in conducting risk assessments for any type of microbial hazard. Exposure data derived from detection methodologies that are both sensitive and specific, address pathogenicity, and are quantitative (rather than presence/absence tests), are critical to conduct meaningful risk assessments. Such occurrence data for microbial risk assessments are usually obtained from published surveillance studies of water, for example, but are more difficult to obtain for the food-borne route. Current methods for the detection of *Cryptosporidium parvum* and *Giardia lamblia* in water, for example, are time-consuming and labor-intensive with several opportunities for error (Teunis and Havelaar, 2002), which obviously impacts the integrity of risk assessment results if such data are incorporated. Some information is available as a result of food-borne outbreak investigations, but to a limited extent (Rose et al., 1995) due to the inherent complexities of the investigations. The challenges associated with adequately defining factors related to exposure introduce the greatest amount of variability and uncertainty in the microbial risk assessment process. This will be discussed further in the Risk Characterization section.

10.2.3 Hazard Characterization (Dose-response Assessment)
The hazard characterization describes the relationship between the dose of the microorganism (parasite) and the extent of the adverse human health effect (infection,

illness, and death). In essence, it is predicting the ability of a pathogen to overcome a host's defenses to initiate infection and perhaps disease. Determining the likelihood of infection or disease in an exposed population is determined from a dose-response curve developed from experimental data. Unlike with chemical hazards, most of the dose-response data regarding microbial hazards—including parasites—were obtained from human studies (rather than animal experiments) (Dupont et al., 1995; Rendtorff, 1954; Rendtorff and Holt, 1954; Teunis et al., 2002). In human studies, participants are given a range of doses (through ingestion, inhalation, or direct contact) and a human health endpoint of interest (infection or disease or both) is determined. Infection may be determined through direct microscopic count of cysts or oocysts in stool samples of participants, for example, and disease would be determined through the observation of appropriate clinical symptoms. Limitations of such human studies include the fact that healthy adults were used and a relatively high amount of (low virulent) dose was administered in order to be able to observe the desired health endpoint of interest using as few participants as possible. In a typical food contamination situation, a person may be exposed to a (more virulent) lower dose of microorganisms. (However, dose-response datasets for protozoan parasites do include low doses.) A dose-response assessment will attempt to predict what the human health outcomes would be in such a situation using the information obtained from dose-response studies.

Although it is the exposure assessment that is the greatest source of variability and uncertainty in a microbial risk assessment (as explained above), most of the controversy surrounding the application of microbial risk assessment results to "real-world" situations targets the dose-response assessment. Haas (1983) was the first to evaluate the ability of dose-response models to adequately represent the microorganism-host interaction and concluded that the models representing the best fit (using maximum-likelihood methods) were the following nonthreshold models: the exponential and the beta-Poisson. More recently, other models such as the Weibull-Gamma, log-probit, and Gompertz, have been evaluated although primarily for bacteria (Holcomb et al., 1999; Teunis et al., 1999). An ideal model not only fits the available data but is also flexible, conservative (to be protective of subpopulations), and simple (Holcomb et al., 1999). Currently, the exponential and beta-Poisson models are recommended for food-borne and waterborne pathogens (Haas, 1983; Haas et al., 1999; Teunis and Havelaar, 2000).

The following is the exponential model:

$$P_i = 1 - \exp(-rN)$$

where P_i = the probability of infection from a single-dose exposure, r = a constant that represents the number of microorganisms that survive and are capable of initiating an infection (i.e., microorganism-specific), and N = the number of microorganisms ingested or inhaled. This model assumes a random distribution of pathogenic microbes and a constant microorganism-host interaction. The parameter r is further defined as:

$$-r = \ln(0.5)/N_{50}$$

Table 10.2. Dose-Response Models for Food-borne Protozoa.

Microorganism	Model	Animal	Reference
Cryptosporidium parvum	Exponential $r = 0.0042$	Humans	Haas *et al.*, 1996
Giardia lamblia	Exponential $r = 0.01982$	Humans	Rose *et al.*, 1991

where N_{50} equals the median infectious dose. The beta-Poisson distribution assumes a heterogeneity between the microorganism-host interaction resulting in two parameters, α and β. The following is the beta-Poisson model:

$$P_i = 1 - (1 + N/\beta)^{-\alpha}$$

where P_i = the probability of infection from a single-dose exposure, N = the number of microorganisms ingested or inhaled, and α and β represent the dose-response curve (microorganism-specific). Ninety-five percent confidence limits to the dose-response parameters can be computed. β can be further defined as:

$$\beta = N_{50}/(2^{1/\alpha} - 1),$$

resulting in the following equation for the beta-Poisson model:

$$P_i = 1 - [1 + N/N_{50}(2^{1/\alpha} - 1]^{-\alpha}.$$

Risks of illness and death can be calculated by multiplying the risk of infection (P_i) by the appropriate morbidity ratio (risk of illness) and then by multiplying the risk of illness by the appropriate mortality ratio (risk of death). Morbidity ratios for *Giardia* and *Cryptosporidium* are reportedly not dose-dependent and are approximately 50% (Rose *et al.*, 1991) and 39% (Haas *et al.*, 1996), respectively. Mortality ratios for both protozoa are about 0.1% (Haas *et al.*, 1999).

More data are needed to fully address the dose-response relationship of food-borne parasites, particularly for *T. gondii* and *Cyclospora*. Factors associated with the food vehicle (e.g., composition), the host (e.g., health status and immune status), as well as the protozoan itself (e.g., virulence factors and strain variation) need to be considered. A cell culture approach has recently been applied to address dose-response issues regarding *Cryptosporidium* (Slifko *et al.*, 2002). Dose-response and model selection information are available for two food-borne protozoa (Haas *et al.*, 1996; Rose *et al.*, 1991) (Table 10.2).

10.2.4 Risk Characterization

The objective of the risk characterization step is to integrate all of the information from the first three steps and provide both a qualitative assessment of the potential human health impact from exposure to the parasitic hazard and (ideally) a quantitative estimate (e.g., 1:10, 1:10,000, 3:100) of the probability of certain human health outcomes actually occurring due to that exposure. Therefore, the probability risk estimate reflects both the likelihood that a microorganism will cause adverse health outcomes within a population as well as the severity of those outcomes.

Risk estimates may be computed as a point-estimate that represents perhaps a "worst-case scenario" where a conservative approach was taken during the risk assessment to be protective of subpopulations, such as the immunocompromised. A more realistic, useful approach is to compute a distribution of risk that represents a range of exposure scenarios. Nevertheless, the goal is to provide science-based direction for risk managers in the mitigation of environmental hazards and risks.

A common approach in the regulatory arena is to define an action level—either in the food or water industry—above which predicted risks are unacceptable. When considering risks associated with protozoa and consumable products, Teunis and Havelaar (2002) propose an action level be based on the following factors: exposure (maximum amount of pathogen consumed), effect (maximum incidence of human health consequences), and associated costs (regarding both human health effects and product maintenance). Although an action level would reflect a maximum level of acceptable risk, this risk limit should be determined from a range (distribution) of exposures. An inevitable reality when resources to mitigate risks are limited, is the challenge of risk-risk comparisons and subsequent decision-making, particularly when such comparisons aren't compatible.

Risk assessments of parasites in food have been minimally explored although this approach has been used to address *Giardia lamblia* and *Cryptosporidium parvum* in water (Aboytes *et al.*, 2004; Haas *et al.*, 1996; Regli *et al.*, 1991). The USEPA recommends that yearly risks of microbial infection not exceed 1:10,000 for potable waters so risk assessments addressing waterborne pathogens can use this risk level when interpreting risk outputs. Approaches have involved either determining the water treatment level required to meet USEPA's recommendation (Rose *et al.*, 1991) or determining the dose associated with a specified risk level (Haas *et al.*, 1996). More recently, risk estimates were computed addressing the occurrence of *Cryptosporidium parvum* in water where not only quantitative data were available, but the detection method was able to identify *infectious* oocysts (Aboytes *et al.*, 2004).

Research regarding exposure factors and dose-response relationships of foodborne parasites—such as *Cyclospora* and *T. gondii*—are needed to fully conduct comprehensive risk assessments. Such assessments could provide a tool to identify risk factors or "critical control points" along with a food production chain as well as during food distribution and processing. Specific food matrix factors and characteristics/parameters associated with the host and the specific microbe(s) of interest would all need to be considered (Buchanan *et al.*, 2000).

10.2.5 Assumptions, Assumptions, Assumptions

Variability and uncertainty are inherent to all risk assessments. Variability can result due to heterogeneous parameters such as those factors related to exposure. Uncertainty can have a role when certain parameters are unknown or specific data are lacking. Assumptions may be made in order to forward the risk assessment process and it is critical that all assumptions are stated as such for proper risk assessment interpretation to occur. There are several places within the risk assessment process where factors are introduced that either underestimate or overestimate the calculated human health risks. Issues related to exposure such as detection method inefficiencies, human consumption patterns, secondary/tertiary transmission, and immunity and multiple exposures may all contribute to inaccurate estimations of risk. In addition, microbial-related factors such as the assumption that all detected microbes are infectious to humans, for example, may also inappropriately impact a computed risk estimate.

Computer software programs are available that address variability and uncertainty. A Monte Carlo simulation can be performed to develop a risk distribution from point estimates (Burmaster and Anderson, 1994). Random variables (representing a defined probability distribution) are entered literally thousands of times resulting in a final probability distribution. This final probability distribution reflects many distributions and input combinations providing perhaps a more realistic evaluation of the risks associated with a particular hazard or hazards.

10.2.6 Emerging Applications of Microbial Risk Assessment

The quantitative microbial risk assessment framework has been used by both the water and food industries as a means to provide and incorporate science-based information during regulatory decision-making and is increasingly becoming part of microbial water monitoring studies (Aboytes *et al.*, 2004). Microbial risk assessment gives public health meaning to laboratory data and can provide direction for addressing microbial-contaminated media, particularly where information gaps are apparent. Although HACCP programs are actually a risk management tool, the risk assessment approach can greatly enhance such program development, particularly during the hazard identification and exposure assessment steps. In addition, with the increasing global food market, risk assessments can provide a means of standardizing the food production process and/or the evaluation of such processes. Recently, the risk assessment framework has been applied in a water/food combined approach to assess the role of microbial—(such as parasites) laden waters (irrigation water, produce wash water, etc.) in contaminating fresh fruits and vegetables. This distinctive application incorporates issues related to both water and food routes of transmission, provides crucial information to enhance (or develop) effective HAACP programs for fruit and vegetable production, and has the potential to have global implications—both for the produce industry as well as human health.

Quantitative microbial risk assessment provides an adaptable, flexible framework for evaluating the public health impacts associated with exposure to a variety of pathogens in different settings. The limited data on food-borne parasites—especially data related to exposure and dose-response—currently restrict the applicability of the framework to adequately address human health risks associated with parasites in food; however, it has also had a role in identifying emerging threats, such as *Cyclospora* (Jaykus, 1996). Quantitative microbial risk assessment—particularly for protozoa—in the food safety arena has yet to be fully realized.

REFERENCES

Aboytes, R., Di Giovanni, G. D., Abrams, F. A., Rheinecker, C., Mcelroy, W., Shaw, N., and Lechevallier, M. W., 2004, Detection of infectious *Cryptosporidium* in filtered drinking water, *J. Am. Water Works Assoc.* **96**(9):88–98.

Buchanan, R. L., Smith, J. L., and Long, W., 2000, Microbial risk assessment: Dose-response relations and risk characterization, *Int. J. Food Microbiol.* **58**:159–172.

Burmaster, D. E., and Anderson, P. D., 1994, Principles of good practice for the use of Monte Carlo techniques in human health and ecological risk assessment, *Risk Anal.* **14**:477–481.

CAC (Codex Alimentarius Commission), 1999, Principles and Guidelines for the Conduct of a Microbiological Risk Assessment, FAO, Rome, CAC/GL-30.

Cassin, M. H., Lammerding, AM. M., Paoli, G. M., and McColl, R. S., 1996, Population response and immunity to *Toxoplasma gondii.* In *Proceedings of the Annual Conference of the Society for Risk Analysis and International Society for Exposure Analysis.* New Orleans, LA, December 8–12, p. 129.

Dupont, H., Chappell, C., Sterling, C., Okhutsen, P., Rose, J., and Jakubowski, W., 1995, Infectivity of *Cryptosporidium parvum* in healthy volunteers, *N. Engl. J. Med.* **332**(13):855.

Haas, C. N., 1983, Estimation of risk due to the doses of microorganisms: A comparison of alternative methodologies, *Am. J. Epidemiol.* **188**:573–582.

Haas, C. N., Crockett, C. S., Rose, J. B., Gerba, C. P., and Fazil, A. M., 1996, Assessing the risk posed by oocysts in drinking water, *J. Am. Water Works Assoc.* **88**:131–136.

Haas, C. N., Rose, J. B., and Gerba, C. P. (eds.), 1999, *Quantitative Microbial Risk Assessment*, John Wiley and Sons, New York.

Holcomb, D. L., Smith, M. A., Ware, G. O., Hung, Y.-C., Brackett, R. E., and Doyle, M.P., 1999, Comparison of six dose-response models foruse with food-borne pathogens, *Risk Anal.* **19**(6):1091–1100.

Jaykus, L.-A., 1996, The application of quantitative risk assessment to microbial food safety risks, *Crit. Rev. Microbiol.* **22**:279–293.

Lammerding, A. M., and Paoli, G. M., 1997, Quantitative risk assessment: An emerging tool for emerging food-borne pathogens, *Emerg. Infect. Dis.* **3**:483–487.

Marks, H. M., Coleman, M. E., Lin, C.-T. J., and Roberts, T., 1998, Topics in microbial risk assessment: Dynamic flow tree process, *Risk Anal.* **18**:309–328.

McMeekin, T. A., Olley, J. N., Ross, T., and Ratkowsky, D. A, 1993, *Predictive Microbiology: Theory and Application*, John Wiley & Sons, New York.

Mead, P. S., Slutsker, L., Dietz, V., McCaig, L. F., Bresee, J. S., Shapiro, C., Griffin, P. M., and Tauxe, R. V., 1999, Food-related illness and death in the United States, *Emerg. Infect. Dis.* **5**:607–625.

NRC (National Research Council), 1983, *Risk Assessment in the Federal Government: Managing the Process*, National Academy Press, Washington, DC.

NRC (National Research Council), 1994, *Science and Judgment in Risk Assessment*. National Academy Press, Washington, DC.

Regli, S., Rose, J. B., Haas, C. N., and Gerba, C. P., 1991, Modeling the risk from *Giardia* and viruses in drinking water, *J. Am. Water Works Assoc.* **83**:76–84.

Rendtorff, R. C., 1954, The experimental transmission of human intestinal protozoan parasites. 2. *Giardia lamblia* cysts given in capsules, *Am. J. Hyg.* **59**:209–220.

Rendtorff, R. C., and Holt, C. J., 1954, The experimental transmission of human intestinal protozoan parasites. 4. Attempts to transmit *Entamoeba coli* and *Giardia lamblia* by water, *Am. J. Hyg.* **60**:327–328.

Rose, J. B., and Sobsey, M. D., 1993, Quantitative risk assessment for viral contamination of shellfish and coastal waters, *J. Food Prot.* **56**:1042–1050.

Rose, J. B., Haas, C. N., and Regli, S., 1991, Risk assessment and control of waterborne giardiasis. *Am. J. Pub. Health* **81**:709–713.

Rose, J. B., Haas, C. N., and Gerba, C. P., 1995, Linking microbiological criteria for foods with quantitative risk assessment, *J. Food Saf.* **15**:121–132.

Slifko, T. R., Huffman, D. E., Bertrand, D., Owens, J. H., Jakubowski, W., Haas, C. N., and Rose, J. B., 2002, Comparison of animal infectivity and cell culture systems for evaluation of *Cryptosporidium parvum* oocysts, *Exper. Parasitol.* **101**:97–106.

Teunis, P. F., and Havelaar, A. H., 2000, The Beta Poisson dose-response model is not a single-hit model, *Risk Anal.* **20**:513–520.

Teunis, P. F., and Havelaar, A. H., 2002, Risk assessment of protozoan parasites, *Int. Biodeteriorat. Biodegradat* **50**:185–193.

Teunis, P. F. M., Nagelkerke, N. J. D., and Haas, C. N., 1999, Dose response models for infectious gastroenteritis, *Risk Anal.* **19**(6):1251–1260.

Teunis, P. F. M., Chappell, C. L., and Okhuysen, P. C., 2002, *Cryptosporidium* dose response studies: Variation between isolates, *Risk Anal.* **22**(1):175–183.

Index

A
AIDS
  associated complications 42
albendazole 177, 220, 222
amoeba 1
  Acantamoeba 1
  Balamuthia 1
  Endolimax 1
  Entamoeba 1
  Iodamoeba 1
  Naegleria 1
amoeba, free living 8
  Acanthamoeba spp 8
  Balamuthia madrillaris 8
  Naegleria fowleri 8
amoeba, nonpathogenic
  Endolimax nana 1, 7
  Entamoeba coli 1, 7
  Entamoeba chattoni 1
  Entamoeba dispar 1
    incidence of amebiasis 2–3
  Entamoeba hartmanni 1, 7
  Entamoeba invadens 1
  Entamoeba moshkovskii 1
  Entamoeba polecki 1
  Iodamoeba butschlii 1, 7
amoeba, pathogenic
  Entamoeba histolytica 2
    clinical significance of 3–5
    incidence of amebiasis 2–3
    morphology of 3
    pathogenesis and immunity 5–6
    therapy 6
amoebic colitis 4
amoebic dysentery 3
amoebicides
  albendazole 177, 220, 222
  chloroquine 6
  dehydroemetine 6
  diloxanide furoate 6
  furazolidone 27
  iodoquinol 6–7, 11
  ketoconazole cream 8
  mebendazole 7, 177
  metronidazole 6, 11
  ornidazole 27
  secnidazole 27
  triclabendazole 186
Angiostrongylus cantonensis 148–150
Angiostrongylus costaricensis 148–150
anisakid worms 146
Anisakis Simplex 135, 145
  background 145
  clinical manifestations 147
  control 148
  diagnosis 147–148
  epidemiology 146–147
  life cycle 146
  prevention 148
  treatment 147–148
apicomplexan parasites, *see Cryptosporidium* spp

B
BioTechnology Frontiers, (BTF) 255
bithionol 167, 181, 186
Blastocystis 1

C
Catostomus commersoni 173
cestodes, *see* tapeworms
cholangiocarcinoma 172
ciliates 1, 8
  Balantidium coli 1
    clinical significance 9–11
    diagnosis and treatment 11
    epidemiology and prevention 11
    life cycle and morphology 9
ciprofloxacine 46
CLB, *see* coccidian parasites
Clonorchis eggs 161
coccidian parasites, *see also Toxoplasma gondii*
  background of 33–35
  differences between species 35
  biology 36–38
    human infections 36–38
    life cycle of 36
  clinical significance 38–39
  diagnosis methods 45–46
    enzyme-linked immunosorbent assay (ELISA) 46
    indirect fluorescent antibody technique (IFAT) 46
    oligonucleotide-ligation assay (OLA) 46
    real-time polymerase chain reaction 46
  foodborne outbreaks 39–40
  transmission and epidemiology
    Cyclospora cayetanensis 39–42
    Isospora belli 42
    Sarcocystis spp 42–45
  treatment and control 46–47
Cryptosporidium spp. 232–235
  control of contamination 82–86
    chlorination 82
    ozone and ultra violet radiation 82, 84
    physical removal 82
    snap freezing 84
  cryptosporidiosis
    transmission 57
  cryptosporidiosis outbreaks 233
  currently recognized 59
  diagnosis and detection
    antigen detection by immunoassays 73–74
    EPA method 80–81
    indirect fluorescent antibody technique (IFAT) 77–78
    microscopic detection of stool specimens 70–73
    molecular method 74–77
    polymerase chain reaction method 78–80
    serologic method 69
  epidemiology
    cryptosporidiosis in immunocompetent persons 61–62
    cryptosporidiosis in immunocompromised persons 62–63
  life cycle and developmental biology 60–61
  taxonomy of 57
    C. andersoni 58
    Cryptosporidium canis 58

Cryptosporidium (cont.)
　Cryptosporidium felis 58
　Cryptosporidium hominis 58
　Cryptosporidium meleagridis 58
　Cryptosporidium muris 58
　Cryptosporidium suis 58
　Cryptosporidium wrairi 58
　transmission routes and infection sources
　　anthroponotic versus zoonotic 63–64
　　food 66–69
　　water 64–66
　treatment
　　paramomycin 81
　　spiramycin 81
Cyclospora cayetanensis 33, 39–42
cysteine proteinases 5
cysticercosis 201–202

D
4,6 diamidino-2′-phenylindole 239
Dientamoeba fragilis
　morphology and transmission 6–7
　therapy 7
digenetic trematodes 161
　Clonorchis sinensis 168
　　clinical signs 171–173
　　diagnosis 171–173
　　epidemiology 170–171
　　hosts for 170
　　life cycle 169–170
　　treatment 171–173
　comparison of food-borne 162
　Fasciolopsis buski
　　clinical signs 188–189
　　diagnosis 188–189
　　epidemiology 188
　　life cycle 187–188
　　treatment 188–189
　Fasciola gigantica
　　clinical signs 184–186
　　diagnosis 184–186
　　epidemiology 184
　　life cycle 182–184
　　treatment 184–186
　Fasciola hepatica
　　clinical signs 184–186
　　diagnosis 184–186
　　epidemiology 184
　　life cycle 182–184

　　treatment 184–186
　Nanophyetus salmincola
　　clinical signs 180–182
　　diagnosis 180–182
　　life cycle 177–179
　　salmon poisoning disease 179–180
　　treatment 180–182
　Opisthorchis felineus 174
　　clinical signs 176–177
　　diagnosis 176–177
　　epidemiology 175
　　life cycle 174–175
　　treatment 176–177
　Opisthorchis viverrini 174
　　clinical signs 176–177
　　diagnosis 176–177
　　epidemiology 175
　　life cycle 174–175
　　treatment 176–177
　Paragonimus spp 161
　　clinical signs 166–168
　　diagnosis 166–168
　　epidemiology 165–166
　　life cycle 163–165
　　treatment 166–168
　prevention and control of 189–191
　　agricultural practices 190
　　cooking time 190
　　drying 190
　　freezing 190
　　pickling 190
　　salting 190
　　smoking 190
Diphyllobothrium spp
　Diphyllobothrium latum
　　biology 211–212
　　clinical aspects 213–214
　　control 214
　　diagnosis 213–214
　　epidemiology 212
　　transmission 213
　　treatment 214
　Diphyllobothrium pacificum
　　clinical presentation 215
　　control 215
　　diagnosis 215
　　epidemiology 214
　　life cycle 214
　　transmission 214
DMSO-carbolfuchsin 238

E
Echinococcus spp
　Echinococcus granulosus
　　biology 216–217

　　clinical aspects 218–219
　　control 220–221
　　diagnosis 219–220
　　epidemiology 217–218
　　treatment 220
　Echinococcus multilocularis
　　biology 221
　　clinical relevance 222
　　control 223
　　diagnosis 222
　　epidemiology 221–222
　　transmission 222
　　treatment 222–223
EhPgp1 6
Entamoeba gingivalis 1
enteric tract infection, of humans, *see* coccidian parasites
eosinophilic meningitis, *see* *Angiostrongylus cantonensis*; *Angiostrongylus costaricensis*

F
flagyl, *see* metronidazole
fluconazole 8
fluorescence *in situ* hybridization (FISH) 251
fluorescent dye stains
　acridine orange 238
　diamidino-2′-phenylindole (DAPI) 238, 250
　propidium iodide (PI) and 4, 6, 238
food borne parasites, risk assessment of
　framework 276
　　assumptions 281–282
　　defining hazard 276
　　exposure assessment 277–278
　　hazard characterization 278–280
　　risk assessment 282
　　risk characterization 280–281
　by National Research Council 275
furazolidone 27

G
Gal/GalNAc lectin 5
Giardia intestinalis 15
　biology of 16
　classification methods 19–20
　　polymerase chain reaction (PCR) method 19

Pulsed field gel
    electrophoresis
    (PFGE) 19
  restriction fragment
    length polymorphisms
    (RFLPs) 19
  zymodeme analysis
    19
control and treatment
  25–27
detection methods 17
  enzyme immunoassays
    (EIA) 17
  immunofluorescent
    antibody microscopy
    (IFA) 17
  US EPA method
    19
genotyping of 20–23
TPI genes 20
host range of 16
taxonomy of 17
transmission and
  epidemiology
  environmental
    24–25
  human 23–24
Gnathostoma spp.
  150–151
Gnathostomiasis, *see*
  Gnathostoma spp
Gongylonema pulchrum
  151–152
GRA4 genes 114
granulomatous amoebic
  encephalitis (GAE) 8

H

hazard analysis critical control
  point (HACCP) programs
  275, 278, 282
Hymenolepis spp
  Hymenolepis nana
    biology 223
    clinical aspects 225
    control 225
    diagnosis 225
    epidemiology 224
    transmission 224
    treatment 225

I

immunomagnetic separation
  (IMS) 237
Invader® technology
  248
Isospora belli 33, 42

K

ketoconazole cream 8

L

Laboratory Multi-Analyte
  Profiling 248
Ligase Chain Reaction
  248

M

macrogametocytes 111
mebendazole 177
merogony 60
meronts, types of 60
metastrongyle nematodes, *see*
  Angiostrongylus cantonensis;
  Angiostrongylus costaricensis
metronidazole 27
monoxenus parasites, *see*
  Cyclospora cayetanensis;
  Isospora belli

N

nematode infection parasites,
  food-borne 135
  *Angiostrongylus
    cantonensis* 148–150
  *Angiostrongylus
    costaricensis* 148–150
  Anisakis simplex, *see*
    Anisakis simplex
  Gnathostoma spp 150–151
  Gongylonema pulchrum
    151–152
  Trichinella, *see Trichinella*
neurocysticercosis, criteria for
  diagnosis of 205
niclofolan 168
niclosamide 181, 206, 209,
  211, 214
nitoxozanide 27
Nomarski differential
  interference contrast
  microscopy (DIC) 237
nucleic acid liberating schemes
  242–243
nucleic acid sequence-based
  amplification (NASBA) 248
nurse cell 137

O

ornidazole 27

P

PAIR treatment 220
parasite concentration
  and isolation technique
  235–237
parasite detection
  methodologies
    advantages of 240–241
    disadvantages 240–241
    evaluation of 253–255

good laboratory
  practices (GLP) 253
lower limit of detection
  (LLOD) 253
thrombospondin-related
  anonymous protein
  (TRAP) gene, use of
  254
immunological-based
  techniques
  speciation and
    genotyping
    techniques 249–250
  strengths and
    weaknesses 248–249
microscopic techniques
  dye stains 238
  fluorescent labeling
    238
  strengths and
    weaknesses 237–238
  USEPA approved 239
nucleic acid techniques
  DNA based protocols
    243–247
  nucleic acid recovery
    techniques 242–243
  RNA based protocols
    248
  strengths and
    weaknesses 239–242
viability techniques
  cell culture and animal
    infectivity models 252
  dye permeability assays
    250
  fluorescence *in situ*
    hybridization (FISH)
    251
  in vitro excystation 251
  reverse transcription
    PCR 251
  strengths and
    weaknesses 250
Percoll/Percoll step gradient
  237
Percoll/sucrose flotation 237
phosphate buffered saline
  (PBS) 235
polymerase chain reaction
  (PCR) 6, 243–244
  multiplex 247
  nested 244
  real-time 246
  reverse transcription 248
praziquantel 167, 173, 177,
  181, 186, 189, 206, 209, 211,
  214, 220, 225
protozoan parasites
  amoeba 1
    axenic cultivation of 2

protozoan parasites (*cont.*)
  life cycle of 2
  nonpathogenic, *see*
    amoeba,
    nonpathogenic
  pathogenic, *see* amoeba,
    pathogenic
  ciliates 1
  food-borne, of human
    concern 277
pyrimethamine therapy 46

Q
Q-beta replicase 248
quinacrine 27

R
18S rRNAgene 254

S
Safe DrinkingWater Act 18–19
safranin methylene blue stain 238
SAG1-4 genes 114
SAG5A genes 114
SAG5B genes 114
SAG5C genes 114
Sanitary and Phyto-Sanitary (SPS) Agreement, of WTO and GATT 275
Sarcocystis hominis 33, 42–45
Sarcocystis spp infection, in countries
  Belgium 44
  Brazil 43, 45
  Canada 44
  Czechoslovakia 43–44
  Ethiopia 44
  Ghana 44
  Hanoi-Haiphong areas 43
  India 44
  Iran 44
  Japan 44
  Malaysia 43
  Mongolia 44
  New Zealand 44
  Thailand 42
  Tibet 43
Sarcocystis suihominis 33, 42–45
Schistosoma eggs 161
secnidazole 27
selective centrifugation techniques *see* parasite concentration and isolation technique
Spirometra spp
  Spirometra mansoides
    biology 215

    clinical aspects 216
    control 216
    diagnosis 216
    epidemiology 215
    transmission 215–216
    treatment 216
SSU-rRNA gene 46
sulfadiazine 8

T
Taenia spp.
  T. asiatica 197, 209
    biology 210
    clinical aspects 210–211
    control 211
    diagnosis 210–211
    epidemiology 210
    transmission 210
    treatment 211
  T. saginata 197
    biology 207–208
    clinical aspects 208–209
    control 209
    diagnosis 208–209
    epidemiology 208
    transmission 208
    treatment 209
  T. solium 197
    biology of 198–200
    clinical aspects 202–204
    control 207
    diagnosis 204–206
    epidemiology 200–201
    transmission 201–202
    treatment 206–207
tapeworms 197
  Diphyllobothrium spp., *see Diphyllobothrium* spp
  Echinococcus spp., *see Echinococcus* spp
  Hymenolepis spp., *see Hymenolepis* spp
  Spirometra spp., *see Spirometra* spp
  Taenia spp., *see Taenia* spp.
tetracycline 11
Toxoplasma gondii
  epidemiology
    in humans 115–118
    in other animal species 122–123
    in poultry 120–121
    in sheep and goats 121–122
    in swine 118–120
  identification 112
    molecular assays 113–114

    PCR-RFLP method 115
    riboprinting 114–115
  inactivation 124–125
  infection 109
  life cycle of 110–112
  pathogenicity 115
  transmission of 112
  treatment
    azithromycin 124
    nifurtimox 124
    pyrimethamine 123
    sulfadiazine 123
  types of 110
Toxoplasma infection, in humans
  in crowded conditions 116
  in Ethiopian immigrants 117
  in Netherlands 117
  in Slovenia 118
  in South America 118
  in Spain 117
  in Tanzania 117
  women 115
    in Catania 116
    in India 116
    in Korea 116
    in Sweden 117
transcription mediated amplification (TMA) 248
Trichina spiralis, *see Trichinella*
Trichinella 135
  background 135
  clinical manifestations 142–143
  control 143–145
  diagnosis 143
  epidemiology
    Trichinella brivoti 138
    Trichinella murrelli 139
    Trichinella native 138
    Trichinella nelsoni 139
    Trichinella papuae 139
    Trichinella pseudospiralis 139
    Trichinella spiralis 138
    Trichinella T6 140
    Trichinella T8 and T9 140
    Trichinella zimbabwensis 140
  human trichinellosis infection 140–142
    due to globalization 141
    ecological modification 142
    human behavior 141
    literacy level 141
    misdiagnosis 142

life cycle 136–137
prevention 143–145
speciation 136
transmission routes 140
treatment 143
triclabendazole 186
trimethoprim sulfamethoxazole (TMPSMX) 46
triose phosphate isomerase (TPI) gene 17
trophozoites 1, 16

U

ulcers 3

US Environmental Protection Agency (USEPA)-mandated Information Collection Rule (ICR) 236
USEPA Method 1623 236

W

waterborne parasites
Ascaris lumbricoides 232
Ascaris suum 232
Balantidium coli 232
Blastocystis hominis 232
Cyclospora cayetanensis 232
Cryptosporidium parvum 231–232
Fasciolopsis buski 232
Fasciola hepatica 232
Giardia lamblia 231–232, 235
Toxoplasma gondii 232
Wisconsin State Laboratory of Hygiene Flow Cytometry Unit (WSLH) 255

Z

zymodeme analysis 6